AF614697

Neuromethods

Series Editor
Wolfgang Walz
University of Saskatchewan
Saskatoon, SK, Canada

For further volumes:
http://www.springer.com/series/7657

Controlled Genetic Manipulations

Edited by

Alexei Morozov

National Institute of Mental Health, National Institutes of Health, Bethesda, MD, USA

Humana Press

Editor
Alexei Morozov
National Institute of Mental Health
National Institutes of Health
Bethesda, MD, USA

ISSN 0893-2336 e-ISSN 1940-6045
ISBN 978-1-61779-532-9 e-ISBN 978-1-61779-533-6
DOI 10.1007/978-1-61779-533-6
Springer New York Dordrecht Heidelberg London

Library of Congress Control Number: 2011944358

Cover illustration: Genes make the brain work. "Red Brain Blue Gene" by John Fossella.

Printed on acid-free paper

Humana Press is part of Springer Science+Business Media (www.springer.com)

Preface to the Series

Under the guidance of its founders Alan Boulton and Glen Baker, the *Neuromethods* series by Humana Press has been very successful since the first volume appeared in 1985. In about 17 years, 37 volumes have been published. In 2006, Springer Science + Business Media made a renewed commitment to this series. The new program will focus on methods that are either unique to the nervous system and excitable cells or which need special consideration to be applied to the neurosciences. The program will strike a balance between recent and exciting developments like those concerning new animal models of disease, imaging, in vivo methods, and more established techniques. These include immunocytochemistry and electrophysiological technologies. New trainees in neurosciences still need a sound footing in these older methods in order to apply a critical approach to their results. The careful application of methods is probably the most important step in the process of scientific inquiry. In the past, new methodologies led the way in developing new disciplines in the biological and medical sciences. For example, Physiology emerged out of Anatomy in the nineteenth century by harnessing new methods based on the newly discovered phenomenon of electricity. Nowadays, the relationships between disciplines and methods are more complex. Methods are now widely shared between disciplines and research areas. New developments in electronic publishing also make it possible for scientists to download chapters or protocols selectively within a very short time of encountering them. This new approach has been taken into account in the design of individual volumes and chapters in this series.

Wolfgang Walz

Preface

Techniques for controlled genetic manipulations have become universal and are used in numerous biomedical fields. Meanwhile, studies of the nervous system prompted development of perhaps the most sophisticated variants of this methodology. The demand for such development is driven by the anatomical and physiological complexity of the brain and by the need for experimental models that can address this complexity through selective manipulation of defined components of the system: specific neuronal populations or selected synapses. The ideal model, which would allow very precise temporal and spatial control of such manipulations, had been considered an unattainable dream just few years ago, but has almost become a reality following recent technical advances, which include the development of genetic tools for inducing or suppressing neuronal activity not only by the chemicals but also by the optical signals.

The general use of "conditional knockouts" or "regulated expression" became a routine, following broad implementation of the Cre-loxP and similar systems. Meanwhile it is getting increasingly important to learn how these already well-established techniques can be combined with continuously generated tools, like new lines of mice with cell-type specific promoters, advanced methods for delivering genetic material into the brain, and new molecules that allow control of neuronal firing or intracellular signaling pathways. As combinations of these techniques should be uniquely crafted to address specific biological questions, the objective of this book is twofold: to supply basic technical information about controlled genetic manipulations and to provide examples of creative implementation of this methodology when addressing a unique biological problem. Subsequently, some chapters of the book describe the most recent developments in the basic methodology, which includes use of Cre-recombinase, and methods for delivery of genetic material into the brain, whereas other chapters focus on applying these techniques to addressing particular biological questions like structural and functional mapping of neuronal circuits, analysis of specific synaptic connections, and modeling or gene therapy of neurological disorders.

Based on this book objective, the contributing authors represent a broad range of expertise in molecular biology, neurophysiology, and behavioral models and pioneered many approaches described in the book.

Bethesda, MD, USA *Alexei Morozov*

Contents

Contributors

ANTOINE ADAMANTIDIS • *Department of Psychiatry, Douglas Mental Health University Institute, McGill University, Montreal, QC, Canada*
KIMI ARAKI • *Institute of Resource Development and Analysis, Kumamoto University, Kumamoto, Japan*
SEBASTIAN AUER • *Molecular Neurobiology Group, Department of Neuroscience, Max Delbrück Center for Molecular Medicine, Berlin, Germany*
VEERLE BAEKELANDT • *Laboratory for Neurobiology and Gene Therapy, Katholieke Universiteit Leuven, Leuven, Flanders, Belgium*
MATHIAS BÄHR • *Department of Neurology, Georg-August-Universität Göttingen, Göttingen, Germany*
LE CONG • *Program in Biological and Biomedical Sciences, Harvard Medical School, Cambridge, Cambridge, MA, USA*
LEAH R. DEBLANDER • *Institute of Neuroscience, University of Oregon, Eugene, OR, USA*
KARL DEISSEROTH • *Department of Biological Engineering, Department of Psychiatry and Behavioral Sciences, Stanford University, Palo Alto, CA, USA*
GREGORY A. DISSEN • *Division of Neuroscience, Oregon National Primate Research Center, Oregon Health & Science University, Beaverton, OR, USA*
MAURICIO DORFMAN • *Division of Neuroscience, Oregon National Primate Research Center, Oregon Health & Science University, Beaverton, OR, USA*
FRANCESCO GALIMI • *Department of Biomedical Sciences/INBB, University of Sassari Medical School, Sassari, Italy*
JEAN-PAUL HERMAN • *CRN2M-UMR 6231, CNRS and Aix-Marseille University, Marseille Cedex 20, France*
INÉS IBAÑEZ-TALLON • *Department of Neuroscience, Max Delbrück Center for Molecular Medicine, Berlin, Germany*
NICOLAS JULLIEN • *CRN2M-UMR 6231, CNRS and Aix-Marseille University, Marseille Cedex 20, France*
NOBUYUKI KAI • *Department of Molecular Genetics, Institute of Biomedical Sciences, Fukushima Medical University School of Medicine, Fukushima, Japan*
CLIFFORD G. KENTROS • *Institute of Neuroscience, University of Oregon, Eugene, OR, USA*
KAZUTO KOBAYASHI • *Department of Molecular Genetics, Institute of Biomedical Sciences, Fukushima Medical University School of Medicine, Fukushima, Japan*
JAN CHRISTOPH KOCH • *Department of Neurology, Georg-August-Universität Göttingen, Göttingen, Germany*
PAUL LINGOR • *Department of Neurology, Georg-August-Universität Göttingen, Göttingen, Germany*
ALEJANDRO LOMNICZI • *Division of Neuroscience, Oregon National Primate Research Center, Oregon Health & Science University, Beaverton, OR, USA*

JOSEPH J. LOTURCO • *Department of Physiology and Neurobiology, University of Connecticut, Storrs, CT, USA*
BRADY J. MAHER • *Department of Physiology and Neurobiology, University of Connecticut, Storrs, CT, USA*
VALERIE MATAGNE • *Division of Neuroscience, Oregon National Primate Research Center, Oregon Health & Science University, Beaverton, OR, USA*
JODI MCBRIDE • *Division of Neuroscience, Oregon National Primate Research Center, Oregon Health & Science University, Beaverton, OR, USA*
TANAYA L. NEFF • *Division of Neuroscience, Oregon National Primate Research Center, Oregon Health & Science University, Beaverton, OR, USA*
SERGIO R. OJEDA • *Division of Neuroscience, Oregon National Primate Research Center, Oregon Health & Science University, Beaverton, OR, USA*
KANA OKADA • *Department of Molecular Genetics, Institute of Biomedical Sciences, Fukushima Medical University School of Medicine, Fukushima, Japan*
GARRET D. STUBER • *Departments of Psychiatry & Cell and Molecular Physiology, UNC Neuroscience Center, University of North Carolina School of Medicine, Chapel Hill, NC, USA*
MASAHIKO TANAKA • *Department of Cellular Biophysics, Graduate School of Pharmaceutical Sciences, Nagoya City University, Nagoya, Japan*
JEAN-MARC TAYMANS • *Laboratory for Neurobiology and Gene Therapy, Katholieke Universiteit Leuven, Leuven, Flanders, Belgium*
JAAN TOELEN • *Laboratory for Neurobiology and Gene Therapy, Katholieke Universiteit Leuven, Leuven, Flanders, Belgium*
ANKE VAN DER PERREN • *Laboratory for Neurobiology and Gene Therapy, Katholieke Universiteit Leuven, Leuven, Flanders, Belgium*
ALDIS P. WEIBLE • *Institute of Neuroscience, University of Oregon, Eugene, OR, USA*
KEN-ICHI YAMAMURA • *Institute of Resource Development and Analysis, Kumamoto University, Kumamoto, Japan; Institute of Molecular Embryology and Genetics, Kumamoto University, Kumamoto, Japan*
FENG ZHANG • *Broad Institute of MIT and Harvard, Harvard University, Cambridge, MA, USA; Department of Brain and Cognitive Sciences, McGovern Institute for Brain Research, Massachusetts Institute of Technology, Cambridge, MA, USA*

Chapter 1

Regulation of Cre Recombinase: Use of Ligand-Regulated and Dimerizable Cre for Transgenesis

Jean-Paul Herman and Nicolas Jullien

Abstract

Drug-regulatable site-specific recombinases have become, during the last decade, a central tool for transgenesis. They allow, depending on the constructs used, the activation or inactivation of target genes in a time-dependent or, if combined with the use of a specific promoter, in a time- and cell-type-dependent manner. The most widely used approach to obtain this regulation relies on the fusion of the recombinase, mostly Cre recombinase, with the ligand-binding domain (LBD) of a steroid receptor, with the resulting Cre-LBD molecule regulated by the cognate steroid. Another, still experimental, method is based on the dimerization of two complementing inactive fragments of Cre, with the drug-induced or spontaneous association of the fragments leading to a reinstatement of the recombinase activity. These different approaches will be described, with a special emphasis on the practical aspects: choice and efficacy of the different variants, drug regimens, and potential pitfalls.

Key words: Cre recombinase, Cre-ER, CrePR1, DiCre, FKBP12, FRAP, Ligand-binding domain, Ligand-induced dimerization, Rapamycin, RU486, Tamoxifen, Transgenesis

1. Introduction

Site-specific recombinases (SSR) have become, during the last decade, a central tool for transgenesis, and this chapter discusses the regulation of Cre recombinase, the SSR most widely used today. Note, however, that several excellent methodological reviews have dealt recently with SSR and, more specifically, with regulatable Cre forms (1–3), and our intention is not to duplicate information given in these reviews. The aim is rather to provide a comprehensive and critical reappraisal of these approaches and their theoretical basis, including some new tools still under development, and to address some specific practical aspects that were

Alexei Morozov (ed.), *Controlled Genetic Manipulations*, Neuromethods, vol. 65,
DOI 10.1007/978-1-61779-533-6_1,

less extensively dealt with previously. Neither did we wish to touch upon aspects that are only indirectly relevant, such as transgenic techniques, histological procedures, breeding strategies, animal husbandry, etc., as these have been dealt with in some of these reviews or others.

Site-specific recombinases are enzymes that have in common the ability to catalyze the exchange (breaking and rejoining) of DNA strands defined by the presence of a DNA recognition sequence, specific for the given enzyme. Most of these recombinases are of prokaryotic origin and belong to either of two main enzyme families, the tyrosine or the serine recombinases, defined on the basis of the amino acid residue involved in the nucleophilic attack of the DNA backbone and the resulting strand breakage (4). Depending on the arrangement of the recognition sites, this recombination can lead to the integration, excision, or inversion of DNA strands, and it was rapidly realized that this capacity to induce rearrangements in DNA makes these enzymes potential tools to induce modifications in mammalian cells and in transgenic organisms. Of the several hundreds SSR known today, a few have been indeed tested and shown to be useful for such purposes: the tyrosine recombinases Cre (5), Flp (6, 7), and Dre (8, 9); the serine recombinases ΦC31 (7, 10, 11), and ß-recombinase (12). Of these, Cre recombinase is the one that is by far the best characterized, the most efficient, for which there exist also the widest array of tools and which, as a consequence, has become the most widely used.

Cre recombinase was described first by Sternberg and his colleagues as a system important in the life cycle of bacteriophage P1 (13, 14). It is a 343-aa-long protein that catalyzes, without requiring energy or cofactors, the recombination between two asymmetric 34-bp DNA sequences called LoxP, and this process leads, when the LoxP sequences flanking a DNA segment are in the same orientation, to the excision of the intervening sequence. The mechanism of action of Cre involves the formation of a highly ordered protein-DNA complex between the two LoxP sequences and four enzyme molecules interacting with each other, followed by a succession of cleavage-religation steps (15, 16). It was soon discovered that Cre has the ability to induce recombination also in the genome of eukaryotic cells (17, 18) and mice (19, 20). With the parallel development of mouse transgenesis, these reports opened up the possibility of using Cre to remodel the genome in transgenic animals, and indeed, Cre soon became a major tool for transgenesis. In vivo, it is used primarily to excise a stretch of DNA flanked ("floxed") by two LoxP sequences. This excision can lead to the inactivation of a gene, when the excision of a critical floxed exon disrupts the transcription of the gene. It can also lead to the activation of a gene, the transcription of which was blocked by the insertion of a floxed stop sequence between the promoter and the coding sequence of that gene.

While Cre with ubiquitous and constitutive expression can be used to achieve excision, regulating Cre activity presents definite advantages. Indeed, restricting the expression of Cre recombinase to a time window defined by the experimenter (temporal regulation) avoids problems linked to the inactivation of a gene early in development: appearance of compensatory processes that may obscure the functional consequences of the loss of the gene, unwanted pleiotropic effects, difficulties to address multiple and subsequent roles of the gene, etc. Moreover, a high-level constitutive activity of Cre may lead to adverse physiological effects on its own, related to the existence of LoxP-like sequences in the eukaryotic genome (21–28), and expressing Cre for only a short period, necessary and sufficient to lead to the desired recombination, will limit this risk. On the other hand, cell/tissue-type-specific expression of Cre (spatial regulation) may help to study the role of a given gene selectively in a given population and address physiological questions more easily.

The spatial regulation of Cre relies on cell-specific promoters that are used to drive the expression of Cre in the cell population or tissue of interest (29–32). This approach is relatively straightforward and relies on classical transgenic techniques. Thus, the theoretical and methodological problems associated with it are those pertaining to the use of specific promoters in transgenesis and as such will not be dealt with here.

Temporal regulation can be obtained in two ways. One can put Cre under the control of a regulatable promoter, such as the tetracycline-regulated promoter (33, 34). This option is, nevertheless, relatively cumbersome, as it requires to maintain three transgenic lines and, for the actual experiment, have the three transgenes (the tTA or rtTA transactivator, Cre, and the floxed target) brought into the same animal. This approach is discussed extensively in the first chapter, and we will not come back to it here. The second option is to regulate the activity of Cre pharmacologically, using a small-molecule inducer. Such an inducer can regulate the access of Cre to its substrate, through the fusion of Cre and a ligand-binding domain of steroid receptors. It can also regulate the presence of active protein, when it is used to dimerize inactive fragments of Cre. These two approaches will be discussed in this chapter.

2. General Overview

2.1. Cre Fusion with the Ligand-Binding Domain of a Steroid Receptor

Since the early 1970s of the last century, Keith Yamamoto and his colleagues were studying the structure and mechanisms of action of steroid receptors. While analyzing the modular composition of these receptors, they found that attaching the ligand-binding domain (LBD) of the glucocorticoid or estrogen receptors to an unrelated protein (the E1A or myc oncogenes) will make this latter regulatable by the cognate steroid (35, 36). These results opened up a new way to achieve regulation of foreign proteins introduced into cells, and during the next few years, this approach was adapted to the regulation of the activity of a large number of proteins (transcription factors, oncogenes, kinases, etc.) that were fused to the LBD of the various vertebrate steroid receptors: the estrogen (ER), glucocorticoid (GR), progesterone (PR), androgen (AR), and mineralocorticoid (MR) receptors (37, 38).

Adaptation of this method to the regulation of recombinases was attempted soon after its initial description. The first demonstrations of the feasibility of the general approach were published almost simultaneously by the Stewart and the Chambon groups in 1995. They showed that when Flp was fused to the LBD of ER, AR, or GR (39), or Cre to the LBD of ER (40), recombinase activity could be regulated by the corresponding steroid. While these reports produced a proof of principle, it was clear that it would be difficult to use these particular fusions, as they relied on the use of the native receptor and would have been activated in vivo by the circulating steroids and in vitro by the steroids present in the serum or agonists such as phenol red (41). However, within 2 years, different mutated forms of these LBD, responsive to artificial analogs but insensitive to the natural steroid ligands, were tested and found working, paving the way for the in vivo use of these fusions (see below).

Note that, besides Cre recombinase, steroid-mediated regulation of other SSR has also been successfully tested. Thus, Flp have been fused to the LBD domains of the ER, the GR, the PR, or the AR (39, 42–44); Dre and ΦC31 to the LBD of PR (9, 44); and ß-recombinase to the AR LBD (45). However, except for Flp-ER^{T2} (43), there exist yet no data concerning the in vivo performance of these systems.

2.1.1. Cre-ER

The first Cre-LBD fusion to be published (40) was based on the human ERα (designated in the following simply as ER), which at that time was the only ER known. It used the C-terminal fragment (aa 282–595) of the receptor, comprising a portion of its D region, the LBD, and the C-terminal F region, that was fused to the C-terminal end of Cre, resulting in a 660-aa protein. This form was activated by the physiological ligand of the receptor, 17ß-estradiol (E_2), but also by the mixed agonist–antagonist tamoxifen. It was validated on cell cultures but has never been used in vivo. Subsequently, several improved versions have been developed.

Cre-ER^{TM} and Cre-ER^{T}

Previous studies conducted on the mouse ER (mER) have shown that the G525R mutation leads to a loss of binding of E_2, while binding of tamoxifen was retained (46). This has led to the development of Cre-ER^{TM}, based on the mouse ER LBD G525R mutant

(47), and of Cre-ER^T using the human ER LBD G521R mutant, analogous to the G525R mutation in mER (48). These recombinases were no longer activated by E_2 but were sensitive to tamoxifen or 4-hydroxytamoxifen (OHT, the active metabolite of tamoxifen; see later). Note that OHT was about tenfold more potent than tamoxifen to activate Cre-ER^{TM} (47), and that the sensitivity of Cre-ER^T to tamoxifen seemed to be higher than that of Cre-ER^{TM} (49, 50).

Cre-ER^{T1} and Cre-ER^{T2}

To obtain a Cre-ER fusion that would be more sensitive to tamoxifen, different mutations, described previously for the human or mouse ER LBD (46, 51, 52), were tested. This led to the development of Cre-ER^{T1}, based on the human ER LBD G400V/L539A/L540A triple mutant, and Cre-ER^{T2}, containing the human ER LBD G400V/M543A/L544A triple mutant (49). While both were more sensitive to OHT than Cre-ER^T, Cre-ER^{T2} had the highest sensitivity, with an EC_{50} of 6 nM in cell cultures. The higher sensitivity of Cre-ER^{T2} relative to Cre-ER^T was found also in transgenic animals (53).

mer-Cre-mer and ER^{T2}-iCre-ER^{T2}

In all the above fusions, the ER LBD had been fused to the C-terminal end of Cre. While fusing it to the N-terminal end of Cre has not been tested, except for one anecdotal report (50), fusing the LBD to both end of Cre has been tested and used several times since. Thus, the G525R mutant of the mouse ER LBD has been added to Cre to obtain mer-Cre-mer (47), while the fusion of the ER^{T2} form of the human ER LBD to the N- and C-terminal end of Cre resulted in ER^{T2}-iCre-ER^{T2} (54, 55). Note that in the latter, the prokaryotic coding sequence for Cre was replaced by a humanized sequence (hence iCre, for "improved Cre"), with human codon-usage preferences, elimination of cryptic splice sites, and minimization of CpG content. While the sensitivity to OHT of these double fusions is 30–50% less than that of the corresponding single C-terminal fusions, their background activity is considerably lower, leading to a much higher induction ratio and a more stringent control (50, 55).

In Vivo Uses of Cre-ER

The approach was validated initially using Cre-reporter mice (56, 57), and the different Cre-ER fusions (Cre-ER^{TM}, Cre-ER^T, and Cre-ER^{T2}; note that Cre-ER^{T1} has not been tested in transgenic animals) controlled either by a ubiquitous promoter, such as the CMV promoter or that of the *Rosa26* locus (48, 58–61), or by various cell-type/tissue-specific promoters. The general conclusions of these studies were that (1) background recombination mediated by the uninduced Cre-ERs is generally low and (2) recombination can be induced in vivo by tamoxifen with good efficacy. It should be stressed at this point, however, that with any of the Cre versions, recombination is rarely total following postnatal induction in the target population ("mosaicism"), and, depending on the tissue, the degree of mosaicism can be more or less important, the reported estimations varying between 15% and 90% recombination. Conversely, when the induction takes place early during embryonic development, especially during early development (E9–E12), the degree of induction is much higher and uniform, and often reaches 100% with each of these versions. These studies helped also to define the optimal conditions of use of the inducer, tamoxifen or OHT. Note that, except for the study of Indra et al. (53), there has been no systematic comparison of the different versions. Thus, drawing a conclusion as to their relative performance is not an easy task and has to rely on the comparison of the results reported by different groups and obtained in different conditions.

Similar conclusions hold for the versions with the double fusion (mer-Cre-mer or ER^{T2}-iCre-ER^{T2}). However, while these versions seem to have a lower background, they also seem to have a lower efficacy, in terms of the extent of induced recombination, than that of the corresponding single fusions (54, 55, 62, 63). On the other hand, when mer-Cre-mer was expressed under a cell-type-specific promoters rather than a ubiquitous one as in the previous report, a high level of induced recombination could be obtained in the target organs (64–66).

In the last 10 years, these inducible Cre variants, especially Cre-ER^{T2} that has been used by the majority of studies, have been largely used to address various biological questions, such as lineage development in different tissues, developmental role of various genes, etc., through cell-type-specific and temporally regulated conditional gene inactivation. In fact, the different tamoxifen-regulated Cre-ER fusions are today by far the predominant version used to obtain regulated recombination in vivo. These studies are too numerous to be cited here, as searching today a bibliographic database for Cre-ER fusions yields hundreds of references, but they do show that these Cre fusions represent powerful tools to obtain data through conditional gene inactivation. The Cre-LBD mouse lines that have been developed in the course of these studies, expressing a tamoxifen-regulated Cre under a general or a cell-type-specific promoter, can be found in the different Cre databases existing today on the Web, the common portal at http://www.creline.org giving an access to most of them: the collaborative database curated by the Nagy lab in Toronto (http://nagy.mshri.on.ca/cre_new/index.php), the Cre publication database (http://ssrc.genetics.uga.edu/), the Cre/CreERT2Zoo database of the Mouse Clinical Institute in Strasbourg (http://www.ics-mci.fr/mousecre), and the Cre database of the Jackson laboratory (http://www.creportal.org) from which lines can also be purchased (http://jaxmice.jax.org/findmice/index.html).

2.1.2. CrePR1

Cre has also been fused with the LBD of the human progesterone receptor (50, 67). For this development, a mutant form of the receptor, hPR891, has been used that has a 42-aa C-terminal deletion. This mutation abolishes binding of progesterone but does not modify the binding of the synthetic anti-progesterone steroid, RU486 (68). Of the different fusion variants tested (C- or N-terminal fusion, variations of the linker peptide, etc.), the best turned out to be the fusion, to the C-terminal end of Cre, of the C-terminal end of hPR891 (aa 641–891) comprising part of the D region and the LBD till aa 891. Note that this variant has a nuclear localization signal (NLS) added to the N-terminal end of Cre that could lead to an increase in the background activity of this fusion Cre (54). Interestingly, the PR-Cre-PR variant (analogous to the mer-Cre-mer version, see above) turned out to be basically inactive. A direct in vitro comparison of CrePR1 with Cre-ER^{T}, Cre-ER^{TM}, and Cre-ER^{T2} showed that background activity of CrePR1 was somewhat lower than that of the best Cre-ER variant (Cre-ER^{T2}) with an equivalent level of induced recombination (50).

An improved version, Cre*PR, contains several modifications relative to CrePR1: extension of the LBD to aa 914 of hPR to increase its sensitivity to RU486 (69), shortening its N-terminal end to decrease accessibility to proteases (these two modifications leading to the use of hPR650–914), use of iCre (see above), removal of the NLS of CrePR1, and truncation of the first 18 aa of Cre, suppression of a cryptic splicing site. These modifications led to a variant that has a tenfold higher sensitivity to RU486 and a greatly reduced background activity (70).

In Vivo Uses

Early studies have shown, using tissue-specific promoters, that CrePR1 can be used to obtain RU486-regulated recombination in adult animals, and, in particular, achieve a good level of recombination in the CNS that was comparable to that published, with the same promoter and reporter lines, for Cre-ER^{T2} lines (50, 71–73). In the followings, several transgenic lines, expressing CrePR1 under the control of specific cytokeratin promoters, have been established (74–77). These lines have subsequently been used repeatedly, by different groups, to study various epidermal pathologies arising following cell-specific gene deletion by in utero or postnatal induction of CrePR1. CrePR1 has also been used to study cardiac physiology, cerebellar development, or kill selected striatal population to study behavioral processes (78–81). On the whole, however, use of CrePR1 has been markedly less widespread than that of the different tamoxifen-regulated Cre-ER forms.

To this day, there have been few reports concerning the in vivo use of Cre*PR. It has been used, with good results, to create conditional KO mice to study development of various neural lineages during embryonic development (82) or development of epidermal tumors (76).

2.1.3. Other Cre-LBD Fusions

Regulation of Cre recombinase through fusion to LBD of other receptors has also been reported and characterized. Thus, Cre fused to the LBD of GR (I747T) has been reported to be regulated, as expected on the basis of the known specificity of this mutant GR, by dexamethasone but not by the natural CR agonist corticosterone (83). Fusion of Cre with the LBD of AR has also been reported (84). However, these Cre variants have been used neither in vitro nor in vivo since.

2.1.4. Mechanisms of Action of Recombinase-LBD Fusions

Diverse hypothesis have been invoked to explain the regulation conferred onto the fusion proteins by the LBD: masking of the active region(s) of the protein by the LBD itself or by Hsp90 bound to it in the absence of steroid (35, 36), sequestration of the fusion protein in the cytoplasm by Hsp90 that would mask the NLS signals, and transfer to the nucleus following the binding of the ligand (85). However, it seems that there is no unifying single explanation that would hold for all cases, and depending on the protein and/or LBD, diverse mechanisms can exist. Concerning the case that is of interest here, a cytoplasmic localization of the fusion protein in the absence of the steroid and a translocation to the nucleus upon addition of the hormone have been observed using immunohistochemistry for the fusion of the LBD of the ER, the PR, or the GR, and Cre recombinase (54, 59, 71, 83, 86–88), no data being available for the other LBD-recombinase fusion variants. Thus, in this specific case, the sequestration of Cre-LBD in the cytoplasm in the absence of ligand and its translocation to the nucleus when the ligand is bound seem to be a valid explanation. Consequently, what is regulated in this case is the access of the recombinase to its substrate and not its enzymatic activity per se. Note, however, that the fact that the three fusions Cre-ER, CrePR, and CreGR behave the same way, i.e., are cytoplasmic in the absence of the ligand and translocate to the nucleus upon addition of the ligand, is in a way unexpected and not in agreement with what is known concerning the localization of the native ER, PR, or GR. Indeed, while the native GR is cytoplasmic in the absence of hormone and translocates to the nucleus upon binding of the hormone, both the ER and the PR have been reported to be localized in the nucleus even in the absence of hormone binding, and, in the case of ER, this is true also for the mutant forms used for the LBD fused with Cre (52, 89–93). Moreover, at least in the case of the GR and the PR, it has been shown, using GFP-LBD fusions, that the nuclear (PR) or cytoplasmic (GR) localization of the receptor in the absence of ligand is determined by the LBD (91). Thus, it could be hypothesized that the Cre-ER or CrePR fusions, but not the CreGR, are localized in the nucleus independently of the presence of the ligand. The fact that their behavior is similar and that they are localized in the nucleus only in the presence of the ligand indicates that the fusion partner of the LBD must also influence its mechanism of action and underlines the limits of our understanding of the precise way the LBD acts in the context of the fusions.

Finally, we can note that the fact that these fusion proteins are able to efficiently catalyze recombination is somewhat puzzling in view of what is known of the crystal structure of the active Cre recombinase. Indeed, the work of van Duyne and his colleagues has shown that the four Cre molecules and two LoxP sequences involved in the reaction form a highly ordered DNA-protein complex, with intricate protein-protein interactions, suggesting that these latter interactions are at the very base of the cooperativity involved in the recombination reaction (15). In particular, the C-terminal helix of one Cre molecule is buried in a pocket of an adjacent Cre molecule. The fact that the fusion of a protein module of the size of Cre itself on the C-terminal end of the latter—or on its both ends, as in ER^{T2}-iCre-ER^{T2}—still allows an efficient recombination seems difficult to reconcile with the hypothesis that these structural constraints play a crucial role and suggest that these interactions might be functionally less important than supposed.

2.1.5. Conclusions

Cre-LBD fusions have proved to offer a reliable way to regulate Cre recombinase in vivo. Of the different existing variants, Cre-ER^{T2} is the one used in majority, and which seems to be the most reliable, offering a low background and a high level of recombination following induction. It does have, however, some limitations as well. Thus, while induction during embryonic development is usually important and can concern close to 100% of a given cell population, adult-stage induction leads usually to more mosaic recombination, with the degree of mosaicism depending on the promoter driving Cre-ER^{T2} expression and on the target gene. Moreover, the inducer tamoxifen has numerous, primarily endocrine, side effects (see later), and this may limit the usefulness of this approach in particular experimental paradigms. Despite these limitations, as of today it seems to be the best option to achieve regulated recombination in transgenic animals. Nevertheless, other options, in particular Cre*PR, that rely on a drug, RU481, with less side effects, or mer-Cre-mer/ERT2-iCre-ERT2, that are devoid of background activity, may represent interesting alternatives.

2.2. Regulation of Cre Recombinase by Dimerization of Inactive Fragments

Regulation of Cre recombinase by tamoxifen, via a fused ER LBD, offers several advantages such as good efficacy and simplicity. It has, nevertheless, also some drawbacks such as the existence of background activity or reliance on tamoxifen, a drug with non-negligible side effects. Moreover, the ability to obtain a cell-type- or tissue-specific expression supposes the use of single promoters with an expression strictly restricted to the given cell type, something that is not always easy to find. These disadvantages motivated the development of an alternative approach, the regulation of Cre

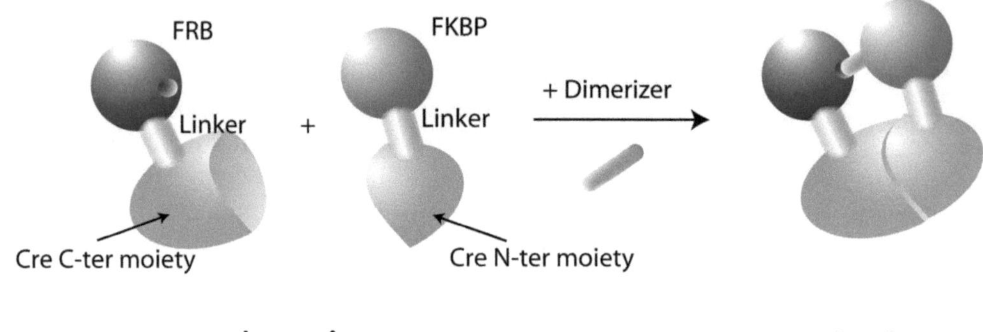

Fig. 1. Principle of the DiCre system. The Cre moieties, taken separately or together, have no recombinase activity. Following the addition of a dimerizer (rapamycin or a rapamycin analog), the drug is bound with high affinity to the dimerization modules FKBP12 and FRB, leading to the physical association of the two Cre moieties and a reinstatement of recombinase activity.

recombinase by ligand-induced complementation of inactive fragments or DiCre (94) and, more recently, non-regulated dimerization systems (Split-Cre) (95–97).

The major advantage of DiCre (and, consequently, of Split-Cre) is to make possible combinatorial recombination. Indeed, being based on the dimerization of two separate peptide fragments of Cre (Fig. 1), these two fragments can be expressed under the control of two different promoters. Thus, in transgenic animals, dimerization, and consequently recombination, will take place only in cells that are at the intersection of the expression domains of these two promoters. This opens the way for more intricate experimental paradigms. It also makes possible a better definition of the cell types in which recombination will occur than what can be obtained with a single promoter. Moreover, DiCre, in contrast to the non-regulated dimerization systems, adds to this spatial definition a temporal regulation, as recombination will depend on the presence of an inducer. Note that such a combinatorial recombination system has been used before, but using two different recombinases (Cre and Flp) (98, 99), a somewhat more complicated setup than fragment dimerization. Note also that besides DiCre, other approaches, using transcription-factor-based ternary systems rather than Cre recombinase, are currently evaluated to achieve combinatorial control of gene expression (100–102).

2.2.1. Ligand-Induced Dimerization of Cre Recombinase (DiCre)

The possibility of reconstituting catalytic activity by complementation through the reassociation of inactive fragments of an enzyme has been known for a long time for bacterial ß-galactosidase (103). Based on this example, different methods have been developed that had in common to use the reconstitution of an enzymatic activity as an index to monitor interaction of modules attached to the enzyme fragments: using ß-galactosidase itself (104), split ubiquitin (105), and protein complementation assay (106, 107).

The success of these methods led us to test whether complementation of inactive fragments could work for Cre and made a new method to regulate its activity. However, in that case, the perspective had to be reversed. Indeed, reinstatement of enzymatic activity was to be considered not as an index revealing the existence of reassociation of attached modules but as the very aim of the reassociation of these modules. Thus, there was a need to have modules with a confirmed capacity to reassociate. An example for such modules is the GCN4 coiled-coil leucine zipper domains (106, 108). A second requirement was to be able to regulate this reassociation, something that was not possible with the leucine zipper domains that reassociate constitutively. The possibility for that was offered by the use of the ligand-induced dimerization procedure (109), based on the heterodimerization, by rapamycin, of the immunophilin FK506-binding protein (FKBP12) (110), and the FKBP12-rapamycin-associated protein (FRAP) (111).

Based on a rational analysis of the structure of Cre recombinase (15, 16), two alternative points were chosen in the enzyme to cut it into two fragments that, taken in themselves, would be devoid of enzymatic activity. These points, between aa 59 and 60 on one hand, between aa 104 and 106 on the other hand, were located in exposed loop regions between two α-helices. Moreover, their orientation was compatible with the attachment of modules that would point above the planar Holliday structure during recombination and not interfere with the interactions between the Cre molecules. FKBP12 and FRB, the FKBP12-rapamycin-binding domain of FRAP (112), were attached to the N-terminal end of the complementing N- and C-terminal fragments of Cre.

DiCre was validated in vitro on cell cultures, and both variants, based on the different sites of cutting Cre, were shown to work reliably (94). There was no recombination at all following the separate transfection of the individual fragments, confirming their lack of activity. Following transient cotransfection of the two complementing fragments into indicator cells, there was some low level recombination that was markedly stimulated following the addition of the dimerizer. The level of induced activity, depending on the variants tested, reached 15–50% of that of intact Cre. In stable transformants expressing the two complementing fragments, background activity was low, and recombinase activity could be induced with rapamycin or the rapamycin analog AP23102 with nanomolar affinities, and recombination could be achieved in 100% of the cells by exposure of the cells to the inducer for only 24 h.

DiCre is a heterodimerization system as the two dimerizing modules (FKBP12 and FRB) are different. This option was chosen to avoid production of non-productive dimers upon addition of the dimerizer. However, it turned out (see below) that the performance of the dimerizer drugs is not satisfactory in vivo. To be able to use small-molecule dimerizer drugs with good in vivo performance, DiCre has been recently modified and changed into a homodimerization system, based on the fusion of the same ligand-regulated dimerization module to both the N- and C-terminal fragments of Cre (Herman et al., in preparation). This DiCre2 system has a somewhat better efficacy than DiCre in terms of induction, with also a lower background activity, and thus represents an improvement of DiCre.

An interesting recent variation on this approach has been the replacement of the modules dependent on a pharmacological agent for dimerization by CIB1 and cryptochrome 2 (CRY2) from *Arabidopsis thaliana*. CRY2 is a blue light absorbing protein that binds CIB1 in its photoexcited state. Fusing these two proteins to the complementing N- and C-terminal fragments of Cre led to a system in which recombination can be induced by illumination of the cells expressing it, with the degree of recombination proportional to the duration of the illumination (113). Given its characteristics (no need for drugs, rapid and reversible activation, potential for induction of recombination in whole organism using two-photon microscopy), this method of light-dependent control of recombination should be a good option for a number of applications.

2.2.2. Non-regulated Dimerization of Cre Recombinase

Overlapping inactive fragments of Cre, obtained by cutting Cre within its C-terminal globular domain (between aa 182 and 196), can complement each other when expressed in the same cell and reconstitute recombinase activity (114). However, the efficacy of the system is relatively poor in vitro (96, 114); moreover, its in vivo evaluation gave disappointing results (E. Casanova, personal communication). To ameliorate this system, artificial antiparallel leucine zipper–forming peptides were fused to the adjacent ends of the fragments obtained by cutting in the same region (aa 190 and 191), resulting in a non-regulated aided dimerization (96). The presence of these peptides ameliorates the efficacy of functional complementation. Efficacy was further ameliorated by fusing nuclear localization sequences to the fragments and reached, in a transient transfection assay and using strong promoters, close to 30% of that of intact Cre.

DiCre has recently been modified by replacing the ligand-dimerizable modules FKBP12 and FRB by GCN4 coiled-coil leucine zipper domains. Similar to the system above, with this system that was called Split-Cre, the level of recombination in vitro could reach 30–50% of that of intact Cre (95). In both of these systems, reconstitution of Cre activity is non-regulated and will occur whenever the two complementing fragments are expressed in the same cells. They have, nevertheless, the advantage of simplicity, as there is no need of pharmacological treatment, and they might certainly be useful when the experimental question can be addressed using simple spatial combinatorial recombination only, without a temporal regulation.

A recent modification of Split-Cre was to fuse the ER^{T2} LBD to the two Cre fragments that can dimerize through the same coiled-coil leucine zipper domains (Split-Cre-ER^{T2}) (97). In this case, although dimerization itself is non-conditional, recombinase activity is regulated by tamoxifen. The potential advantage of this system, relative to DiCre, is to rely, for the induction of Cre activity, on a drug that has been well characterized in vivo. At the same time, relying on tamoxifen means also to have to deal with the problems inherent to this drug (toxicity, endocrine side effects, etc., see below), and that can become a real disadvantage, especially if DiCre2 turns out to have good in vivo performances. A first in vitro evaluation, using transient transfection assays, of Split-Cre-ER^{T2} has shown that it can work and that recombinase activity can be induced by tamoxifen, somewhat surprisingly given the size and position (on the adjacent ends of the Cre fragments) of the ER^{T2} modules relative to the Cre fragments. However, while its reported efficacy seems to be quite low, it is somewhat difficult to evaluate the real performances of the system from the data already published, and a more extensive characterization will be needed to assess its real potential.

2.2.3. In Vivo Uses of DiCre and Split-Cre

To obtain a proof of principle for its in vivo activity, DiCre has been evaluated in an experimental setting (knock-in of a bicistronic expression cassette into the *Rosa26* locus) simplified relative to its projected use, i.e., the utilization of two separate cell-type-specific promoters, and compared to ER^{T2}-iCre-ER^{T2} introduced into the same locus (62, 115). Following crossing with R26R mice, no background recombination could be observed in double transgenic postnatal animals, and rapamycin treatment induced widespread recombination. However, the degree of recombination was highly variable among tissues and was, in most places, quite limited. Moreover, recombinase activity could not be induced during embryonic development. The reasons of this latter limitation are not clear but could be linked to a high endogenous FKBP12 expression during embryonic development (116) or limited passage of the inducer through the placental barrier. Equally disappointing is that the rapamycin analog AP23102, supposed to be devoid of side effects, was basically inactive in vivo, presumably due to unfavorable pharmacokinetic characteristics. Consequently, induction was dependent on the use of rapamycin, an immunosuppressive drug with numerous side effects in vivo. Thus, DiCre seems to be ill-adapted for in vivo usage in its present form. A new version, DiCre2, based on a more performant homodimerization system (see above), is more promising, as we have observed that recombination can be induced in vivo by the non-toxic rapalog AP20187 with a good efficiency (C. Monetti, N. Jullien, A. Nagy and J.P. Herman, unpublished results).

The non-regulated dimerization systems, based on the presence of leucine zippers, were also tested in vivo, using either the same or two different cell-type-specific promoters to drive the expression of the complementing Cre moieties in the pancreas (Pdx1 promoter) (96) or the CNS (PLP, GFAP or GAD67 promoters) (95) and crossing the animals with a Cre indicator strain. For both systems, recombination was observed in triple transgenic animals in the expected cell types. Recombination was, in all cases, mosaic, and concerned around 30% of potential target cells in the pancreas (96), and possibly a lower proportion in glial cells (95), although a quantitative estimation is difficult to reach on the basis of the available micrographs in this latter case. More recently, Split-Cre was used successfully to label neural stem cells in the adult CNS that express simultaneously two glial promoters, the GFAP and prominin promoters (117). Although the reported proportion of the labeled cells was very low, this could be due both to the low frequency of such cells and the route of administration of the dimerizable Cre (intracerebroventricular injection of viral vectors expressing it).

2.2.4. Conclusions

While still largely experimental, dimerization represents a novel and exciting approach to the in vivo regulation of Cre recombinase, and it certainly possesses the potential to become an important tool for conditional transgenesis and help to address sophisticated biological questions. However, before becoming a routine tool, this approach has still to be improved to make it more reliable, decrease mosaicism, and improve the methods for temporal regulation. Two key factors will have to be optimized for that. The first is the dimerizer drug that has to be devoid of side effects and have good pharmacokinetic characteristics (long half-life in plasma, limited/slow metabolism, passage through physiological barriers). The second factor is the expression level of the components of the system. Indeed, this approach requires the concurrent expression of the complementing moieties at a relatively high level to have a reasonable probability of successful combination. Thus, the choice of the promoters to be used will have a crucial role in determining the success of this approach: they will have to ensure a high-level expression, be as specific as possible, and be expressed simultaneously in the target cell population.

3. Methodological Aspects

As showed above, dimerization-regulated Cre recombinase is presently still under development and not adapted for routine use. For this reason, this approach will not be dealt with in this chapter that will be restricted to the Cre fusions with the ER LBD or the PR LBD, the two regulatable forms that are currently in use.

3.1. Cre-ER

3.1.1. Choice and Expression of the Cre-ER Variant

Among the different existing Cre-ER LBD fusion variants, Cre-ER^{T2} seems as of today to be the best choice, associating a low background activity with the highest sensibility to tamoxifen leading to a high level of induced recombination (53, 87). Note that using the humanized version of Cre, rather than the prokaryotic sequence, may increase the efficacy of recombination, although the background activity level may also slightly increase (54, 55). While background activities for Cre-ER^{T} or Cre-ER^{T2} are reported usually to be low (around 1% or 2% of the target population), there are cases in the literature with relatively high levels of background recombination, with one report documenting unregulated recombination in up to 50% of the target cell population (60, 118–123). This high-level background activity could be related to the influence of the genomic environment of the transgene, the promoter, and/or a high-level expression of Cre-ER^{T2}. Background activity could be lowered by using the double LBD ER^{T2}-iCre-ER^{T2} or mer-Cre-mer that in vitro displays a lower background activity (50, 55, 62), although that may come at the price of a somewhat lower efficacy of the tamoxifen-induced recombination. Alternatively, the use of a dual control strategy, placing Cre-ER^{T2} under the control of an inducible promoter, may provide a tight control and eliminate background recombinase activity (122, 124).

3.1.2. Use of Tamoxifen

General Observations

Tamoxifen is a mixed antagonist–agonist of the ER receptor, with an affinity around 10 nM, that has been used for decades for the treatment and prevention of breast cancer. Given its clinical use, its pharmacologic and toxicological properties have been well characterized (125, 126), and these should be taken into account when determining the parameters (dosage, frequency, ways of administration, possible side effects) used for the in vivo induction of recombinase activity.

First, when determining the dose and dosage regimen used, it is important to keep in mind that the LD_{50} of tamoxifen in mice is 200 mg/kg when given intraperitoneally (ip) but 5 g/kg when given orally (125). The safety margin is therefore considerably higher with oral administration, and this strongly argues in favor of administering tamoxifen by oral gavage, especially during pregnancy, when higher doses are often used. In fact, induction of recombinase activity was documented to be, at the same dosage, of equivalent magnitude and more rapid with oral administration of tamoxifen when compared to its ip administration (66, 128). Thus, the recommended route is by oral gavage, a simple, quick, and reliable method. Oral gavage presents the additional advantage of not inducing intraperitoneal inflammation, in contrast with the ip administration of the drug (127). Note that other ways of administration have been occasionally reported by various investigators but do not seem to be very effective. Given its poor water solubility (<0.1 mg/ml), adding it to water is not a real option.

Subcutaneous pellets have been tested but seem to lead to ulceration of the skin (M.M. Peña, personal communication). Incorporation of tamoxifen citrate into food is an option (129, 130), but in that case, the dose ingested is less well controlled, and the animals have to be exposed to the tamoxifen-containing food for a longer period.

Following its administration, tamoxifen is metabolized and gives rise to various active metabolites, among them to 4-hydroxytamoxifen (OHT) that has a much higher affinity to ER than tamoxifen and is considered to be the main active form. Thus in vitro it is about tenfold more active to induce recombinase activity than tamoxifen (47). This provides a rationale for the utilization of OHT rather than tamoxifen itself. It should be noted, however, that its plasma concentration stays considerably lower than that of tamoxifen and that its half-life in plasma is also shorter (6 vs. 12 h) (131), and this may explain that the two drugs are used usually at the same dose. With repeated administrations, tamoxifen accumulates progressively in tissues that act like a reservoir and contributes to the gradual increase of plasma levels upon repeated injections over several days. Tamoxifen passes the placental or blood–brain barrier, thus peripheral administration can ensure Cre induction in the CNS or, in case of pregnant females, in the embryo. Likewise, it passes into the milk of lactating females, making possible to treat pups by giving the drug to the mother.

It is its binding to the ER LBD that makes tamoxifen useful for the induction of the Cre-ER LBD fusion. However, its antiestrogenic properties, linked to this binding, are also at the origin of various physiological side effects that complicate the life of the experimenter but that have to be taken into account. These side effects are impacting primarily the endocrine system (125, 126, 132). First of all, and the most significant from the point of view of its use to address developmental questions, there are those that are affecting pregnancy and in utero development. Thus, tamoxifen can induce a transient infertility when given to adult animals. Given early during pregnancy, tamoxifen blocks embryo implantation, while, when given around E9–11, it leads to significant impairment of feto-placental development: reduction of decidual weight and of weight of the fetus, and higher rate of intrauterine fetal death. Finally, it also impairs parturition. Diminishing the number of injections, using oral administration, and delivering pups by cesarean section allow one to limit these side effects when administering tamoxifen to pregnant mice. Second, there are the endocrine effects linked to its early postnatal administration. Indeed, even a single injection to pups can abolish their sexual activity once they become adult and provoke hypogonadism, but also lasting morphological alterations in the sexually dimorphic area of the brain (133, 134). Thus, tamoxifen is a "dirty" drug, and its side effects may preclude using it for studies for which the endocrine status of the animals is important. In all case, using it requires the inclusion of a proper control group to check for the influence of tamoxifen itself on the parameter of interest.

Practical Aspects

Working solution (10 mg/ml) of tamoxifen or OHT (refs T5648 and H7904, obtained both from Sigma) is obtained by suspending the drug first in ethanol (optional) at 100 mg/ml and diluting it tenfold in sunflower oil to obtain the desired concentration. At this step, the solution is sonicated to disperse the drug. The solution can be kept, protected from light, at 4°C for a week or at –20°C for several weeks, resonicating it before use. Note that given that OHT is considerably more expensive, while its in vivo efficacy in practice, at comparable doses and as related to its pharmacokinetic properties (see above), is not really higher than that of tamoxifen, most studies rely on the utilization of tamoxifen.

Several general remarks can be made concerning the dosage of tamoxifen. First, increasing its doses leads to an increase in levels of recombination. Note, however, that a recent publication could not confirm such a proportionality, at least in early embryos (127). For embryos, this effects plateau around 10 mg per injection, while the plateau is reached with 2–3 mg in adults (59, 135). On the other hand, the efficacy can be increased not only by increasing the dose per injection, but by increasing the frequency of administration to twice per day (112, 120, 136, 137). Finally, for a given dose of tamoxifen, efficacy is increased if Cre-ER^{T2} is used rather than Cre-ER^{T} or Cre-ER^{TM} (53, 87).

In practice, for treatment of adult animals, the drug regimen most often used is 1 mg tamoxifen given daily for 5 days, with lower doses leading to less widespread recombination (128). On the other hand, to induce Cre within the CNS, doubling this dose, and giving 1 mg (or, for some authors, even 2 mg) twice a day for 5 days, is more effective (112, 120, 136, 137). The same drug regimen can be used for lactating females, when the aim is to treat pups (120, 138), although for older pups (P12–16), use of 3 mg twice daily seems to be better adapted (138). Treatment during embryonic development (E7–E14) usually involves a single high dose (4–10 mg) of tamoxifen, lower (4–5 mg) at early stages, gradually increasing toward later stages, although at later stages (E13–E15), repeated treatments—over 2 days—can also be utilized (59, 139, 140). Note that during late embryonic development (E16–E18), use of the same drug regimen as that used for adults, i.e., 1 mg twice daily for 3 days, results in a good level of recombination in the embryo or newborn pup (138). Oral gavage is especially recommended for treatment during gestation. Indeed, this route of administration seems, besides offering a higher safety margin, to decrease loss of embryos. Thus, Danielian et al. reported loss of embryos when 2 mg tamoxifen was injected ip at E8.5 while Santagati et al. used 6 mg by oral gavage at the same age, without embryo loss (135, 139). Note, however, that higher doses may compromise embryonic development and are better adapted for short survival times. If extended survival and birth of the embryos is required, a lower dose of tamoxifen may have to be used. It is also reminded that tamoxifen, especially when given later during pregnancy, may compromise parturition, and in these case, delivery by cesarean section should be considered. Finally, for local epidermal induction, tamoxifen can also be applied topically to the shaved skin as a 100 or 200 mg/ml solution in DMSO (60, 141). Note, finally, that it has been reported that housing together animals treated by tamoxifen with animals that are not treated may lead to induction of the transgene also in the latter animals, supposedly through coprophagous behavior or skin contact (142). Thus, non-treated control animals should be caged separately from treated ones.

The kinetics of induction after a single tamoxifen injection has been found to be quite rapid. Following injection at E8, nuclear translocation of Cre was observed in the embryos as soon as 6-h postinjection, a time when the first traces of recombination can also be observed, peaks at 24 h and subsides by 36–48 h (59, 87).

3.2. CrePR

3.2.1. Choice and Expression of the CrePR Variant

As mentioned in the previous chapter, the in vivo use of CrePR has been very limited, and we have few specific information concerning this form of regulatable Cre. In vitro, Cre*PR has better characteristics (lower background, higher sensitivity to RU486) than the original CrePR1 version (70). Although there has been up to now no direct comparison concerning the in vivo performance of these two CrePR forms, Cre*PR has been shown to work also in vivo and is probably the better choice for use in animals.

3.2.2. Use of RU486

RU486 (mifepristone) is an antiprogesterone that binds to the PR receptor and induces abortion. Under 2 mg/kg, it does not affect fetoplacental development (132). However, doses higher than that are used to induce recombination during embryonic development, and it is recommended therefore to coadminister progesterone (between 0.5 and 3 mg/day) to avoid induction of abortion (74, 82). Progesterone does not bind to CrePR1 or Cre*PR and will not affect the induction of Cre by RU486.

RU486 can be bought from Sigma. For injection to animals, it can be suspended in water containing 0.25% carboxymethylcellulose and 0.5% Tween80 (50), or dissolved in 100% ethanol to a concentration of 100 mg/ml, and then diluted to the desired concentration in sesame oil (82). The doses that have been reported for the in vivo induction of CrePR1 or Cre*PR are highly variable. Indeed, they vary, when treating adult animals, between 1 g/kg for 2 days (78, 81), 80 mg/kg for 8 days (50), and 8 mg/kg for 5 days (71) and, when administered during embryonic development (E9.5–E14), between 7 mg/kg (82) and 100 μg/kg (74, 76). Thus, at this stage, and given the limited number of published reports, it is difficult to define a minimal effective dose. Topical application of RU486 mixed to a cosmetic hand cream at a concentration of 1% has also been used to induce recombination in the skin (75).

3.3. Outlines of the Basic Strategies for the Use of Cre-LBD Mice

This final section gives just a cursory outline of the general approaches when using Cre-LBD constructs in vivo. Indeed, these aspects being not really specific for these constructs, we will not go into the details.

3.3.1. Cre Lines

Using Cre-LBD fusions for conditional transgenesis involves various multistep breeding schemes, starting with mice of a line harboring the modulatable Cre construct with mice that have the target gene of interest floxed. Examples for such breeding schemes are given in Feil et al. (3) or Seibler et al. (58). Establishing Cre-LBD fusion mouse lines is based on the techniques of transgenesis or homologous recombination and will not be discussed here. The number of already established Cre-LBD fusion lines has grown considerably during the last few years. Most of them express Cre-ER^{T} or Cre-ER^{T2} under a tissue/cell-type-specific promoter and have been generated by classical transgenesis or by homologous recombination. There are also some lines in which Cre-ER^{T2} is controlled by a general and ubiquitously expressed promoter. These promoters are the endogenous promoter of the Rosa26 locus (58, 62), the artificial strong CAG promoter (139), and the human ubiquitin C (UbiC) promoter (143). The different existing tamoxifen- or RU486-regulated Cre-LBD lines can be found in the various existing Cre databases (http://nagy.mshri.on.ca/cre_new/index.php; http://www.creportal.org; http://www.ics-mci.fr/mousecre/browser), and many of them are available through the Jackson laboratories (http://jaxmice.jax.org/findmice/index.html). A recently established consortium, CREATE (coordination of resources for conditional expression of mutated mouse alleles), offers a unified portal for access to many of these sites (http://www.creline.org). Note, however, that the expression pattern of these lines has been rarely investigated extensively. In fact, this pattern is often simply inferred from the known pattern of the expression of the promoter that has been used. However, the full expression pattern of a promoter is rarely known. Moreover, this expression pattern may not be faithfully reproduced when the promoter Cre-LBD construct has been introduced by classical transgenesis, as opposed to knock-ins or BAC transgenesis, as the promoter fragment used for the construct may lack crucial enhancers to ensure a fully specific pattern, or the genomic environment of the transgene may modify the expression of the latter (position effect). This can lead to ectopic expressions of the transgene ("leakage") or incomplete expression in the target cell population (mosaicism). A clear example is the divergent recombination pattern obtained in the CNS when using the different existing CamKIIa-Cre lines (144). Therefore, it is advisable to control, following the acquisition of a new Cre line, that the expression pattern corresponds to that expected and important for the given experiments.

As noted before, induction of recombinase activity by tamoxifen is usually incomplete, even when using a specific promoter, leading to tissue mosaicism. Increasing the level of expression of the Cre-LBD fusion may increase the level of induced recombination (48, 59, 128, 135, 137). However, high recombinase expression may also increase background activity (121). Moreover, strong induced Cre expression may exert toxic effects on its own, related possibly to the existence of cryptic LoxP-like sites in the mammalian genome and leading to cell death and various developmental defects (21, 23–28). The optimal Cre-LBD dosage is therefore a compromise that allows efficient recombination while minimizing adverse and non-specific effects. This dosage will depend on the copy number of the Cre and locus of insertion (when classical transgenesis is used), on its mono- or biallelic character, on the strength of the promoter, etc. Nevertheless, it is difficult to predict beforehand the level of Cre expression and the resulting adverse effects. Thus, one would think that knocking in the Cre-LBD fusion gene to put it under the control of an endogenous promoter would decrease the likelihood of adverse effects, as in this case, Cre-LBD is present in a single copy, and endogenous promoters are usually not as strong as viral or artificial promoters. However, toxicity of the induced Cre in the hematopoietic system has been documented even with a Cre-ER^{T2} line in which the fusion gene was knocked in under the control of the endogenous promoter of the *Rosa26* locus (27, 58). Thus, the presence of adverse effects has to be controlled in all cases with the final Cre line that is to be used.

3.3.2. Indicator Lines

Characterizing the pattern of recombination brought about following induction of Cre activity frequently involves the use of Cre-reporter lines. These lines are based on the visualization of a reporter gene, the expression of which is dependent on the Cre-mediated excision of a stop sequence placed between a general and ubiquitous promoter and the CDS of the reporter. The most widely used line is the R26R line that expresses ß-galactosidase in cells where recombination has taken place (56). In this case, the indicator construct has been knocked in downstream of the ubiquitously expressed promoter of the *Rosa26* locus. The R26-EGFP, R26-EYFP, and R26-ECFP lines, generated by the same procedure, express, after recombination, the green, yellow, or cyan fluorescent proteins, respectively, from the same locus (145, 146). Other lines are the transgenic Z/EG and Z/AP lines, that express, under the control of the strong CAG promoter, ß-galactosidase before recombination and EGFP or alkaline phosphatase after (147, 148). Finally, lines that express, after recombination, the new and especially bright fluorescent proteins ZsGreen or dtTomato (resulting in an enhanced sensitivity) under the control of the CAG promoter have recently been developed, the whole indicator construct being inserted into the *Rosa26* locus (123). The advantage of these latter lines is that expression of the reporter gene depends on a strong artificial promoter, rather than on the weaker Rosa26 promoter, but at the same time the construct is inserted into a locus that stays open throughout life, preventing silencing of the transgene. It must be stressed, however, that in fact, neither of these promoters is really ubiquitous, i.e., active in every cells of the organism all the time. This explains that the recombination patterns one can observe when comparing several reporter lines are not necessarily the same (60, 77, 115, 123, 124, 149, 150). For example, the Z/EG or Z/AP lines seem to have a lower sensitivity and to show a more limited recombination than other lines. In general, this pattern depends not only on the reporter line used, the localization of the floxed sequence within the genome, but also the age at which Cre recombination is induced (recombination during early development being much more widespread than in adults) or the age at which the animals are sacrificed (the Rosa26 promoter seems to be less active for example in the adult CNS), and even the mouse strains. Thus, it is a good practice, when characterizing the recombination pattern obtained with a new Cre line, to do this characterization in parallel in several Cre-reporter lines and, moreover, vary the conditions of recombination. The final characterization will be done, however, with the floxed target gene of interest.

3.3.3. Target Lines

The target lines have a critical stretch of the genome floxed, the aim being the controlled excision of this stretch to achieve inactivation, or conversely, the activation of the gene of interest. Thus, the floxed DNA sequence may correspond to a critical exon of the gene, with its excision leading to a missense or prematurely stopped mRNA. It can also be a stuffer sequence containing a stop codon that blocks the transcription of the downstream gene of interest. This latter can be a reporter gene, as for the Cre-indicator mice, or it can be a toxic gene, the expression of which leads to the death of the expressing cells (81). The procedures to establish these lines are outside of the scope of this review.

3.3.4. General Aspects of Induced Recombination

The pattern of recombination of an endogenous gene may not reproduce that observed with a general Cre-reporter line (see above), and, moreover, the pattern of recombination of different endogenous genes also differs, even when using the same Cre line. This can be related to different accessibility of the floxed gene, as determined by their localization in the genome within domains with different chromatin states that can be open or only partially open or silenced. Moreover, as already noted for indicator lines, the degree of recombination depends also on the age of the animals. Thus, recombination is more extensive when Cre is induced during embryonic development and can reach, in this case, 100%. In all cases, the definitive proof of the recombination of the target gene and its extent in the given experimental situation should be obtained by direct controls, done, if possible and depending on the tools available, using in situ hybridization, Southern blots, qRT-PCR, or immunohistochemistry.

4. Conclusions

Although regulation of Cre recombinase by a steroid hormone receptor LBD fused to it has some drawbacks, such as side effects of the inducer, mosaicism, and existence of some background, it is today a mature technology that can be used for conditional transgenesis. Indeed, these potential problems can be controlled or, at least, taken into account provided the use of adequate controls. Moreover, presently, there is no other technology that would have a comparable efficacy and simplicity, even if other approaches, such as the regulation by dimerization, could emerge as interesting alternatives in the future.

References

1. Birling M, Gofflot F, Warot X (2009) Site-specific recombinases for manipulation of the mouse genome. Methods Mol Biol 561: 245–263.
2. Feil R (2007) Conditional somatic mutagenesis in the mouse using site-specific recombinases. Handb Exp Pharmacol: 3–28.
3. Feil S, Valtcheva N, Feil R (2009) Inducible Cre mice. Methods Mol Biol 530: 343–363.
4. Grindley ND, Whiteson KL, Rice PA (2006) Mechanisms of Site-Specific Recombination. Annu Rev Biochem 75: 567–605.
5. Nagy A (2000) Cre recombinase: the universal reagent for genome tailoring. Genesis 26: 99–109.
6. Dymecki SM (1996) Flp recombinase promotes site-specific DNA recombination in embryonic stem cells and transgenic mice. Proc Natl Acad Sci USA 93: 6191–6196.
7. Raymond CS, Soriano P (2007) High-Efficiency FLP and ΦC31 Site-Specific Recombination in Mammalian Cells. PLoS ONE 2: e162.
8. Sauer B, McDermott J (2004) DNA recombination with a heterospecific Cre homolog identified from comparison of the pac-c1 regions of P1-related phages. Nucleic Acids Res 32: 6086–6095.
9. Anastassiadis K, Fu J, Patsch C, Hu S, Weidlich S, et al. (2009) Dre recombinase, like Cre, is a highly efficient site-specific recombinase in E. coli, mammalian cells and mice. Disease Models & Mechanisms 2: 508–515.
10. Belteki G, Gertsenstein M, Ow DW, Nagy A (2003) Site-specific cassette exchange and germline transmission with mouse ES cells expressing (phgr)C31 integrase. Nature Biotech 21: 321–324.
11. Sangiorgi E, Shuhua Z, Capecchi MR (2008) In vivo evaluation of PhiC31 recombinase activity using a self-excision cassette. Nucleic Acids Res 36: e134–e134.
12. Díaz V, Rojo F, Martínez-A C, Alonso JC, Bernad A (1999) The prokaryotic beta-recombinase catalyzes site-specific recombination in mammalian cells. J Biol Chem 274: 6634–6640.
13. Sternberg N, Hamilton D (1981) Bacteriophage P1 site-specific recombination. I. Recombination between loxP sites. J Mol Biol 150: 467–486.
14. Sternberg N, Hamilton D, Hoess R (1981) Bacteriophage P1 site-specific recombination. II. Recombination between loxP and the bacterial chromosome. J Mol Biol 150: 487–507.
15. Van Duyne G (2001) A structural view of cre-loxp site-specific recombination. Annu Rev Biophys Biomol Struct 30: 87–104.
16. Guo F, Gopaul D, van Duyne G (1997) Structure of Cre recombinase complexed with

DNA in a site-specific recombination synapse. Nature 389: 40–6.

17. Sauer B, Henderson N (1988) Site-specific DNA recombination in mammalian cells by the Cre recombinase of bacteriophage P1. Proc Natl Acad Sci USA 85: 5166–5170.
18. Sauer B (1987) Functional expression of the cre-lox site-specific recombination system in the yeast Saccharomyces cerevisiae. Mol Cell Biol 7: 2087–2096.
19. Orban PC, Chui D, Marth JD (1992) Tissue- and site-specific DNA recombination in transgenic mice. Proc Natl Acad Sci USA 89: 6861–6865.
20. Lakso M, Sauer B, Mosinger B, Lee EJ, Manning RW, et al. (1992) Targeted oncogene activation by site-specific recombination in transgenic mice. Proc Natl Acad Sci USA 89: 6232–6236.
21. Naiche LA, Papaioannou VE (2007) Cre activity causes widespread apoptosis and lethal anemia during embryonic development. Genesis 45: 768–775.
22. Buerger A, Rozhitskaya O, Sherwood MC, Dorfman AL, Bisping E, et al. (2006) Dilated cardiomyopathy resulting from high-level myocardial expression of Cre-recombinase. J. Card. Fail 12: 392–398.
23. Forni P, Scuoppo C, Imayoshi I, Taulli R, Dastru W, et al. (2006) High levels of Cre expression in neuronal progenitors cause defects in brain development leading to microencephaly and hydrocephaly. J Neurosci 26: 9593–602.
24. Schmidt E, Taylor D, Prigge J, Barnett S, Capecchi M (2000) Illegitimate Cre-dependent chromosome rearrangements in transgenic mouse spermatids. Proc Natl Acad Sci USA 97: 13702–7.
25. Loonstra A, Vooijs M, Beverloo H, Allak B, van Drunen E, et al. (2001) Growth inhibition and DNA damage induced by Cre recombinase in mammalian cells. Proc Natl Acad Sci USA 98: 9209–14.
26. Thyagarajan B, Guimaraes M, Groth A, Calos M (2000) Mammalian genomes contain active recombinase recognition sites. Gene 244: 47–54.
27. Higashi AY, Ikawa T, Muramatsu M, Economides AN, Niwa A, et al. (2009) Direct hematological toxicity and illegitimate chromosomal recombination caused by the systemic activation of CreERT2. J Immunol 182: 5633–5640.
28. Lee J, Ristow M, Lin X, White MF, Magnuson MA, et al. (2006) RIP-Cre revisited, evidence for impairments of pancreatic beta-cell function. J Biol Chem 281: 2649–2653.
29. Zinyk D, Mercer E, Harris E, Anderson D, Joyner A (1998) Fate mapping of the mouse midbrain-hindbrain constriction using a site-specific recombination system. Curr Biol 8: 665–8.
30. Sclafani A, Skidmore J, Ramaprakash H, Trumpp A, Gage P, et al. (2006) Nestin-Cre mediated deletion of Pitx2 in the mouse. Genesis 44: 336–44.
31. Tsien J, Chen D, Gerber D, Tom C, Mercer E, et al. (1996) Subregion- and cell type-restricted gene knockout in mouse brain. Cell 87: 1317–26.
32. Gu H, Marth J, Orban P, Mossmann H, Rajewsky K (1994) Deletion of a DNA polymerase beta gene segment in T cells using cell type-specific gene targeting. Science 265: 103–106.
33. Schonig K, Schwenk F, Rajewsky K, Bujard H (2002) Stringent doxycycline dependent control of CRE recombinase in vivo. Nucleic Acids Res 30: e134.
34. St-Onge L, Furth P, Gruss P (1996) Temporal control of the Cre recombinase in transgenic mice by a tetracycline responsive promoter. Nucleic Acids Res 24: 3875–7.
35. Picard D, Salser SJ, Yamamoto KR (1988) A movable and regulable inactivation function within the steroid binding domain of the glucocorticoid receptor. Cell 54: 1073–1080.
36. Eilers M, Picard D, Yamamoto KR, Bishop JM (1989) Chimaeras of myc oncoprotein and steroid receptors cause hormone-dependent transformation of cells. Nature 340: 66–68.
37. Picard D (2000) Posttranslational regulation of proteins by fusions to steroid-binding domains. Meth. Enzymol 327: 385–401.
38. Picard D (1994) Regulation of protein function through expression of chimaeric proteins. Curr. Opin. Biotechnol 5: 511–515.
39. Logie C, Stewart AF (1995) Ligand-regulated site-specific recombination. Proc Natl Acad Sci USA 92: 5940–5944.
40. Metzger D, Clifford J, Chiba H, Chambon P (1995) Conditional site-specific recombination in mammalian cells using a ligand-dependent chimeric Cre recombinase. Proc Natl Acad Sci USA 92: 6991–6995.
41. Berthois Y, Katzenellenbogen JA, Katzenellenbogen BS (1986) Phenol red in tissue culture media is a weak estrogen: implications concerning the study of estrogen-responsive cells in culture. Proc Natl Acad Sci USA 83: 2496–2500.
42. Angrand PO, Woodroofe CP, Buchholz F, Stewart AF (1998) Inducible expression based on regulated recombination: a single vector

strategy for stable expression in cultured cells. Nucleic Acids Res 26: 3263–3269.

43. Hunter NL, Awatramani RB, Farley FW, Dymecki SM (2005) Ligand-activated Flpe for temporally regulated gene modifications. Genesis 41: 99–109.

44. Sharma N, Moldt B, Dalsgaard T, Jensen TG, Mikkelsen JG (2008) Regulated gene insertion by steroid-induced {Phi}C31 integrase. Nucleic Acids Res 36: e67.

45. Servert P, Garcia-Castro J, Díaz V, Lucas D, Gonzalez MA, et al. (2006) Inducible model for beta-six-mediated site-specific recombination in mammalian cells. Nucleic Acids Res 34: e1.

46. Danielian PS, White R, Hoare SA, Fawell SE, Parker MG (1993) Identification of residues in the estrogen receptor that confer differential sensitivity to estrogen and hydroxytamoxifen. Mol Endocrinol 7: 232–240.

47. Zhang Y, Riesterer C, Ayrall A, Sablitzky F, Littlewood T, et al. (1996) Inducible site-directed recombination in mouse embryonic stem cells. Nucleic Acids Res 24: 543–8.

48. Feil R, Brocard J, Mascrez B, LeMeur M, Metzger D, et al. (1996) Ligand-activated site-specific recombination in mice. Proc Natl Acad Sci USA 93: 10887–90.

49. Feil R, Wagner J, Metzger D, Chambon P (1997) Regulation of Cre Recombinase Activity by Mutated Estrogen Receptor Ligand-Binding Domains. Biochemical and Biophysical Research Communications 237: 752–757.

50. Kellendonk C, Tronche F, Casanova E, Anlag K, Opherk C, et al. (1999) Inducible site-specific recombination in the brain. J Mol Biol 285: 175–82.

51. Tora L, Mullick A, Metzger D, Ponglikitmongkol M, Park I, et al. (1989) The cloned human oestrogen receptor contains a mutation which alters its hormone binding properties. EMBO J 8: 1981–1986.

52. Mahfoudi A, Roulet E, Dauvois S, Parker MG, Wahli W (1995) Specific mutations in the estrogen receptor change the properties of antiestrogens to full agonists. Proc Natl Acad Sci USA 92: 4206–4210.

53. Indra A, Warot X, Brocard J, Bornert J, Xiao J, et al. (1999) Temporally-controlled site-specific mutagenesis in the basal layer of the epidermis: comparison of the recombinase activity of the tamoxifen-inducible Cre-ER(T) and Cre-ER(T2) recombinases. Nucleic Acids Res 27: 4324–7.

54. Shimshek D, Kim J, Hubner M, Spergel D, Buchholz F, et al. (2002) Codon-improved Cre recombinase (iCre) expression in the mouse. Genesis 32: 19–26.

55. Casanova E, Fehsenfeld S, Lemberger T, Shimshek D, Sprengel R, et al. (2002) ER-based double iCre fusion protein allows partial recombination in forebrain. Genesis 34: 208–214.

56. Soriano P (1999) Generalized lacZ expression with the ROSA26 Cre reporter strain. Nature Genet 21: 70–1.

57. Akagi K, Sandig V, Vooijs M, Van der Valk M, Giovannini M, et al. (1997) Cre-mediated somatic site-specific recombination in mice. Nucleic Acids Res 25: 1766–1773.

58. Seibler J, Zevnik B, Kuter-Luks B, Andreas S, Kern H, et al. (2003) Rapid generation of inducible mouse mutants. Nucleic Acids Res 31: e12.

59. Hayashi S, McMahon A (2002) Efficient recombination in diverse tissues by a tamoxifen-inducible form of Cre: a tool for temporally regulated gene activation/inactivation in the mouse. Dev Biol 244: 305–18.

60. Vooijs M, Jonkers J, Berns A (2001) A highly efficient ligand-regulated Cre recombinase mouse line shows that LoxP recombination is position dependent. EMBO Reports 2: 292–7.

61. Guo C, Yang W, Lobe C (2002) A Cre recombinase transgene with mosaic, widespread tamoxifen-inducible action. Genesis 32: 8–18.

62. Jullien N, Goddard I, Selmi-Ruby S, Fina J, Cremer H, et al. (2008) Use of ERT2-iCre-ERT2 for conditional transgenesis. Genesis 46: 193–199.

63. Bugeon L, Danou A, Carpentier D, Langridge P, Syed N, et al. (2003) Inducible gene silencing in podocytes: A new tool for studying glomerular function. Journal of the American Society of Nephrology 14: 786–791.

64. Tannour-Louet M, Porteu A, Vaulont S, Kahn A, Vasseur-Cognet M (2002) A tamoxifen-inducible chimeric Cre recombinase specifically effective in the fetal and adult mouse liver. Hepatology 35: 1072–1081.

65. Birbach A, Casanova E, Schmid JA (2009) A Probasin-MerCreMer BAC allows inducible recombination in the mouse prostate. Genesis 47: 757–764.

66. Park EJ, Sun X, Nichol P, Saijoh Y, Martin JF, et al. (2008) System for tamoxifen-inducible expression of cre-recombinase from the Foxa2 locus in mice. Developmental Dynamics 237: 447–453.

67. Kellendonk C, Tronche F, Monaghan A, Angrand P, Stewart F, et al. (1996) Regulation

of Cre recombinase activity by the synthetic steroid RU 486. Nucleic Acids Res 24: 1404–11.

68. Vegeto E, Allan GF, Schrader WT, Tsai MJ, McDonnell DP, et al. (1992) The mechanism of RU486 antagonism is dependent on the conformation of the carboxy-terminal tail of the human progesterone receptor. Cell 69: 703–713.

69. Wang Y, Xu J, Pierson T, O'Malley BW, Tsai SY (1997) Positive and negative regulation of gene expression in eukaryotic cells with an inducible transcriptional regulator. Gene Ther 4: 432–441.

70. Wunderlich F, Wildner H, Rajewsky K, Edenhofer F (2001) New variants of inducible Cre recombinase: a novel mutant of Cre-PR fusion protein exhibits enhanced sensitivity and an expanded range of inducibility. Nucleic Acids Res 29: E47.

71. Minamino T, Gaussin V, DeMayo FJ, Schneider MD (2001) Inducible gene targeting in postnatal myocardium by cardiac-specific expression of a hormone-activated Cre fusion protein. Circ. Res 88: 587–592.

72. Kitayama K, Abe M, Kakizaki T, Honma D, Natsume R, et al. (2001) Purkinje Cell-Specific and Inducible Gene Recombination System Generated from C57BL/6 Mouse ES Cells. Biochemical and Biophysical Research Communications 281: 1134–1140.

73. Tsujita M, Mori H, Watanabe M, Suzuki M, Miyazaki J, et al. (1999) Cerebellar Granule Cell-Specific and Inducible Expression of Cre Recombinase in the Mouse. J Neurosci 19: 10318–10323.

74. Zhou Z, Wang D, Wang X, Roop DR (2002) In utero activation of K5.CrePR1 induces gene deletion. Genesis 32: 191–192.

75. Morris RJ, Liu Y, Marles L, Yang Z, Trempus C, et al. (2004) Capturing and profiling adult hair follicle stem cells. Nature Biotech 22: 411–417.

76. Caulin C, Nguyen T, Longley MA, Zhou Z, Wang X, et al. (2004) Inducible Activation of Oncogenic K-ras Results in Tumor Formation in the Oral Cavity. Cancer Res 64: 5054–5058.

77. Berton TR, Wang XJ, Zhou Z, Kellendonk C, Schütz G, et al. (2000) Characterization of an inducible, epidermal-specific knockout system: differential expression of lacZ in different Cre reporter mouse strains. Genesis 26: 160–161.

78. Takeuchi T, Miyazaki T, Watanabe M, Mori H, Sakimura K, et al. (2005) Control of Synaptic Connection by Glutamate Receptor {delta}2 in the Adult Cerebellum. J Neurosci 25: 2146–2156.

79. Baurand A, Zelarayan L, Betney R, Gehrke C, Dunger S, et al. (2007) Beta-catenin downregulation is required for adaptive cardiac remodeling. Circ Res 100: 1353–1362.

80. Mishina M, Sakimura K (2007) Conditional gene targeting on the pure C57BL/6 genetic background. Neuroscience Research 58: 105–112.

81. Kishioka A, Fukushima F, Ito T, Kataoka H, Mori H, et al. (2009) A Novel Form of Memory for Auditory Fear Conditioning at a Low-Intensity Unconditioned Stimulus. PLoS ONE 4: e4157.

82. Rose MF, Ahmad KA, Thaller C, Zoghbi HY (2009) Excitatory neurons of the proprioceptive, interoceptive, and arousal hindbrain networks share a developmental requirement for Math1. Proc Natl Acad Sci USA 106: 22462–22467.

83. Brocard J, Feil R, Chambon P, Metzger D (1998) A chimeric Cre recombinase inducible by synthetic, but not by natural ligands of the glucocorticoid receptor. Nucleic Acids Res. 26: 4086–4090.

84. Kaczmarczyk S, Green J (2003) Induction of cre recombinase activity using modified androgen receptor ligand binding domains: a sensitive assay for ligand-receptor interactions. Nucleic Acids Research 31: e86.

85. Scherrer LC, Picard D, Massa E, Harmon JM, Simons SS, et al. (1993) Evidence that the hormone binding domain of steroid receptors confers hormonal control on chimeric proteins by determining their hormone-regulated binding to heat-shock protein 90. Biochemistry 32: 5381–5386.

86. Brocard J, Warot X, Wendling O, Messaddeq N, Vonesch J, et al. (1997) Spatio-temporally controlled site-specific somatic mutagenesis in the mouse. Proc Natl Acad Sci USA 94: 14559–14563.

87. Zervas M, Millet S, Ahn S, Joyner A (2004) Cell behaviors and genetic lineages of the mesencephalon and rhombomere 1. Neuron 43: 345–57.

88. Indra A, Warot X, Brocard J, Bornert J, Xiao J, et al. (1999) Temporally-controlled site-specific mutagenesis in the basal layer of the epidermis: comparison of the recombinase activity of the tamoxifen- inducible Cre-ER(T) and Cre-ER(T2) recombinases. Nucleic Acids Res. 27: 4324–4327.

89. King WJ, Greene GL (1984) Monoclonal antibodies localize oestrogen receptor in the nuclei of target cells. Nature 307: 745–747.

90. Welshons WV, Lieberman ME, Gorski J (1984) Nuclear localization of unoccupied oestrogen receptors. Nature 307: 747–749.
91. Wan Y, Coxe KK, Thackray VG, Housley PR, Nordeen SK (2001) Separable features of the ligand-binding domain determine the differential subcellular localization and ligand-binding specificity of glucocorticoid receptor and progesterone receptor. Mol. Endocrinol 15: 17–31.
92. Picard D, Kumar V, Chambon P, Yamamoto KR (1990) Signal transduction by steroid hormones: nuclear localization is differentially regulated in estrogen and glucocorticoid receptors. Cell Regul 1: 291–299.
93. Perrot-Applanat M, Logeat F, Groyer-Picard MT, Milgrom E (1985) Immunocytochemical study of mammalian progesterone receptor using monoclonal antibodies. Endocrinology 116: 1473–1484.
94. Jullien N, Sampieri F, Enjalbert A, Herman J (2003) Regulation of Cre recombinase by ligand-induced complementation of inactive fragments. Nucleic Acids Res 31: E131.
95. Hirrlinger J, Scheller A, Hirrlinger PG, Kellert B, Tang W, et al. (2009) Split-cre complementation indicates coincident activity of different genes in vivo. PLoS ONE 4: e4286.
96. Xu Y, Xu G, Liu B, Gu G (2007) Cre reconstitution allows for DNA recombination selectively in dual-marker-expressing cells in transgenic mice. Nucleic Acids Res 35: e126.
97. Hirrlinger J, Requardt RP, Winkler U, Wilhelm F, Schulze C, et al. (2009) Split-CreERT2: temporal control of DNA recombination mediated by split-Cre protein fragment complementation. PLoS ONE 4: e8354.
98. Awatramani R, Soriano P, Rodriguez C, Mai J, Dymecki S (2003) Cryptic boundaries in roof plate and choroid plexus identified by intersectional gene activation. Nature Genetics 35: 70–75.
99. Farago AF, Awatramani RB, Dymecki SM (2006) Assembly of the brainstem cochlear nuclear complex is revealed by intersectional and subtractive genetic fate maps. Neuron 50: 205–218.
100. Potter CJ, Tasic B, Russler EV, Liang L, Luo L (2010) The Q system: a repressible binary system for transgene expression, lineage tracing, and mosaic analysis. Cell 141: 536–548.
101. Luan H, Peabody NC, Vinson CR, White BH (2006) Refined spatial manipulation of neuronal function by combinatorial restriction of transgene expression. Neuron 52: 425–436.
102. Zhang S, Ma C, Chalfie M (2004) Combinatorial marking of cells and organelles with reconstituted fluorescent proteins. Cell 119: 137–144.
103. Ullmann A, Jacob F, Monod J (1967) Characterization by in vitro complementation of a peptide corresponding to an operator-proximal segment of the beta-galactosidase structural gene of Escherichia coli. J Mol Biol 24: 339–343.
104. Rossi F, Charlton C, Blau H (1997) Monitoring protein-protein interactions in intact eukaryotic cells by beta-galactosidase complementation. Proc Natl Acad Sci USA 94: 8405–10.
105. Johnsson N, Varshavsky A (1994) Split ubiquitin as a sensor of protein interactions in vivo. Proc Natl Acad Sci USA 91: 10340–4.
106. Pelletier J, Campbell-Valois F, Michnick S (1998) Oligomerization domain-directed reassembly of active dihydrofolate reductase from rationally designed fragments. Proc Natl Acad Sci USA 95: 12141–6.
107. Michnick SW, Ear PH, Manderson EN, Remy I, Stefan E (2007) Universal strategies in research and drug discovery based on protein-fragment complementation assays. Nature Rev Drug Discov 6: 569–582.
108. O'Shea EK, Klemm JD, Kim PS, Alber T (1991) X-ray structure of the GCN4 leucine zipper, a two-stranded, parallel coiled coil. Science 254: 539–544.
109. Rivera V, Clackson T, Natesan S, Pollock R, Amara J, et al. (1996) A humanized system for pharmacologic control of gene expression. Nature Med 2: 1028–32.
110. Siekierka J, Hung S, Poe M, Lin C, Sigal N (1989) A cytosolic binding protein for the immunosuppressant FK506 has peptidyl-prolyl isomerase activity but is distinct from cyclophilin. Nature 341: 755–7.
111. Chiu M, Katz H, Berlin V (1994) RAPT1, a mammalian homolog of yeast Tor, interacts with the FKBP12/rapamycin complex. Proc Natl Acad Sci USA 91: 12574–8.
112. Chen J, Zheng X, Brown E, Schreiber S (1995) Identification of an 11-kDa FKBP12-rapamycin-binding domain within the 289-kDa FKBP12-rapamycin-associated protein and characterization of a critical serine residue. Proc Natl Acad Sci USA 92: 4947–51.
113. Kennedy MJ, Hughes RM, Peteya LA, Schwartz JW, Ehlers MD, et al. (2010) Rapid blue-light-mediated induction of protein interactions in living cells. Nature Meth 7: 973–975.

114. Casanova E, Lemberger T, Fehsenfeld S, Mantamadiotis T, Schutz G (2003) alpha Complementation in the Cre recombinase enzyme. Genesis 37: 25–9.

115. Jullien N, Goddard I, Selmi-Ruby S, Fina J, Cremer H, et al. (2007) Conditional Transgenesis Using Dimerizable Cre (DiCre). PLoS ONE 2: e1355.

116. Yazawa S, Obata K, Iio A, Koide M, Yokota M, et al. (2003) Gene expression of FK506-binding proteins 12.6 and 12 during chicken development. Comp Biochem Physiol A Mol Integr Physiol 136: 391–9.

117. Beckervordersandforth R, Tripathi P, Ninkovic J, Bayam E, Lepier A, et al. (2010) In Vivo Fate Mapping and Expression Analysis Reveals Molecular Hallmarks of Prospectively Isolated Adult Neural Stem Cells. Cell Stem Cell 7: 744–758.

118. Ji B, Song J, Tsou L, Bi Y, Gaiser S, et al. (2008) Robust acinar cell transgene expression of CreErT via BAC recombineering. Genesis 46: 390–395.

119. Casper KB, Jones K, McCarthy KD (2007) Characterization of astrocyte-specific conditional knockouts. Genesis 45: 292–299.

120. Leone D, Genoud S, Atanasoski S, Grausenburger R, Berger P, et al. (2003) Tamoxifen-inducible glia-specific Cre mice for somatic mutagenesis in oligodendrocytes and Schwann cells. Molecular and Cellular Neuroscience 22: 430–440.

121. Imayoshi I, Ohtsuka T, Metzger D, Chambon P, Kageyama R (2006) Temporal regulation of Cre recombinase activity in neural stem cells. Genesis 44: 233–238.

122. Kemp R, Ireland H, Clayton E, Houghton C, Howard L, et al. (2004) Elimination of background recombination: somatic induction of Cre by combined transcriptional regulation and hormone binding affinity. Nucleic Acids Res 32: e92.

123. Madisen L, Zwingman TA, Sunkin SM, Oh SW, Zariwala HA, et al. (2010) A robust and high-throughput Cre reporting and characterization system for the whole mouse brain. Nature Neurosci 13: 133–140.

124. Badea TC, Hua ZL, Smallwood PM, Williams J, Rotolo T, et al. (2009) New mouse lines for the analysis of neuronal morphology using CreER(T)/loxP-directed sparse labeling. PLoS ONE 4: e7859.

125. Furr BJ, Jordan VC (1984) The pharmacology and clinical uses of tamoxifen. Pharmacol. Ther 25: 127–205.

126. Jordan VC, Murphy CS (1990) Endocrine Pharmacology of Antiestrogens as Antitumor Agents. Endocr Rev 11: 578–610.

127. Ellisor D, Zervas M (2010) Tamoxifen dose response and conditional cell marking: Is there control? Molecular and Cellular Neuroscience 45: 132–138.

128. Kühbandner S, Brummer S, Metzger D, Chambon P, Hofmann F, et al. (2000) Temporally controlled somatic mutagenesis in smooth muscle. Genesis 28: 15–22.

129. Andersson KB, Winer LH, Mørk HK, Molkentin JD, Jaisser F (2009) Tamoxifen administration routes and dosage for inducible Cre-mediated gene disruption in mouse hearts. Transgenic Res 19: 715–725.

130. Kiermayer C, Conrad M, Schneider M, Schmidt J, Brielmeier M (2007) Optimization of spatiotemporal gene inactivation in mouse heart by oral application of tamoxifen citrate. Genesis 45: 11–16.

131. Robinson SP, Langan-Fahey SM, Johnson DA, Jordan VC (1991) Metabolites, pharmacodynamics, and pharmacokinetics of tamoxifen in rats and mice compared to the breast cancer patient. Drug Metab. Dispos 19: 36–43.

132. Sadek S, Bell SC (1996) The effects of the antihormones RU486 and tamoxifen on feto-placental development and placental bed vascularisation in the rat: a model for intrauterine fetal growth retardation. Br J Obstet Gynaecol 103: 630–641.

133. Vancutsem P, Roessler M (1997) Neonatal treatment with tamoxifen causes immediate alterations of the sexually dimorphic nucleus of the preoptic area and medial preoptic area in male rats. Teratology 56: 220–8.

134. Csaba G, Karabelyos C (2001) The effect of a single neonatal treatment (hormonal imprinting) with the antihormones, tamoxifen and mifepristone on the sexual behavior of adult rats. Pharmacol Res 43: 531–4.

135. Danielian PS, Muccino D, Rowitch DH, Michael SK, McMahon AP (1998) Modification of gene activity in mouse embryos in utero by a tamoxifen-inducible form of Cre recombinase. Curr Biol 8: 1323–1326.

136. Mori T, Tanaka K, Buffo A, Wurst W, Kühn R, et al. (2006) Inducible gene deletion in astroglia and radial glia - A valuable tool for functional and lineage analysis. Glia 54: 21–34.

137. Erdmann G, Schutz G, Berger S (2007) Inducible gene inactivation in neurons of the adult mouse forebrain. BMC Neuroscience 8: 63.

138. Weber P, Schuler M, Gerard C, Mark M, Metzger D, et al. (2003) Temporally controlled site-specific mutagenesis in the germ cell lineage of the mouse testis. Biology of Reproduction 68: 553–559.

139. Santagati F, Minoux M, Ren S, Rijli FM (2005) Temporal requirement of Hoxa2 in cranial neural crest skeletal morphogenesis. Development 132: 4927–4936.
140. Naiche LA, Papaioannou VE (2007) Tbx4 is not required for hindlimb identity or post-bud hindlimb outgrowth. Development 134: 93–103.
141. Vasioukhin V, Degenstein L, Wise B, Fuchs E (1999) The magical touch: genome targeting in epidermal stem cells induced by tamoxifen application to mouse skin. Proc Natl Acad Sci USA 96: 8551–8556.
142. Brake RL, Simmons PJ, Begley CG (2004) Cross-contamination with tamoxifen induces transgene expression in non-exposed inducible transgenic mice. Genetics and Molecular Research 3: 456–462.
143. Gruber M, Hu C, Johnson RS, Brown EJ, Keith B, et al. (2007) Acute postnatal ablation of Hif-2α results in anemia. Proc Natl Acad Sci USA 104: 2301–2306.
144. Gavériaux-Ruff C, Kieffer BL (2007) Conditional gene targeting in the mouse nervous system: Insights into brain function and diseases. Pharmacol. Ther 113: 619–634.
145. Srinivas S, Watanabe T, Lin C, William C, Tanabe Y, et al. (2001) Cre reporter strains produced by targeted insertion of EYFP and ECFP into the ROSA26 locus. BMC Developmental Biology 1: 4.
146. Mao X, Fujiwara Y, Chapdelaine A, Yang H, Orkin SH (2001) Activation of EGFP expression by Cre-mediated excision in a new ROSA26 reporter mouse strain. Blood 97: 324–326.
147. Novak A, Guo C, Yang W, Nagy A, Lobe C (2000) Z/EG, a double reporter mouse line that expresses enhanced green fluorescent protein upon Cre-mediated excision. Genesis 28: 147–55.
148. Lobe CG, Koop KE, Kreppner W, Lomeli H, Gertsenstein M, et al. (1999) Z/AP, a double reporter for cre-mediated recombination. Dev Biol 208: 281–292.
149. Hebert J, McConnell S (2000) Targeting of cre to the Foxg1 (BF-1) locus mediates loxP recombination in the telencephalon and other developing head structures. Dev Biol 222: 296–306.
150. Ellisor D, Koveal D, Hagan N, Brown A, Zervas M (2009) Comparative analysis of conditional reporter alleles in the developing embryo and embryonic nervous system. Gene Expr Patterns 9: 475–489.

Chapter 2

Genetic Manipulations Using Cre and Mutant *LoxP* Sites

Kimi Araki and Ken-ichi Yamamura

Abstract

The bacteriophage P1-derived Cre/*lox* recombination system has been extensively used to engineer the genome of cultured cells and experimental animals. Cre recombinase recognizes the *lox*P site, which is composed of two 13-bp inverted repeats and an 8-bp spacer region, and mediates both intramolecular (excisive) and intermolecular (integrative) recombination between two *lox*P sites. The excision reaction is efficient and can be used in conditional knockout strategies. On the other hand, integrative recombination is inefficient because the integrated DNA retains *lox*P sites at both ends and is easily excised again if the Cre recombinase is still present. However, integrative recombination is expected to be a powerful tool for genome engineering in mouse embryonic stem (ES) cells because it allows precise and repeated knock-in of any DNA into *lox* sites placed in the genome. To promote integrative recombination, two kinds of mutant *lox* systems have been developed and successfully used in ES cells to produce exchangeable (multipurpose) alleles. In this chapter, we describe a Cre/mutant *lox* system for integrative recombination, and we present an application of this system to gene targeting. By incorporating mutant *lox* sites into gene targeting vectors, we can first produce a null allele. Subsequently, any gene of interest, including the Cre recombinase gene itself, fluorescent genes, luciferase genes, mutated cDNAs, and human cDNAs, can be inserted and expressed under the endogenous promoter of the targeted gene. By combining other recombination systems, such as Flp/*FRT*, we can also convert null alleles into conditional alleles.

Key words: Site-specific recombination, Cre, Mutant *lox*, Exchangeable gene targeting, Flp/*FRT*, Embryonic stem cell

1. Introduction: Cre/*Lox* Systems

1.1. Brief Summary of the Cre-LoxP System

The Cre/*lox* recombination system is derived from bacteriophage P1 and arose to prominence as the most powerful tool for genome engineering (1, 2). Cre recombinase catalyzes reciprocal site-specific recombination between two specific 34-bp sites, called *lox*P sites. Each *lox*P site is composed of two 13-bp inverted repeats that serve as Cre binding sites, and an 8-bp central spacer region that participates in strand exchange during recombination (3, 4).

Alexei Morozov (ed.), *Controlled Genetic Manipulations*, Neuromethods, vol. 65,
DOI 10.1007/978-1-61779-533-6_2, © Springer Science+Business Media, LLC 2012

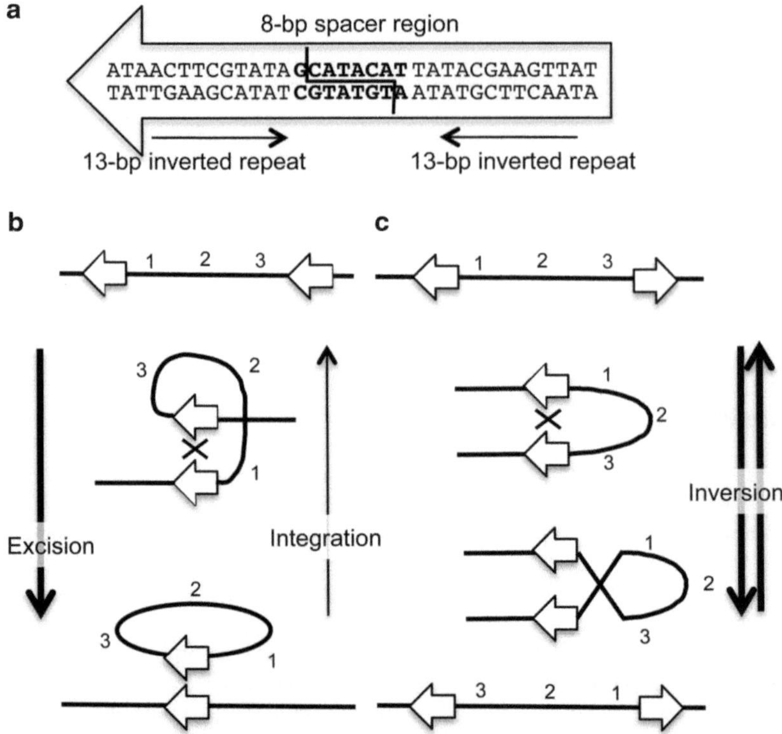

Fig. 1. Cre/*lox* recombination system. (**a**) Wild-type *lox*P sequence. The 34-bp *lox*P site is composed of two 13-bp inverted repeat regions and an 8-bp asymmetric spacer region. The asymmetric spacer region dictates the orientation of the *lox*P sequence, as indicated by the *large arrow*. (**b**) Recombination between two *lox*P sites with the same orientation. Excisive intermolecular recombination occurs efficiently; however, integrative intermolecular recombination rarely occurs due to reinsertion. (**c**) Recombination between two *lox*P sites in the inverse orientation. In this case, recombination results in inversion of floxed DNA.

The sequence of the spacer region is asymmetric and confers directionality to the *lox*P site (Fig. 1a) (5). Depending on the orientation of the *lox*P sites with respect to one another, the recombination can result in excision, inversion, or integration (Fig. 1b, c). Integrative recombination is very powerful for genome engineering in mouse embryonic stem (ES) cells, because it allows precise and repeated knock-in of any exogenous DNA into chromosomally located *lox* sites that have been introduced by gene targeting. However, integrative recombination between wild-type *lox*P sites is inefficient due to re-excision through intramolecular recombination in the presence of Cre recombinase (6). Studies of mutated *lox*P sites have revealed that two classes of mutations can promote Cre-mediated integration or replacement.

1.2. Left Element/Right Element (LE/RE) Mutant Strategy

One class of mutant *lox*, originally reported by Albert et al. (7), is the left element/right element (LE/RE) mutant strategy that uses an LE mutant *lox* site carrying mutations in the left-inverted repeat region and RE mutant *lox* site carrying mutations in the right-inverted repeat region. Recombination between an LE mutant *lox* site and an RE mutant *lox* site results in the production of a double mutant *lox* site having mutations in both ends and a wild-type *lox*P site (Fig. 2a). Since the binding affinity of the double mutant *lox* site for Cre recombinase is severely decreased, the integrated DNA is stably retained.

Albert et al. introduced five nucleotide changes into the left 13-bp element (*lox*71) or into the right 13-bp element (*lox*66) (Table 1) and succeeded in producing *lox*-site-specific integration

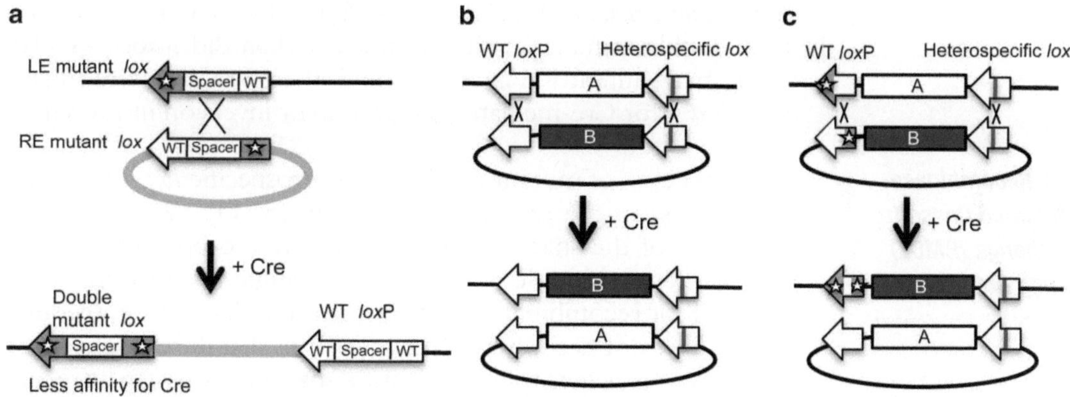

Fig. 2. Cre/mutant *lox* recombination system. (**a**) Integrative recombination using LE/RE mutant *lox* sites. Recombination between LE mutant *lox* and RE mutant *lox* sites results in the production of a double mutant *lox* site having mutations in both ends and a wild-type *lox*P site. *Asterisk* represents mutation. (**b**) RMCE using heterospecific *lox* sites. The heterospecific *lox* site is indicated by an *arrow* with a *center line* representing a mutation in the spacer region. A chromosomal cassette (*open box*) flanked by two heterospecific *lox* sites is replaced through Cre-mediated recombination with another cassette (*solid box*) located on a plasmid. (**c**) Combinational RMCE with LE/RE mutant *lox* and heterospecific *lox* sites. Although the recombination efficiency is almost identical to that of *lox*P and heterospecific *lox*, the replaced cassette is quite stable since it is never recombined by Cre.

Table 1
Mutant *lox* sites

LE mutant *lox* site	Sequence of the left-inverted repeat region	RE mutant *lox* site	Sequence of the right-inverted repeat region
*lox*71	TACCGTTCGTATA	*lox*66	TATACGAACGGTA
*lox*JT15	AATTATTCGTATA	*lox*JTZ2	TATACGAATACCT
*lox*JT12	AGTTGTTCGTATA	*lox*JTZ17	TATAGCAATTTAT
*lox*JT510	TAACGTTCGTATA	*lox*KR3	TATACCTTGTTAT

using tobacco cells (7). We used ES cells and assessed the frequency of *lox*-site-specific integration over random integration. The efficiency of *lox*-site-specific integration with LE/RE mutant *lox* sites was 2–16%, which was much higher than the 0.2% efficiency of *lox*P-*lox*P recombination (8). Recently, Thomson et al. performed mutational analysis of LE/RE mutant *lox* sites using *Escherichia coli* and identified a novel LE/RE mutant *lox* pair, *lox*JT15 and *lox*JTZ17 (Table 1), that showed approximately 1,500-fold higher integration rates than *lox*71and *lox*66 (9). The combination of *lox*JTZ17 and *lox*71 produced a frequency that was tenfold higher than *lox*71 and *lox*66. To confirm whether such large differences in frequency are also observed in ES cells, we compared six RE mutant *lox* sites, including *lox*JTZ17, focusing on their recombination efficiency with *lox*71. Unlike in *E. coli*, all of the RE mutant *lox* sites showed similar recombination efficiency in ES cells. However, two RE mutant *lox* sites, *lox*JTZ17 and *lox*KR3, produced more stable (inactive) double mutant *lox* sites with *lox*71 than did *lox*66/71 (10). These two mutant RE *lox* sites would, therefore, be more suitable than *lox*66 for Cre-mediated integration or inversion in ES cells.

1.3. Recombinase-Mediated Cassette Exchange (RMCE) Using Heterospecific Lox Sites

A second class of mutant *lox* site is a heterospecific *lox* site carrying mutation(s) in the central spacer region (5, 11, 12). Cre cleaves the DNA of the spacer region to generate a 6-bp staggered cut, and sequence homology in the 6-bp single-stranded region is essential for recombination between *lox* sites. Therefore, recombination does not occur between two *lox* sites that differ in the spacer region, whereas *lox* sites with identical spacer regions can be recombined efficiently. Recombination using heterospecific *lox* sites is termed recombinase-mediated cassette exchange (RMCE), in which a chromosomal cassette flanked by two heterospecific *lox* sites is exchanged for another cassette located on a plasmid by Cre-mediated recombination (Fig. 2b) (13). To date, *lox*511 (14), *lox*2272 (15), and *lox*5171 (16) (Table 2) have been successfully used for RMCE in ES cells.

We compared the recombination efficiencies of insertion by the LE/RE mutant *lox* strategy and RMCE in ES cells and found that RMCE was two- to threefold more efficient (17). In a comparison

Table 2
Heterospecific *lox* sites

Heterospecific *lox* site	Sequence of spacer region
*lox*511	GTATACAT
*lox*2272	GGATACTT
*lox*5171	GTACACAT

between *lox*2272 and *lox*511 sites, the former showed approximately 1.5-fold higher efficiency, probably due to unfavorable recombination between *lox*P and *lox*511 (17).

1.4. Combination of LE/RE Mutant Lox Sites and Heterospecific Lox Sites

LE/RE mutant *lox* and heterospecific *lox* sites can be used simultaneously in RMCE, as shown in Fig. 2c. The recombination efficiency of this combinational RMCE is almost the same as that of usual RMCE with wild-type *lox*P and heterospecific *lox* sites. The merit of combinational RMCE is high stability of the recombined product even in the presence of Cre protein. Taking advantage of this stability, we successfully integrated the *cre* gene under the control of a strong promoter through combinational RMCE using *lox*71-*lox*2272 and *lox*66-*lox*2272 cassettes in ES cells (17). The inserted *cre* gene was strongly expressed and stably maintained. Cre-expressing mouse lines were established from the ES clones, and the inserted *cre* gene was stably maintained in sequential generations. Thus, the *cre* knock-in system using combinational RMCE should be useful for the production of various Cre-driver mice.

2. Vector Designs for Exchangeable Gene Targeting

By incorporating mutant *lox* sites into gene targeting vectors, the selection marker gene retained in the targeted locus can be exchanged with any other gene of interest, for example, the *cre* gene, fluorescent protein genes, luciferase genes, mutated cDNAs, or cDNAs of other species. Thus, a targeted null allele can be converted into various knock-in alleles. In this section, several designs of such exchangeable targeting vectors and replacement patterns using Cre-mediated recombination are presented.

2.1. Disruption of ATG-Containing Exons

The purpose of this targeting vector is to insert and express a cDNA (containing an open reading frame) under the control of the promoter of a targeted gene. Therefore, the ATG codon of the targeted gene is required to be disrupted and replaced with an LE mutant *lox* site. The backbone of the targeting vector is shown in Fig. 3a. Using a *lox*P-polyadenylation (pA) signal-heterospecific *lox* cassette, two patterns of replacement are possible, as described in the following sections. The 5′ arm of the targeting vector should extend until just before the ATG codon of the target gene (Fig. 3b). Homologous recombination (Fig. 3c) will result in a null allele due to deletion of the start codon. The selection marker can be removed by transient expression of Cre (18) or by mating with Cre-deleter mice (19) (Fig. 3d). Using the original targeted ES clone, the allele can be further modified in two ways.

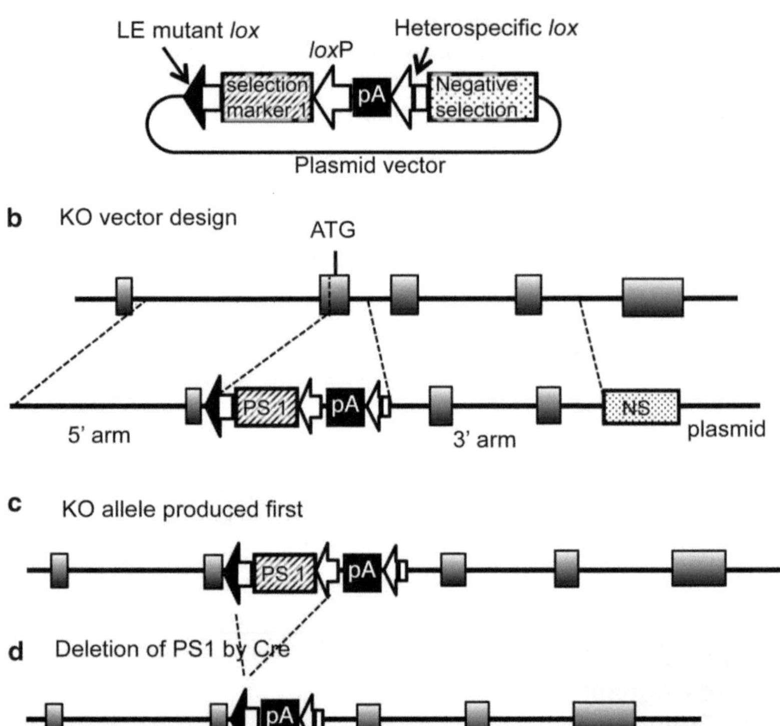

Fig. 3. ATG-containing exon knockout. (**a**) Backbone of KO vector. The transcriptional directions of positive selection marker gene 1 and the negative selection marker gene are not fixed (either direction is acceptable). However, the direction of the polyadenylation signal (pA) should be the same as the direction of the targeted gene. (**b**) Design of KO vector. The positive selection marker cassette (PS1) should be placed just before the ATG codon of the target gene. (**c**) Targeted allele. Since the endogenous ATG codon is deleted, this allele should be null. (**d**) PS1-deleted allele through Cre-mediated recombination.

2.2. Insertion of cDNA

A gene of interest (GoI) can be inserted into the ATG position of a targeted gene (Fig. 4), and the GoI is then driven by the endogenous promoter of the targeted gene. Design of such a replacement vector is shown in Fig. 4a. Importantly, the selection marker in this vector should not contain a pA signal to provide further selection of the recombination event, as described below. The replacement and Cre expression vectors are coelectroporated into the targeted ES clone in their circular forms. The *cre* gene is transiently expressed and mediates recombination. Since the replacement plasmid and the targeting vector in the ES genome both carry two *lox* sites with the same spacer region (RE or LE mutant *lox* and *lox*P), it is expected that intramolecular recombination between the two *lox* sites should occur after coelectroporation, resulting in the production of two intermediate molecules, as shown in Fig. 4b. Integrative recombination then occurs between LE mutant *lox* sites in the genome and RE mutant *lox* sites in the intermediate molecule. ES cells in which

a Replacement vector

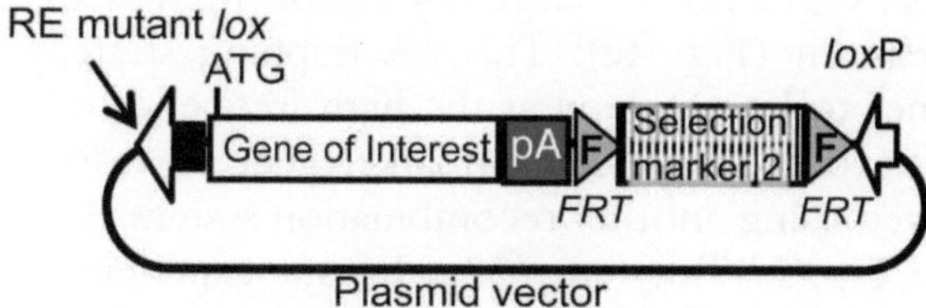

b 2-step recombination by Cre

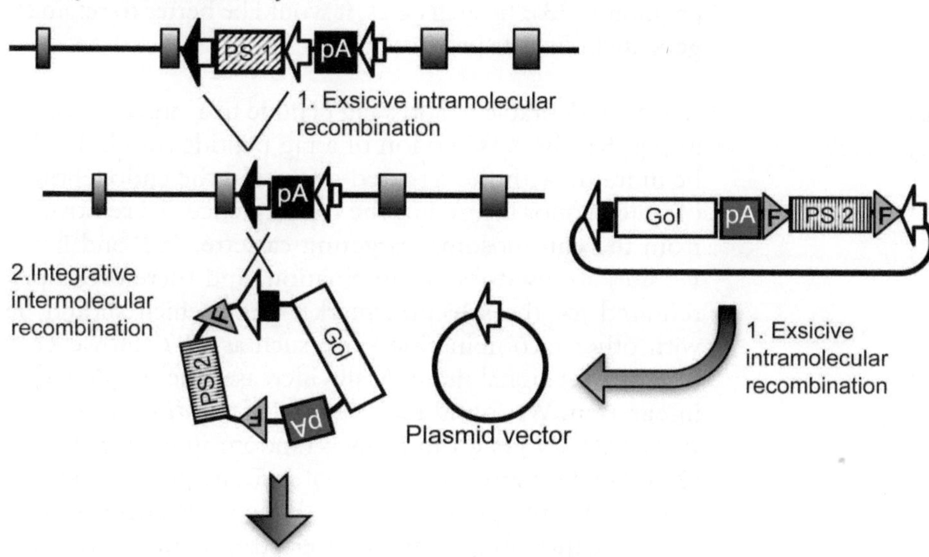

c Replaced allele with gene of interest (GoI) by Cre

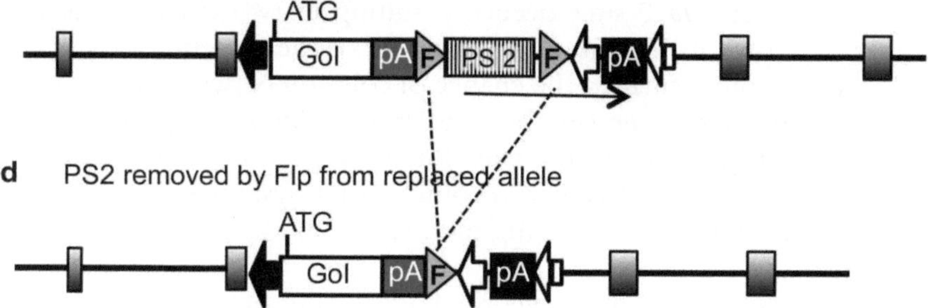

Fig. 4. Knock-in of cDNA through Cre-mediated recombination. (**a**) Replacement vector. The positive selection marker gene 2 (PS2) cassette should not contain a pA signal. (**b**) Intermediate molecules produced through intramolecular recombination. The PS1 cassette is removed from the targeted allele, and the knock-in vector is divided into two circular molecules. Integrative recombination then occurs between LE and RE mutant *lox* sites. GoI, gene of interest. (**c**) Replaced allele. Only upon Cre-mediated site-specific integration does the PS2 gene fuse to the pA signal on the targeting vector, thereby making the cells drug-resistant. (**d**) PS2-deleted allele through Flp-mediated recombination.

the replacement cassette is integrated into the LE mutant *lox* site are selected for with an appropriate drug. Because the selection marker gene in the replacement vector does not contain a pA signal, most of random integrants should be drug sensitive. Only upon

Cre-mediated targeted integration does the selection marker gene fuse to the pA signal on the targeting vector, thereby making the cells drug-resistant (Fig. 4c). This pA trapping strategy enables replaced clones to be obtained at the high frequency of 60–100% (20, 21). If needed, the selection marker gene used for insertion can be removed using another recombination system (Fig. 4d), for example, the Flp/*FRT* system (22). In our experience, inserted genes are expressed efficiently in the presence of the selectable marker gene driven by the mouse *phosphoglycerate kinase-1 (Pgk)* promoter (23). In such cases, it would be better to retain the marker gene including its promoter.

2.3. Insertion of a Tag Peptide

It is often desirable to add a tag peptide to a targeted gene. As shown in Fig. 5a, the ATG codon of a tag peptide should be designed to be in frame with the targeted gene, and the endogenous sequence of splice donor is fused to the tag sequence. To remove a pA signal from the chromosomal targeting cassette, *lox*P and heterospecific *lox* sites are used for recombination, and therefore, a pA signal is required for the selection marker gene which should be flanked with other recombination sites, such as *FRT*. However, the addition of a pA signal dramatically increases the frequency of random integration. We found that the use of the *diphtheria toxin A fragment* (*DT-A*) gene can reduce random integrants by almost half (24). Furthermore, an additional *lox*P site placed between the *DT-A* gene and the plasmid vector sequence is effective in increasing recombination frequency. After electroporation with a Cre expression plasmid, intramolecular recombination between RE mutant *lox* and *lox*P sites occurs, resulting in two circular molecules, as shown in Fig. 5b. Since the size of the targeting DNA molecule becomes smaller, the chance of collision between chromosomal *lox* and plasmid *lox* sites becomes higher. After site-specific recombination (Fig. 5c, d), the selection marker has to be removed, either in ES cells by another recombination system or by mating with recombinase-expressing mice (Fig. 5e).

2.4. Conditional Knockout

Exon replacement, via the engineering of conditional alleles, is a highly effective strategy for the production of mice that model genetic disease. A conditional allele can be produced by flanking a exon of interest with *lox* sites (floxing). The conditional allele can then be exchanged for a mutated exon using a targeting vector, as shown in Fig. 7a. In the exon-exchange targeting vector (Fig. 6a, b), the replacing exon is flanked by *lox*P sites, a heterospecific *lox* site is placed between the coding region of the positive selection marker gene and the pA signal, and the positive selection marker cassette is flanked by other recombination sites such as *FRT*. After homologous recombination (Fig. 6c), the selection marker cassette has to be removed through excising recombination by the other recombination system to produce the conditional allele (Fig. 6d).

a Replacement vector

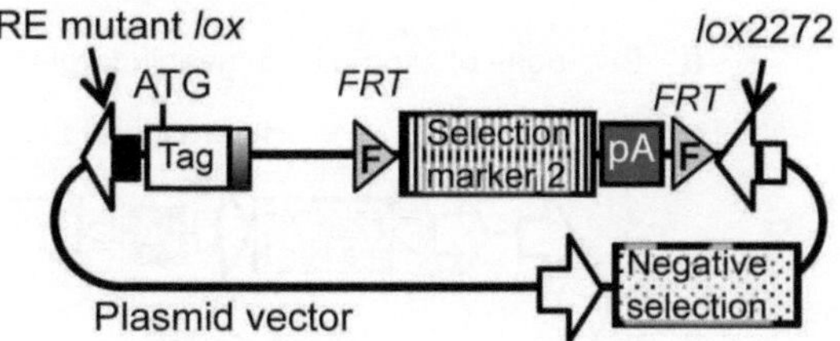

b Intermediate molecule produced by intramolecular recombination

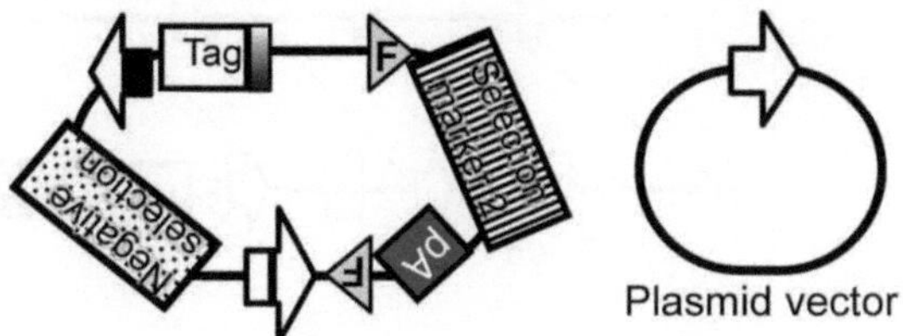

c Replacement by Cre

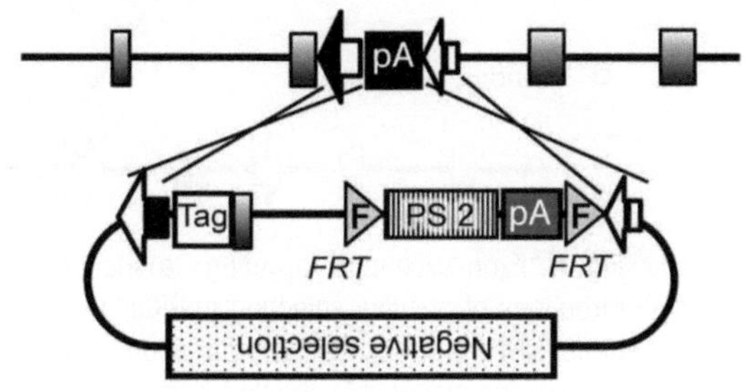

d Replaced allele

e PS2 removed by Flp from replaced allele

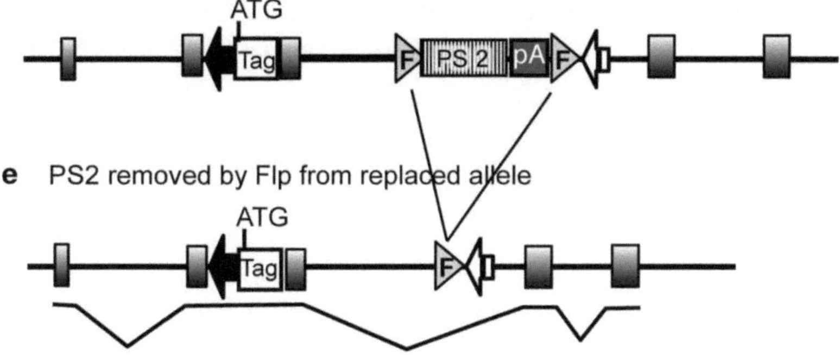

Fig. 5. Production of a tag-fused gene through Cre-mediated recombination. (**a**) Replacement vector. A DNA fragment between RE mutant *lox* and heterospecific *lox* sites is integrated into chromosomal lox sites in the targeting vector. In this case, positive selection marker gene 2 should have a pA signal. To reduce random integrants, addition of a negative selection marker gene and a *lox*P site is effective. (**b**) Intermediate molecules produced through intramolecular recombination. The knock-in vector is divided into two circular molecules. (**c**) Cassette-exchange recombination occurs between LE/RE mutant *lox* sites and heterospecific *lox* sites. PS2, positive selection marker gene 2. (**d**) Replaced allele. Since the PS2 gene exists in the intron, transcription from the endogenous promoter may be disturbed. (**e**) PS2-deleted allele through Flp-mediated recombination. After removal of the PS2 cassette, a fused transcript can be produced.

a Backbone of exon-exchangeable targeting vector

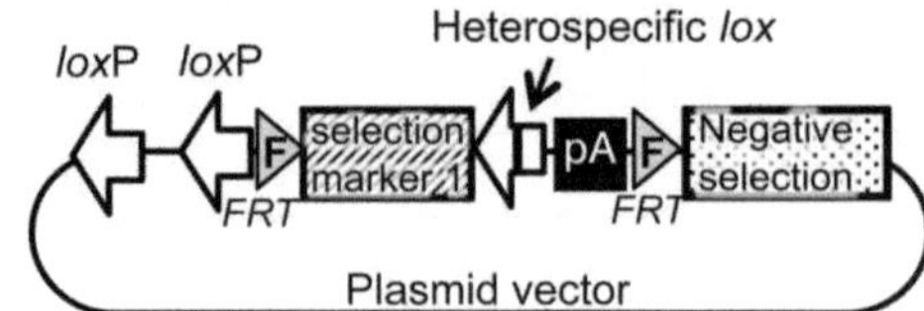

b Targeting vector design

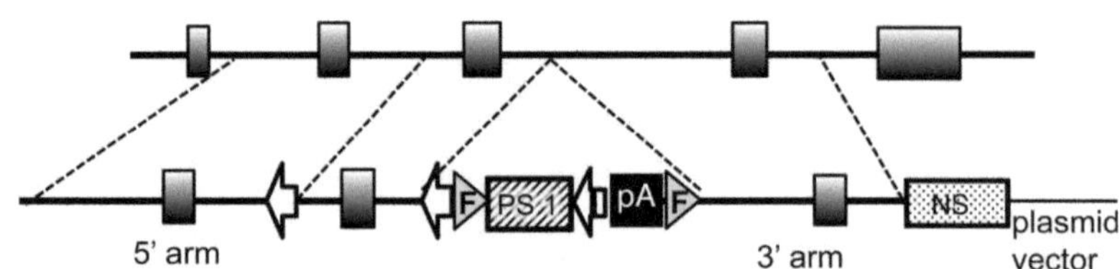

c Targeted allele produced first

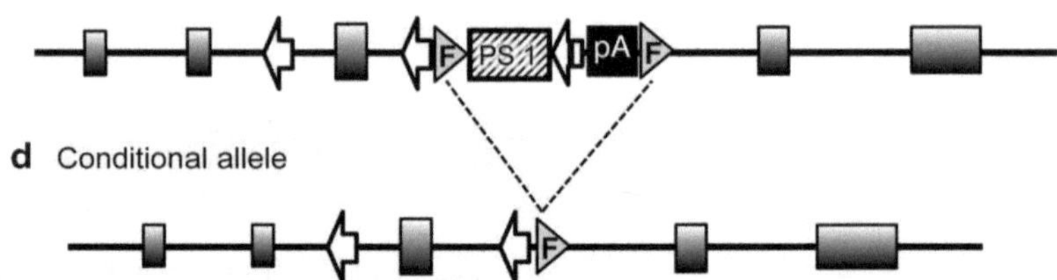

Fig. 6. Exon-exchange targeting. (**a**) Backbone of targeting vector. The transcriptional directions of positive selection marker gene 1 (PS1) and the negative selection marker gene (NS) are not fixed (either direction is acceptable). However, the direction of the polyadenylation signal (pA) should be the same as the direction of targeted gene. (**b**) Design of KO vector. The exon to be replaced in the next step is placed between two *lox*P sites. (**c**) Targeted allele produced first. Since the PS1 gene exists in the intron, transcription of the targeted gene may be disturbed. (**d**) Conditional allele produced by Flp-mediated recombination. After removal of the PS2 cassette, only a *lox*P site and *FRT* site remain in the intron, which should leave a normally transcribed gene.

2.5. Exchange of an Exon

Design of a replacement vector is shown in Fig. 7a. The selection marker in this vector should not contain a pA signal. The negative selection marker gene and the following *lox*P site are not essential, because the recombination efficiency without these elements is sufficiently high due to pA trapping. After coelectroporation of the replacement and Cre expression vectors, two-step site-specific recombination should occur, intramolecular excising recombination followed by cassette replacement recombination (Fig. 7b). Only upon site-specific recombination does the selection marker gene fuse to the pA signal on the targeting vector, thereby making the cells drug-resistant (Fig. 7c). After removal of the selection marker gene with the other recombination system, the mutated exon is incorporated to the mRNA of the targeted gene.

a Replacement vector

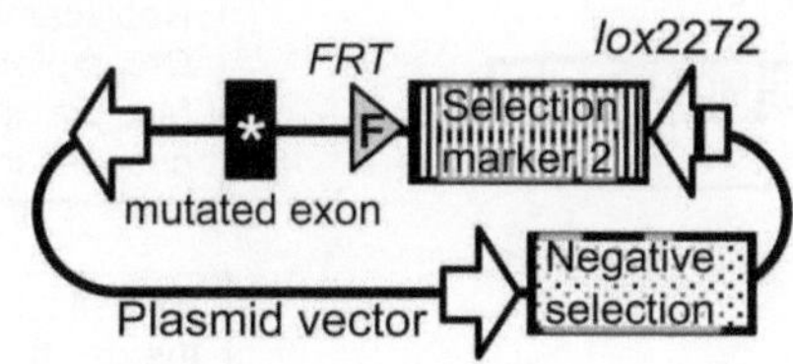

b Recombination by Cre

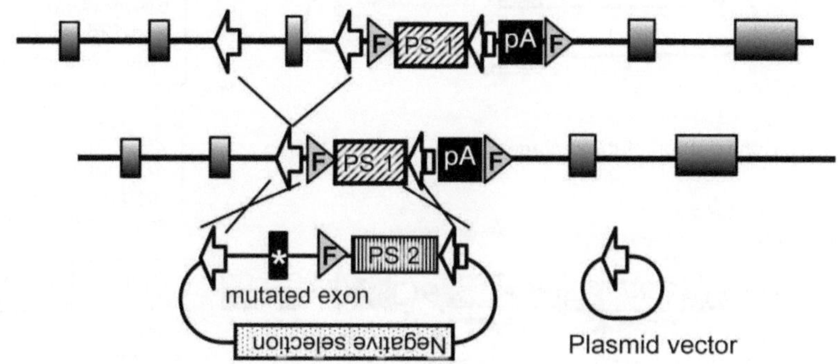

c Allele replaced by Cre

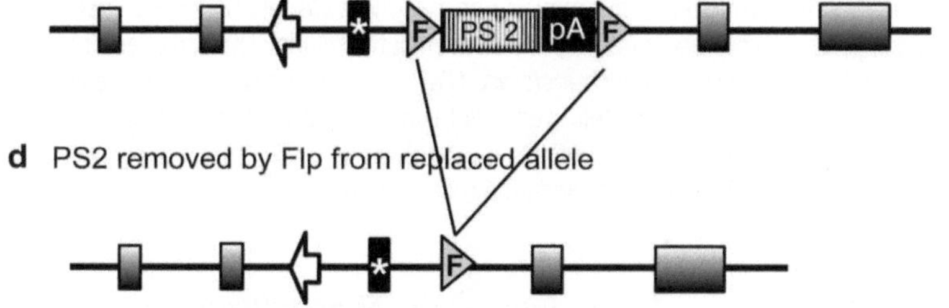

Fig. 7. Exchange of an exon through Cre-mediated recombination. (**a**) Replacement vector. Positive selection marker gene 2 (PS2) should not have a pA signal. A negative selection marker gene and an additional *lox*P site are optional. (**b**) Intermediate molecules produced through intramolecular recombination. First, the floxed exon is removed and the replacement vector is divided into two circular molecules, and then intermolecular cassette-exchange recombination occurs. (**c**) Replaced allele. Since the PS2 gene exists in the intron, transcription from the endogenous promoter may be disturbed. (**e**) PS2-deleted allele through Flp-mediated recombination. After removal of the PS2 cassette, the allele containing the mutated exon is produced.

3. Example of Cre-Mediated Recombination in ES Cells

3.1. Vector Design and Protocol

In this section, we will illustrate how cassette exchange is performed in ES cells by describing a specific example. Figure 8a shows our exchangeable targeting vector containing a *lox*71-*Pgk*-neo^{R}-*lox*P-pA-*lox*2272 cassette (25), and Fig. 8b shows a replacement vector for the insertion of a cDNA fragment. We usually use

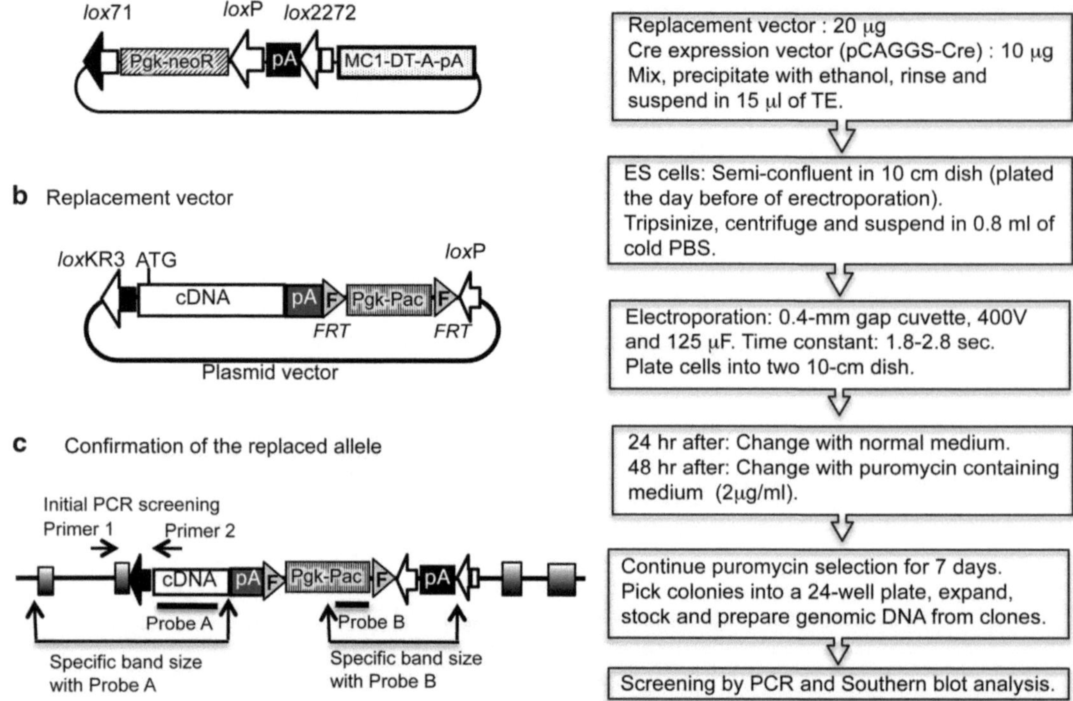

Fig. 8. Example of Cre-mediated recombination. (**a**) Exchangeable KO vector contains a *lox*71-Pgk-neo^R-*lox*P-pA-*lox*2272 cassette. Pgk, the promoter of the mouse *phosphoglycerate kinase-1* gene; neo^R, the open reading frame (ORF) of the *neomycin phosphotransferase* gene; pA, polyadenylation signal. (**b**) Replacement vector to insert a cDNA. The *Pgk* promoter and the *puromycin N-acetyltransferase* (*Pac*) gene are used as a positive selection marker. The *Pac* gene does not have a pA signal. (**c**) Replaced allele. For initial screening, a PCR assay that detects the 5′ junction is used. Then, both junctions should be confirmed by Southern blot analysis. It is also important to confirm that no random integration occurs. (**d**) Electroporation protocol for Cre-mediated replacement.

the *puromycin N-acetyltransferase* (*Pac*) gene as a positive selection marker gene in the replacement vector (26). The *Pac* gene has no pA signal to select recombined clones.

Figure 8d depicts the protocol for coelectroporation and selection. Twenty micrograms of replacement plasmid and 10 μg of pCAGGS-Cre (18), a vector producing strong expression of the *cre* gene, are used in their circular forms. The ES cells are harvested from a 10-cm dish and electroporated with plasmids at 400 V and 125 μF. Selection for 7 days with puromycin at 2 μg/ml is started 48 h after electroporation. Colonies become visible under the microscope from 3 days after starting selection. Since Cre-mediated site-specific recombination fuses the pA signal to the *Pac* gene, we can predict the degree of replacement success from the number of colonies. If the number of colonies is over 50, correctly targeted clones will be successfully obtained, but if the number is under 20, the colonies might all be random integrants.

Genomic DNA from clones has to be examined to determine whether the replacement cassette is correctly inserted. For initial screening, a PCR assay that detects the 5′ junction, as shown in Fig. 8c, is recommended. Next, both junctions should be confirmed through Southern blot analysis using restriction enzyme(s) that digest inside and outside of the inserted cassette to fragments of specific sizes (Fig. 8c). Southern blot analysis is essential to confirm correct targeting with no random integration event.

3.2. Production of Humanized Mice by the Insertion of Human cDNA

Humanized mice can be produced by introducing a homologous human gene or cDNA into the endogenous mouse gene locus. With such an approach, the inserted human gene/cDNA is expressed under the mouse promoter with similar temporal and spatial expression to that of the endogenous mouse gene. Such humanized mice are excellent animal models for human diseases and for genome-based drug discovery.

We conducted humanization of the mouse *transthyretin* (*Ttr*) gene (25). TTR protein is synthesized mainly in the liver, choroid plexus, and retinal pigment cells, and is secreted into plasma, cerebrospinal fluid, and vitreous body, respectively (27–29). It circulates as a tetramer and serves as a transporter for thyroxine and the retinol-binding protein (RBP) (30). The human *transthyretin* (*TTR*) gene spans about 7 kilobases (kb) and is composed of four exons (31). *TTR* has about 110 variants, more than 90 of which are associated with human amyloidosis. To produce humanized mice carrying different TTR variants, we used the exchangeable targeting vector containing a neomycin resistance gene flanked by *lox*71 and *lox*P sites, as shown in Fig. 8a.

The production of the replaced allele was done in two steps. The first step was the production of the targeted KO allele carrying *lox*71 and *lox*P sites by homologous recombination (Fig. 9a). After electroporation with the exchangeable targeting vectors, we isolated and analyzed 98 neo-resistant clones and obtained 5 targeted clones. Two out of the 5 clones produced germ-line chimeras. Therefore, these clones were used for the second step, the Cre-mediated site-specific integration of *TTR* cDNA. We constructed a replacement vector containing *lox*66-exon-intron cassette-hTTR cDNA-pA-*FRT*-Pgk-*Pac*-*FRT*-*lox*P. The targeted ES clones were coelectroporated with the replacement vector and the Cre expression vector, pCAGGS-Cre. Twelve puromycin-resistant colonies were isolated from each clone and were analyzed by PCR and Southern blot analysis for the site-specific recombination event. All 24 puromycin-resistant clones had recombined as expected; the neo cassette was replaced with the *TTR* cDNA and Pgk-Pac cassette (Fig. 9b). Thus, pA trapping upon site-specific integration effectively selected for recombined clones.

Chimeric mice were produced with the humanized ES clones, and mouse lines were established. Humanized mice of all

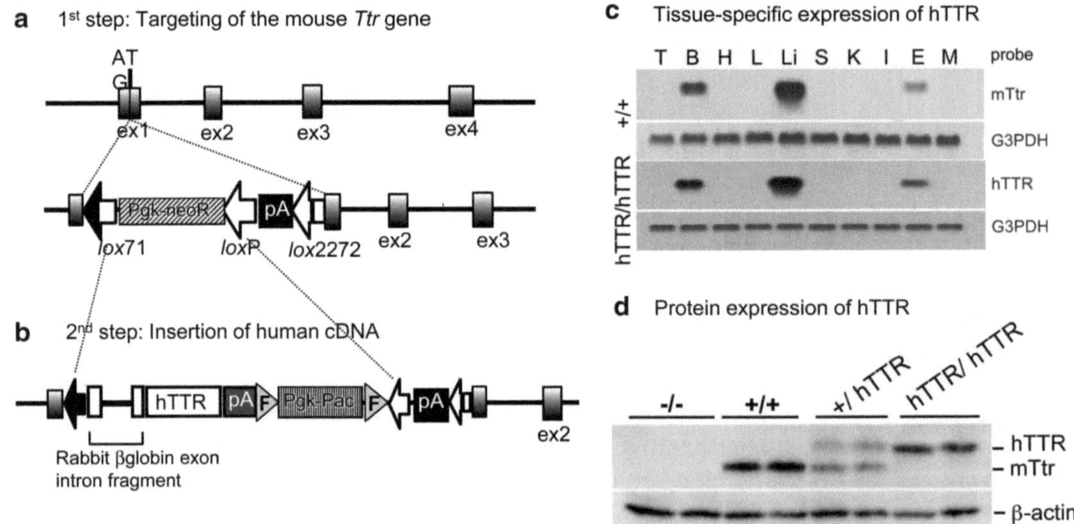

Fig. 9. Production of a humanized allele of the *Ttr* locus. (**a**) Knockout *Ttr* allele produced with a *lox*71-Pgk-neo^R-*lox*P-pA-*lox*2272 cassette. The mouse *Ttr* gene is composed of 4 exons, and the ATG codon resides in exon 1. The ATG codon was removed, and the neo cassette was inserted into the ATG position. (**b**) Replaced allele with a human *TTR* cDNA. To increase expression of the inserted cDNA, an exon-intron cassette was placed before the cDNA. (**c**) Northern blot analysis comparing the expression patterns of the endogenous *Ttr* gene in WT mice and the inserted human *TTR* cDNA in *TTR* allele homozygotes. *T* thymus, *B* brain, *H* heart, *L* lung, *Li* liver, *S* spleen, *K* kidney, *I* intestine, *E* eye, *M* muscle. (**d**) Western blot analysis of liver extracts using an anti-TTR antibody recognizing both human and mouse TTR proteins.

genotypes appeared normal up to at least 6 months of age. To compare the expression patterns of endogenous *Ttr* and the inserted *TTR* cDNA, RNAs were extracted from various tissues of wild-type and humanized mice and analyzed by Northern blotting. In all mice examined, the *TTR* cDNA was consistently expressed in the liver, brain, and eyes (Fig. 9c), which are precisely the same tissues in which endogenous *Ttr* is expressed (32). This suggests that the *TTR* cDNA, located in the *Trt* locus, is correctly regulated under the promoter of the *Ttr* gene.

We analyzed protein levels of human and mouse TTR in KO (−/−), WT (+/+), +/*TTR*, and *TTR/TTR* mice using western blotting of liver extracts. As shown in Fig. 9d, both mouse and human TTR proteins were detected in the liver of +/*TTR* mice, but only human protein was detected in the liver of *TTR/TTR* mice. The level of TTR in the liver of *TTR/TTR* mice was twice that in the livers of +/*TTR* mice. These results showed human TTR protein was expressed in a gene-dose-dependent manner in the humanized mice.

4. Combination of the Cre/*Lox* System with Other Recombination Systems

With the Cre/mutant *lox* system, any DNA of interest can be inserted into the mouse genome; however, re-excision by Cre is impossible, and removal of the inserted DNA requires the use of another recombination system. To date, three recombination systems have been described, although their recombination activities in vivo remain to be examined.

The first is the well-known Flp/*FRT* system (33), which is derived from *Saccharomyces cerevisiae* (34). Flp is a recombinase which recognizes the *FRT* site. Wild-type Flp protein showed limited recombination efficiency in mammalian cells due to thermoinstability (35). Although mutational screening identified Flpe protein, which showed four times higher activity (36), the activity of Flpe is still lower than that of Cre protein. Therefore, the Flp/*FRT* system has been mainly used for deletion of selection marker gene in ES cells. However, in 2007, Raymond and Soriano reported codon-optimized Flp (*Flpo*), which showed high recombination efficiency in ES cells, comparable with that of Cre (37).

The second recombination system is ΦC31 from *Streptomyces lividans.* ΦC31 recombinase recognizes two heterotypic sequences, attB and attP. Raymond and Soriano also improved ΦC31 (ΦC31o) and produced ΦC31o-expressing mice (37). Although the recombination efficiency of ΦC31o in vivo was slightly lower than that of Cre, it would be of use when complete recombination is not required.

The third and most promising recombination system is Dre/*rox*, as reported by Sauer and McDermott in 2004 (38). Dre/rox was identified from the P1-like transducing phage D6, which was isolated from *Salmonella oranienburg.* Dre recombinase recognizes and recombines *rox* sites composed of 32 bp. Recently, Anastassiadis et al. reported that Dre/*rox* is highly efficient in mice as well as in ES cells, similar to Cre/*lox* (39).

By placing two recognition sites of another recombination system on the 5′ and 3′ ends of the inserted DNA, the DNA can be removed conditionally through mating with transgenic mice expressing the recombinase in a tissue-specific manner. Thus, through the combinatorial use of different recombination systems, we can control gene expression in an off-on-off manner.

References

1. Nagy A (2000) Cre recombinase: the universal reagent for genome tailoring. Genesis 26: 99–109
2. Branda CS, Dymecki SM (2004) Talking about a revolution: The impact of site-specific recombinases on genetic analyses in mice. Dev Cell 6:7–28
3. Voziyanov Y, Pathania S, Jayaram M (1999) A general model for site-specific recombination by the integrase family recombinases. Nucleic Acids Res 27:930–941
4. Hoess RH, Ziese M, Sternberg N (1982) P1 site-specific recombination: nucleotide sequence of the recombining sites. Proc Natl Acad Sci USA 79:3398–3402
5. Hoess RH, Wierzbicki A, Abremski K (1986) The role of the *lox*P spacer region in P1 site-specific recombination. Nucleic Acids Res 14:2287–2300
6. Fukushige S, Sauer B (1992) Genomic targeting with a positive-selection *lox* integration vector allows highly reproducible gene expression in mammalian cells. Proc Natl Acad Sci USA 89:7905–7909
7. Albert H, Dale EC, Lee E et al. (1995) Site-specific integration of DNA into wild-type and mutant *lox* sites placed in the plant genome. Plant J 7:649–659
8. Araki K, Araki M, Yamamura K (1997) Targeted integration of DNA using mutant *lox* sites in embryonic stem cells. Nucleic Acids Res 25:868–872
9. Thomson JG, Rucker EB, 3rd, Piedrahita JA (2003) Mutational analysis of *lox*P sites for efficient Cre-mediated insertion into genomic DNA. Genesis 36:162–167
10. Araki K, Okada Y, Araki M et al. Comparative analysis of right element mutant *lox* sites on recombination efficiency in embryonic stem cells. BMC Biotechnol 10:29
11. Lee G, Saito I (1998) Role of nucleotide sequences of *lox*P spacer region in Cre-mediated recombination. Gene 216:55–65
12. Langer SJ, Ghafoori AP, Byrd M et al. (2002) A genetic screen identifies novel non-compatible *lox*P sites. Nucleic Acids Res 30:3067–3077
13. Bouhassira EE, Westerman K, Leboulch P (1997) Transcriptional behavior of LCR enhancer elements integrated at the same chromosomal locus by recombinase-mediated cassette exchange. Blood 90:3332–3344
14. Bethke B, Sauer B (1997) Segmental genomic replacement by Cre-mediated recombination: genotoxic stress activation of the p53 promoter in single-copy transformants. Nucleic Acids Res 25:2828–2834
15. Kolb AF (2001) Selection-marker-free modification of the murine beta-casein gene using a *lox*2272 (correction of *lox*2722) site. Anal Biochem 290:260–271
16. Osipovich AB, Singh A, Ruley HE (2005) Post-entrapment genome engineering: first exon size does not affect the expression of fusion transcripts generated by gene entrapment. Genome Res 15:428–435
17. Araki K, Araki M, Yamamura K (2002) Site-directed integration of the cre gene mediated by Cre recombinase using a combination of mutant *lox* sites. Nucleic Acids Res 30:e103
18. Araki K, Imaizumi T, Okuyama K et al. (1997) Efficiency of recombination by Cre transient expression in embryonic stem cells: comparison of various promoters. J Biochem (Tokyo) 122:977–982
19. Hayashi S, Tenzen T, McMahon AP (2003) Maternal inheritance of Cre activity in a Sox2Cre deleter strain. Genesis 37:51–53
20. Taniwaki T, Haruna K, Nakamura H et al. (2005) Characterization of an exchangeable gene trap using pU-17 carrying a stop codon-beta geo cassette. Dev Growth Differ 47:163–172
21. Miura K, Yoshinobu K, Imaizumi T et al. (2006) Impaired expression of importin/karyopherin b1 leads to post-implantation lethality. Biochem Biophys Res Commun 341:132–138
22. Rodriguez CI, Buchholz F, Galloway J et al. (2000) High-efficiency deleter mice show that FLPe is an alternative to Cre-*lox*P. Nat Genet 25:139–140
23. Li Z, Zhao G, Shen J et al. (2010) Enhanced expression of human cDNA by phosphoglycerate kinase promoter-puromycin cassette in the mouse transthyretin locus. Transgenic Res 20:191–200
24. Araki K, Araki M, Yamamura K (2006) Negative selection with the Diphtheria toxin A fragment gene improves frequency of Cre-mediated cassette exchange in ES cells. J Biochem 140:793–798
25. Zhao G, Li Z, Araki K et al. (2008) Inconsistency between hepatic expression and serum concentration of transthyretin in mice humanized at the transthyretin locus. Genes Cells 13:1257–1268
26. Araki K, Imaizumi T, Sekimoto T et al. (1999) Exchangeable gene trap using the Cre/mutated *lox* system. Cell Mol Biol 45:737–750
27. Liddle CN, Reid WA, Kennedy JS et al. (1985) Immunolocalization of prealbumin: distribution in normal human tissue. J Pathol 146:107–113
28. Herbert J, Wilcox JN, Pham KT et al. (1986) Transthyretin: a choroid plexus-specific transport

protein in human brain. The 1986S. Weir Mitchell award. Neurology 36:900–911

29. Cavallaro T, Martone RL, Dwork AJ et al. (1990) The retinal pigment epithelium is the unique site of transthyretin synthesis in the rat eye. Invest Ophthalmol Vis Sci 31:497–501
30. Vranckx R, Savu L, Maya M et al. (1990) Characterization of a major development-regulated serum thyroxine-binding globulin in the euthyroid mouse. Biochem J 271:373–379
31. Tsuzuki T, Mita S, Maeda S et al. (1985) Structure of the human prealbumin gene. J Biol Chem 260:12224–12227
32. Wakasugi S, Maeda S, Shimada K (1986) Structure and expression of the mouse prealbumin gene. J Biochem 100:49–58
33. McLeod M, Craft S, Broach JR (1986) Identification of the crossover site during FLP-mediated recombination in the Saccharomyces cerevisiae plasmid 2 microns circle. Mol Cell Biol 6:3357–3367
34. Dymecki SM (1996) Flp recombinase promotes site-specific DNA recombination in embryonic stem cells and transgenic mice. Proc Natl Acad Sci USA 93:6191–6196
35. Buchholz F, Ringrose L, Angrand PO et al. (1996) Different thermostabilities of FLP and Cre recombinases: implications for applied site-specific recombination. Nucleic Acids Res 24:4256–4262
36. Buchholz F, Angrand PO, Stewart AF (1998) Improved properties of FLP recombinase evolved by cycling mutagenesis. Nat Biotechnol 16:657–662
37. Raymond CS, Soriano P (2007) High-efficiency FLP and PhiC31 site-specific recombination in mammalian cells. PLoS One 2:e162
38. Sauer B, McDermott J (2004) DNA recombination with a heterospecific Cre homolog identified from comparison of the pac-c1 regions of P1-related phages. Nucleic Acids Res 32:6086–6095
39. Anastassiadis K, Fu J, Patsch C et al. (2009) Dre recombinase, like Cre, is a highly efficient site-specific recombinase in E. coli, mammalian cells and mice. Dis Model Mech 2:508–515

Chapter 3

Using Recombinant Adeno-Associated Viral Vectors for Gene Expression in the Brain

Anke Van der Perren, Jaan Toelen, Jean-Marc Taymans, and Veerle Baekelandt

Abstract

Recombinant AAV vectors currently enjoy an excellent track record in brain applications such as generating preclinical models of neurodegeneration and gene therapy for brain disorders. Indeed, rAAV vectors have been useful in modeling diseases such as Parkinson's disease (discussed below) and have also been tested in various phases of clinical development for Parkinson's disease (Christine et al., Neurology 73:1662–1669, 2009; Kaplitt et al., Lancet 369:2097–2105, 2007) and Alzheimer's disease (Mandel 2010, Curr Opin Mol Ther 12:240–247, 2010). In this review, we will discuss the vectorology of rAAV, rAAV production, and purification of the different rAAV serotypes. We will also describe locoregional transduction of the brain using rAAV vectors and illustrate these techniques with specific examples of applications such as non-invasive imaging of reporter genes and disease modeling in Parkinson's disease.

Key words: Adeno-associated viral vectors, Animal model, Neurodegeneration, Parkinson's disease, Transduction, Reporter gene, Vector production, Gene therapy

1. Introduction

Recombinant gene vectors derived from adeno-associated virus (AAV), or recombinant adeno-associated viral vectors (rAAV vectors), have been an attractive gene delivery vehicle since they were first engineered almost three decades ago (1). The main reason for this is the unique combination of attractive properties: rAAV vectors are relatively safe and ensure efficient gene transfer. rAAV vectors are considered safe because the virus they are derived from, AAV, is a non-pathogenic parvovirus whose replication depends on co-infection with a lytic helper virus, usually a member of the adenovirus (2) or herpes virus family (3). rAAV vectors are also devoid of all wild-type AAV genes, making reversion to replication competent forms virtually impossible. In addition, unlike the administration

Alexei Morozov (ed.), *Controlled Genetic Manipulations*, Neuromethods, vol. 65,
DOI 10.1007/978-1-61779-533-6_3,

of adenoviral vectors, the administration of rAAV vectors in vivo usually does not elicit a host immune response resulting in destruction of the transduced cells as shown in preclinical studies in small animal models.

rAAV vectors are capable of delivering gene cassettes of 4.5–5 kb (4, 5). Transcription of rAAV gene cassettes occurs without integration into the host genome; however they will readily persist for months to years in slowly dividing or non-dividing cells. Indeed, studies have demonstrated that rAAV vectors can efficiently transduce a number of somatic tissues, including muscle (6), liver (7), heart (8), retina (9), and the central nervous system (CNS) (10). To date, the most widely used rAAV vector is based on serotype 2 of AAV. In the past decade, a large number of clinical studies using rAAV2 vectors as gene delivery vehicle have been launched for the treatment of a variety of inherited and acquired human diseases (11, 12). Despite these favorable properties, the first translation of rAAV into a human clinical trial resulted in a rAAV2 capsid specific immune response (13). This failure encouraged the search for novel AAV serotypes that may circumvent the intrinsic characteristics of rAAV2. This search has revealed a diverse family of more than 120 novel AAV variants with unique tissue/cell type tropisms and efficient gene transfer capability (14, 15). This major advance in rAAV vectorology has significantly broadened potential applications of rAAV vectors in clinical gene therapy or disease modeling and now offers more options for the selection of a suitable rAAV variant per specific application.

Recombinant AAV vectors currently enjoy an excellent track record in brain applications such as generating preclinical models of neurodegeneration and gene therapy for brain disorders. Indeed, rAAV vectors have been useful in modeling diseases such as Parkinson's disease (discussed below) and have also been tested in various phases of clinical development for Parkinson's disease (16, 17) and Alzheimer's disease (18). In this review, we will discuss the vectorology of rAAV, rAAV production, and purification of the different rAAV serotypes. We will also describe locoregional transduction of the brain using rAAV vectors and illustrate these techniques with specific examples of applications such as non-invasive imaging of reporter genes and disease modeling in Parkinson's disease.

2. AAV Vectorology

2.1. From Virus to Vector

To generate a recombinant viral vector, the original viral genome has to be dissected and separated into *cis*- and *trans*-acting sequences. As coding sequences for regulatory and structural proteins work in *trans*, they can easily be expressed from heterologous plasmids or even be incorporated into the chromosomal DNA of the production cell. The *cis*-acting sequences are linked to a promoter sequence and to the coding sequence of any desired transgene, the whole usually being referred to as transfer construct. Historically, most recombinant AAV vectors have been based on serotype 2 (AAV2/2) that constitutes the prototype of the genus, and rAAV vectors were produced by means of two plasmids (the transfer and the packaging plasmid) and an infectious adenovirus (Fig. 1). The transfer plasmid carries a transgene expression cassette flanked by the AAV2 inverted terminal repeats (ITRs) which are the only *cis*-acting elements required for rescue, replication, and packaging of the recombinant genome (19). The AAV2 termini consist of a 145-nucleotide-long inverted terminal repeat that, due to the multipalindromic nature of its terminal 125 bases, can fold on itself via complementary base pairing and form a characteristic T-shaped hairpin structure. The AAV2 nonstructural (rep) gene and, depending on the serotype used, a specific structural (cap) gene are supplied in *trans* on the second plasmid, the so-called packaging plasmid. The adenoviral (Ad) helper functions were originally supplied by infection of rAAV producer cells with a wild-type adenovirus. The finding that Ad helper functions are provided by expression of E1A, E1B, E2A, E4ORF6, and VA RNAs enabled subsequent Ad-free production of rAAV vector stocks by incorporating these sequences into a plasmid (referred to as adenoviral helper plasmid) and transfecting it together with two above-mentioned plasmids. Upon introduction of all these constructs into the production cells, vector particles are generated.

2.2. Naturally Occurring and Artificial rAAV Serotypes

An increasingly important area in the development of rAAV as a vector concerns the engineering of altered cell tropisms to narrow or broaden rAAV-mediated gene delivery and to increase its efficiency in tissues refractory to rAAV2/2 infection. The transduction efficiency of cells by rAAV2/2 depends on the cellular receptor content and can be impaired by the mechanisms involved in either intracellular virion trafficking and uncoating (20, 21) or single-to-double strand genome conversion (22–24). As these processes depend either directly or indirectly on capsid conformation, both in vitro and in vivo transduction efficiencies depend on the serotype used. In the context of efficient and specific transduction in the central nervous system, the selection of a tailored rAAV capsid either artificially or by selecting natural occurring serotypes becomes imperative.

The approaches for a rational and targeted design rely on the modification of the rAAV capsid structure by chemical, immunological, or genetic means (25). Another route to find the ideal rAAV tropism exploits the natural capsid diversity of newly isolated serotypes by packaging rAAV2 genomes into capsids derived from other human or non-human AAV isolates (26). To this end, up until now, most researches employ hybrid *trans*-complementing

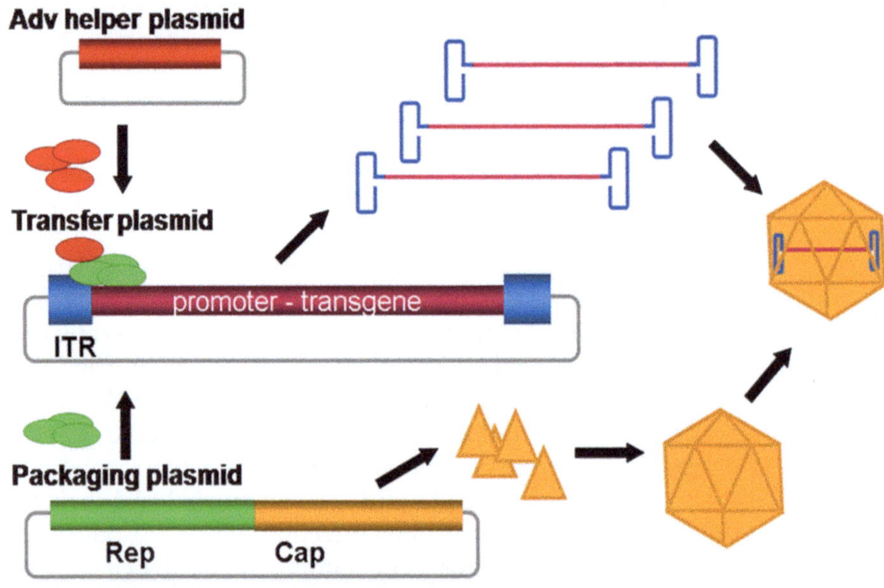

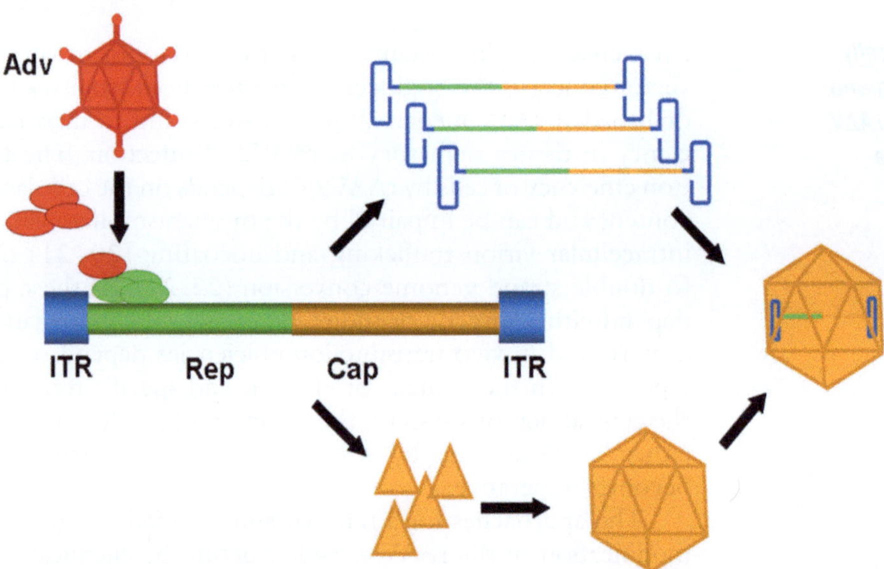

Fig. 1. The adeno-associated virus and a recombinant adeno-associated viral vector.

constructs that encode *rep* from AAV2 whereas *cap* is derived from the serotype displaying the cell tropism of choice. These hybrid vectors are therefore indicated as rAAV2/1, rAAV2/5, etc., where the second number refers to the capsid serotype. Several rAAV serotypes have already been evaluated for their tropism in the central nervous system.

2.3. rAAV Transgene Expression

The rAAV vector technology can be used for both overexpression and silencing of certain genes. In disease models, the transgene encoded by the rAAV can be a reporter protein for optimal visualization of the labeled cells by immunohistochemical or noninvasive imaging techniques such as bioluminescence, nuclear medicine techniques, or magnetic resonance imaging (27). The expressed protein can also be a disease-related protein used for disease modeling (28–31) or a therapeutic protein for gene therapy applications (32–34). On the other hand, rAAV vectors can be used to suppress certain proteins in the transduced cells by making use of the RNAi mechanism (35). If the expression of the transgene has to be limited to a specific cell type, one can make use of tissue-specific transcriptional control elements such as tissue-specific promoters (36) or other regulatory elements.

3. AAV Production and Purification

3.1. Production Systems

Extensive efforts have focused on developing versatile and scalable manufacturing processes for rAAV vector production, with attention to compatibility with good manufacturing practice (GMP) (37–40). At present, several well-developed and validated production platforms include (1) the above-mentioned helper-free 293/triple-transfection method (40) and the two plasmid-based system (41), (2) the stable rep/cap cell line/adenovirus (Adv)–AAV hybrid infection method (42, 43), (3) the recombinant herpes simplex virus (HSV) helper infection-based rAAV production method (44), (4) Ad helper virus infection-dependent rAAV producer cell lines (38), and (5) the recombinant baculovirus/insect cell-based production system (45, 46).

While scalable systems based upon AAV adenovirus, AAV herpesvirus, and AAV baculovirus hybrids hold promise for clinical applications, they require labor-intensive generation of reagents and are not highly suited to small and intermediate scale preclinical studies in animals where several combinations of serotype and genome may need to be tested. Furthermore, several desired outcomes of the downstream process, e.g., the separation of empty from full particles, are difficult to obtain using scaled chromatography systems and must be individually tailored for each serotype.

In our research group, we optimized a scalable and flexible serum-free rAAV vector production system, allowing a swift adaptation for production of different serotypes (47, 48). This production protocol will be described here in detail.

3.2. Production of rAAV Serotypes Based on Isolation from the Supernatant

In a Hyperflask cell culture vessel (1,720-cm^2 growth area, Corning Life Science, USA), 293 T cells are seeded at a concentration of 1E+08 cells in 500 ml of Optimem with 2% FCS. After 24 h, a transient transfection procedure is carried out based on the PEI (polyethylenimine) transfection technique using the AAV transfer, AAV rep/cap, and pAdbDeltaF6 plasmids. The DNA–PEI complex is added to the Optimem medium supplemented with 2% FCS and added to the 293 T cells. The supernatant is collected after 5 days, filtered through a 0.22-μm filter (Stericup-HV filter unit, Millipore NV, Brussels, Belgium), brought to a final concentration of 1 M NaCl, and concentrated using tangential flow filtration (TFF). The TFF is performed using a Minim™; a benchtop TFF machine, with the Centramate™ LV holder (Pall, Ann Arbor, MI, USA); and a 100-kDa cut-off Omega screen channel cassette, according to the manufacturer's protocol (Fig. 2).

3.3. rAAV Purification Using a Single Iodixanol Gradient Centrifugation Step

The concentrated supernatant is purified using an iodixanol step gradient in a Beckman Ti-70 fixed angle rotor at 27, 000 rpm for 0.5, 2, 6, and 24 h. For the different gradient mixtures, the iodixanol (OptiPrep density gradient medium, Sigma-Aldrich, Bornem, Belgium) is diluted with PBS and a 5-M NaCl solution in order to obtain 20%, 30%, and 40% (w/v) solutions with a final concentration of 1 M NaCl. A discontinuous gradient is made by carefully underlayering the concentrated supernatant with 5 ml of 20% iodixanol (1 M NaCl), 3 ml of 30% iodixanol (1 M NaCl), 3 ml of 40% iodixanol (1 M NaCl), and 3 ml of 60% iodixanol. After centrifugation, the rAAV-containing fraction is collected and centrifuged in a Vivaspin 6 (PES, 100 kDa MWCO, Sartorius AG, Goettingen, Germany) using a swinging bucket rotor at 3,000 g.

3.4. Quality Control of rAAV Preparations

The above-mentioned protocol is compatible with multiple rAAV capsids. We have produced rAAV vectors based on the following serotypes: 1, 5, 7, 8, and 9. The obtained rAAV preparations are submitted to a quality control protocol including in vitro transduction assays (to assess the functionality of the vector), qPCR analysis (for determination of genome copies), and silver stain and Western blot analysis (for detection of contaminations after purification) (Fig. 3).

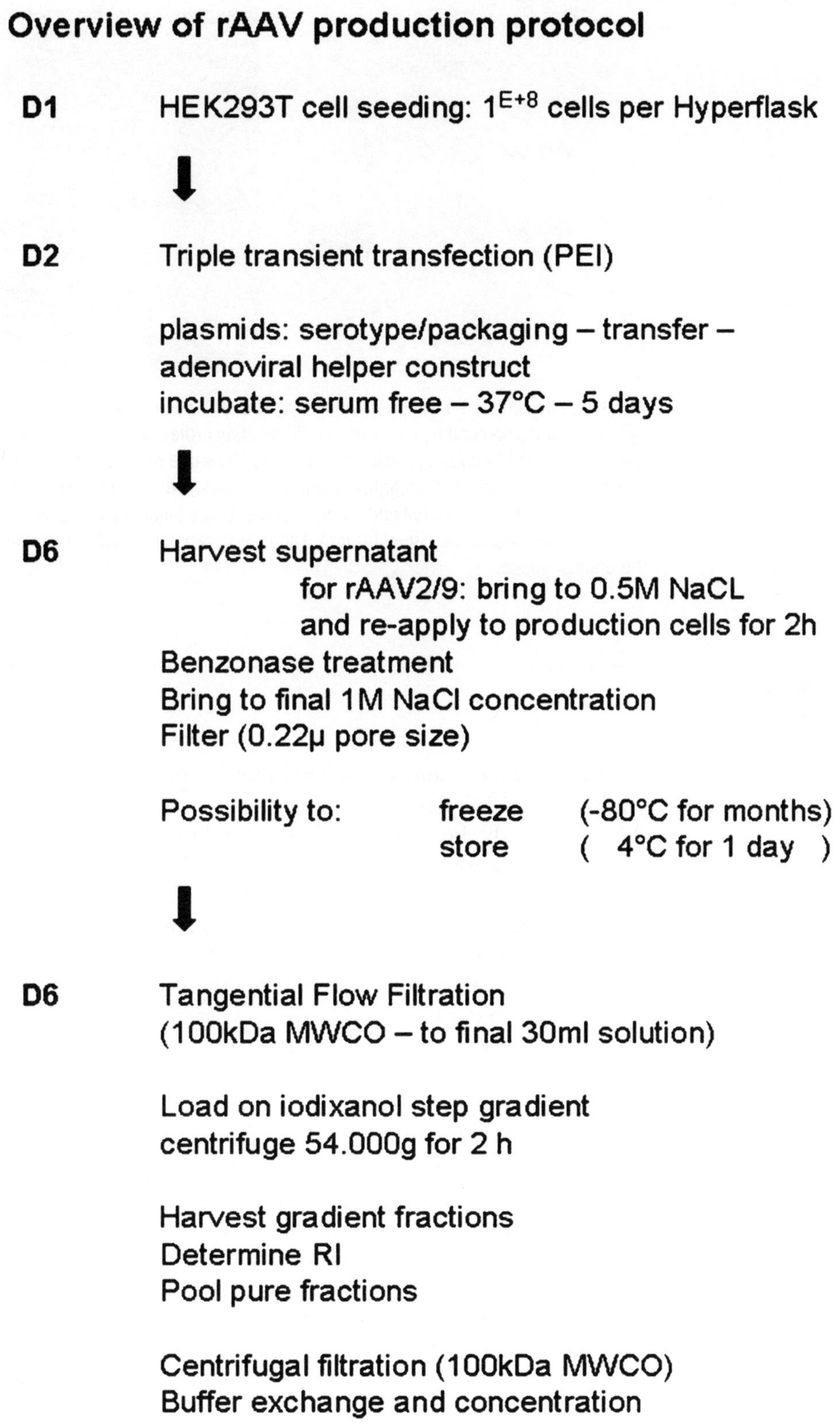

Overview of rAAV production protocol

D1 HEK293T cell seeding: 1^{E+8} cells per Hyperflask

⬇

D2 Triple transient transfection (PEI)

plasmids: serotype/packaging – transfer – adenoviral helper construct
incubate: serum free – 37°C – 5 days

⬇

D6 Harvest supernatant
for rAAV2/9: bring to 0.5M NaCL
and re-apply to production cells for 2h
Benzonase treatment
Bring to final 1M NaCl concentration
Filter (0.22µ pore size)

Possibility to: freeze (-80°C for months)
store (4°C for 1 day)

⬇

D6 Tangential Flow Filtration
(100kDa MWCO – to final 30ml solution)

Load on iodixanol step gradient
centrifuge 54.000g for 2 h

Harvest gradient fractions
Determine RI
Pool pure fractions

Centrifugal filtration (100kDa MWCO)
Buffer exchange and concentration

Fig. 2. Overview of rAAV production and purification protocol.

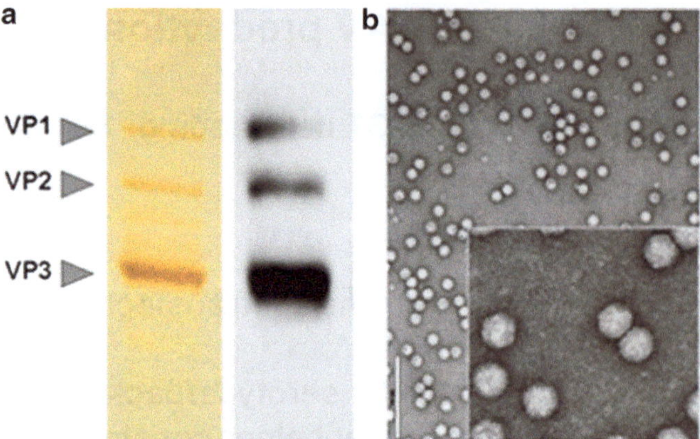

Fig. 3. Quality control of rAAV preparations. (**a**) The rAAV vector preparations are analyzed by silver stain to detect the presence of contaminating protein and a Western blot to detect the presence of AAV capsid proteins VP1, 2, 3. (**b**) The same preparations are also imaged using electron microscopy (negative staining with uranyl acetate) to assess the sample purity and the presence of full and empty capsids. Large images were taken at a magnification of 98, 000 (scale bar = 200 nm). The inserts represent higher magnifications of the original image.

4. Locoregional Transduction

As an alternative and complementary approach to transgenesis, viral vectors have been used to create new models for CNS disorders or to study the function of disease-related proteins (see below). The use of transgenic mouse models has been very instrumental, but these models are, however, mainly restricted to a single species, often limited in specific targeting and in general lacking controlled transgene expression. In contrast, viral vector-mediated locoregional/stereotactic gene delivery provides a novel way of delivering proteins to the CNS as it has several advantages: local transgene delivery allows for specific targeting of brain regions; transgene expression can be induced during adulthood, bypassing the risk of compensatory mechanisms during development; different doses can be applied; furthermore, models can be created in multiple species, ranging from rodents to non-human primates, and finally, different combinations of genes can easily be made. Since the first proof of principle of this technique, a continually growing number of publications have proven the value of viral vector-mediated locoregional transduction (29, 30, 33, 49).

For some gene therapy strategies, e.g., for metabolic disorders that affect the CNS, a more widespread delivery throughout the brain is required. Intraventricular injection of rAAV vectors in neonatal mice can therefore be a good alternative as robust, widespread expression can be achieved in this way (50, 51). rAAV vectors have also been used to target the brain less invasively by intravascular injection (IV) (52). For a long time, the vascular delivery approach to target neurons in the CNS was inefficient, likely due to the inability of the vector to cross the blood brain barrier (BBB) (53, 54). However, recently Foust and colleagues tested a new serotype AAV 9, with unique serological characteristics, for gene delivery to the CNS after intravenous injection in mice. Transduction within the CNS was obtained, and interestingly, the pattern of transduction shifted from primarily neuronal targeting in neonates to astrocytic targeting in adult-treated animals (52). These findings show the unique capacity of AAV2/9 vectors to cross the BBB in varying degrees depending on the time point of vector administration.

5. AAV Serotypes

Of the different viral vector systems, rAAV has become a common vector of choice to deliver genes to the CNS (55–61). To date, the most widely used rAAV vector is based on the AAV2 serotype, which has, however, been shown to contain some drawbacks. rAAV2 transduces neurons efficiently in the immediate vicinity of the injection site (62) but requires multiple injections, convection-enhanced delivery (63, 64), or addition of agents such as mannitol, basic fibroblast growth factor (bFGF), or heparin to transduce larger brain volumes (65–68). Furthermore, the clinical application of rAAV2/2 may be limited by pre-existing immunity to AAV2 which is present in most humans (69). Finally, rAAV2 mainly targets neurons although other cell types in the CNS might also be important. The limitations of rAAV2 have necessitated the evaluation of alternative serotypes with broader cellular targets, higher transduction efficiencies, and potential to evade preexisting immunity to the AAV2 capsid (56–60). Among the more than 100 identified AAV variants that have been isolated from adenovirus stock or from human/non-human primate tissues, AAV 1 to AAV 10 are currently being developed as recombinant vectors for gene therapy applications (26, 70–74). The cellular tropism of different AAV serotypes is diverse partly because they bind different cellular receptors (Table 1). To date, the cellular receptor(s) and co-receptor(s) for most serotypes remain unknown. The transduction pattern can, however, not be completely predicted by receptor binding. The cellular tropism of different AAV serotypes is believed to be related to the cumulative effects of viral binding to multiple cell surface receptors, cellular uptake, intracellular processing, nuclear delivery, uncoating, and second-strand DNA conversion (75, 76). Therefore, the complete molecular mechanism underlying the recognition and entry of AAV serotypes needs to be further dissected. At the moment, one needs to rely on empirical determination of the transduction efficiency and cellular specificity for each serotype.

Table 1
Overview of known cellular receptors for different AAV serotypes

Serotype	Receptor	Co-receptor
AAV 1	Sialic compounds on N-linked glycoproteins (77–80)	
AAV2	Heparin sulfate proteoglycan (81)	Integrin $\alpha V\beta_5$, laminin-R, the fibroblast GFR 1, hepatocyte GFR (82–85)
AAV3	Heparin sulfate proteoglycan (81)	Laminin-R, the fibroblast GFR 1(85–87)
AAV4	O-linked sialic acid (88, 89)	
AAV5	N-linked sialic acid (88, 89)	Platelet-derived growth factor receptor (90)
AAV6	Sialic compounds on N-linked glycoproteins (77–80)	
AAV8	Laminin receptor (85)	
AAV9	Laminin receptor (85)	

Different groups have studied the transduction efficiency and tropism of novel rAAV serotypes in specific brain regions of different species of animals (Table 2). These studies show that more recently discovered serotypes are as good as and often better than AAV2 in transducing the CNS and serve as good alternatives. The discrepancy between some studies in terms of transduction efficiency might be explained by differences in constructs like promoter, transgene, and regulatory elements (36, 91, 92); the titers used; production and purification methods (93); and species specificity. It is also clear that it is not always possible to extrapolate findings between different species (94).

Most rAAV serotypes described mention a specific tropism for neurons and inefficient transduction of other cell types within the brain such as astrocytes, oligodendrocytes, and epithelial cells (57, 95). Recently, preferential transduction of astrocytes and oligodendrocytes was observed with newer serotypes (rh43, hu.32, hu.11, pi.2, hu.48R3, and rh.8) (96, 97), but the overall percentage of transduction was rather low. Neurons, astrocytes, and oligodendrocytes are implicated in many different CNS diseases (98). Therefore, selection of a viral vector that is able to transduce a specific cell type may be of interest for certain CNS diseases. The combination of a certain serotype together with a tissue-specific promoter (36) or other regulatory elements may offer an alternative.

Several studies in the CNS report the ability of different rAAV serotypes to transduce high numbers of neuronal cells in the brain without apparent toxicity (58–60, 99). Limited astrocytosis or

Table 2
Summary of published reports that have compared different rAAV serotypes in different brain regions and species

First author	Species	Site of injection	Serotypes tested	Promoter/transgene	Results
Taymans et al. 2007 (58)	Mouse	STR, SN, HPC	rAAV 2/1, 2/2, 2/5, 2/7, 2/8	CMV-eGFP	rAAV 2/1, 2/5, 2/7, 2/8 > rAAV 2/2 (high titers) rAAV 2/5, 2/7 >2/8 (lower titers) Expression major neuronal
Paterna et al. 2004 (56)	Rat	STR, SN, MFB	rAAV 2/2 and 2/5	CBA-eGFP	rAAV 2/5 > 2/2 rAAV 2/5 higher transduced region rAAV 2/2 higher expression level/cell Majority NeuN+
Burger et al. 2004 (57)	Rat	SN, HPC, STR, GP, SC	rAAV 2/1, 2/2, 2/5	CBA-eGFP	rAAV 2/1 > 2/5 > 2/2 Higher transduction efficiency in all regions Majority NeuN+ rAAV 2/1: highest TD volume and number
Klein et al. 2006 (101)	Rat	HPC	rAAV 2/1, 2/2, 2/5, 2/8	CMV/CBA-eGFP	rAAV 2/8 > rAAV 2/5 > 2/2, 2/1 Expression exclusively neuronal
Reimsnider et al. 2007 (99)	Rat	STR	rAAV 2/1, 2/2, 2/5, 2/8	CBA-eGFP	rAAV 2/1, 2/2: fast expressors rAAV 2/5, 2/8: slow expressors
Mc Farmand et al. 2009 (59)	Rat	SN	rAAV 2/1, 2/2, 2/5, 2/8	CMV-eGFP	rAAV 2/1 > rAAV 2/5, 2/8 > 2/2
Van der Perren et al. 2010 (100)	Rat	SN	rAAV 2/1, 2/2, 2/5, 2/6.2, 2/7, 2/8, 2/9	CMV-eGFP-t2A-fLuc CMVie synapsin-eGFP-T2A-fLuc	rAAV 2/1, 2/7, 2/9 > 2/2, 2/5, 2/6.2, 2/8: TD volume rAAV 2/1, 2/7, 2/8, 2/9 > 2/2, 2/5, 2/6.2: TD cell number

(continued)

Table 2 (continued)

First author	Species	Site of injection	Serotypes tested	Promoter/transgene	Results
Blits et al. 2009 (61)	Rat	Red nucleus	rAAV 2/1–2/6, 2/8	CMV-eGFP	rAAV 2/1: highest expression level rAAV 2/1, 2/6: fast expressors rAAV 2/5 and 2/8: slow expressors Expression exclusively neuronal
Doiya et al. 2009 (60)	Primate	STR	rAAV 2/1, 2/5, 2/8	CBA-eGFP	*STR*: rAAV 2/5 highest TD volume rAAV 2/5, 2/1 > 2/8 highest number of TD cells *SN*: rAAV 2/1 and 2/8 equal efficiency
Markakis et al. 2009 (102)	Primate	STR, SN	rAAV 2/1, 2/2, 2/3, 2/4, 2/5, 2/6	CMV-eGFP	rAAV 2/5 and 2/1: highest TD volume rAAV 2/5: highest number of TD cells Expression was both neuronal and astrocytic (47% GFAP+ –53% NeuN+)
Vite et al. 2003 (103)	Cat	CC, STR, TH, IC	rAAV 2/1, 2/2, 2/5	GUSB-GUSB	rAAV 2/1 > 2/2 rAAV 2/5 no detectable transduction
Kornum et al. 2009 (104)	Mini pig Neonatal rat	STR STR	rAAV 2/1, 2/5 rAAV 2/1, 2/5	CBA-eGFP CBA-eGFP	rAAV 2/5 > 2/1 highest TD volume and number of TD cells rAAV 2/1 exclusively neuronal cells rAAV 2/5 low frequency of nonneuronal cells

STR striatum, *SN* substantia nigra, *CC* cerebral cortex, *HPC* hippocampus, *MFB* medial forebrain bundle, *CMV* cytomegalovirus promoter, *CBA* chicken beta actin promoter, *eGFP* green fluorescent protein, *TD* transduced gliosis adjacent to the needle tract with negligible inflammatory response in view of the widespread transduction and high transgene expression levels has been reported after injecting low (10E10 GC/ml) or high titers (10E12 GC/ml) (59, 100). One study has reported neurotoxicity with high titers of rAAV2/8-eGFP but not with rAAV 2/5 or 2/1 in rodents (101). This study demonstrates that with the development of more potent viral vectors, there is an upper limit to vector dose and/or transgene expression levels that a living organism can tolerate; however, this threshold is relatively high.

6. Specific Example

6.1. Different rAAV Serotypes in the Rat Substantia Nigra: Non-invasive Imaging and Immunohistochemistry

Using multicistronic constructs different transgenes can be efficiently expressed from the same plasmid (105). We used two different reporter genes, firefly luciferase (fLuc) and green fluorescent protein (eGFP), to compare the transduction efficiency and tropism of seven different rAAV serotypes in the adult rat brain. Using bioluminescent imaging (BLI), we could noninvasively investigate the location, magnitude, duration, and the kinetics of transgene expression in the brain. Immunohistochemical analysis for eGFP revealed the identification and number of the labeled cells (Fig. 4).

For all rAAV serotypes tested, gene expression was found to be stable over time. This is in accordance with the literature as stable expression up to 1 year post-injection has been reported (106–108). Differences described in the onset of transgene expression between different serotypes refer to earlier time points (4–7 days) (99). Immunohistochemical analysis at 6 weeks following injection revealed distinct expression patterns for the different serotypes. Further detailed confocal analysis revealed that mainly dopaminergic neurons and almost no astrocytes were transduced. These results show that various rAAV serotypes efficiently transduce nigral dopaminergic neurons, but significant differences in transgene expression pattern and level were observed (100).

6.2. Disease Modeling: Parkinson's Gene

Generation of animal models of disease has been spurred by the identification of disease genes from genetic linkage and association studies (109). These genes are attractive candidates for disease modeling because of their translational quality and because the mode of inheritance of the disease gene offers clues as to how to design the disease model. For instance, at least ten genes have been identified as a cause of familial forms of Parkinson's disease (PD) (110–112). In this section, we will focus on rAAV-mediated disease modeling using one of these genes, alpha-synuclein. Alpha-synuclein is inherited in an autosomal dominant fashion, suggesting a gain of function as toxic mechanism. Alpha-synuclein is also a

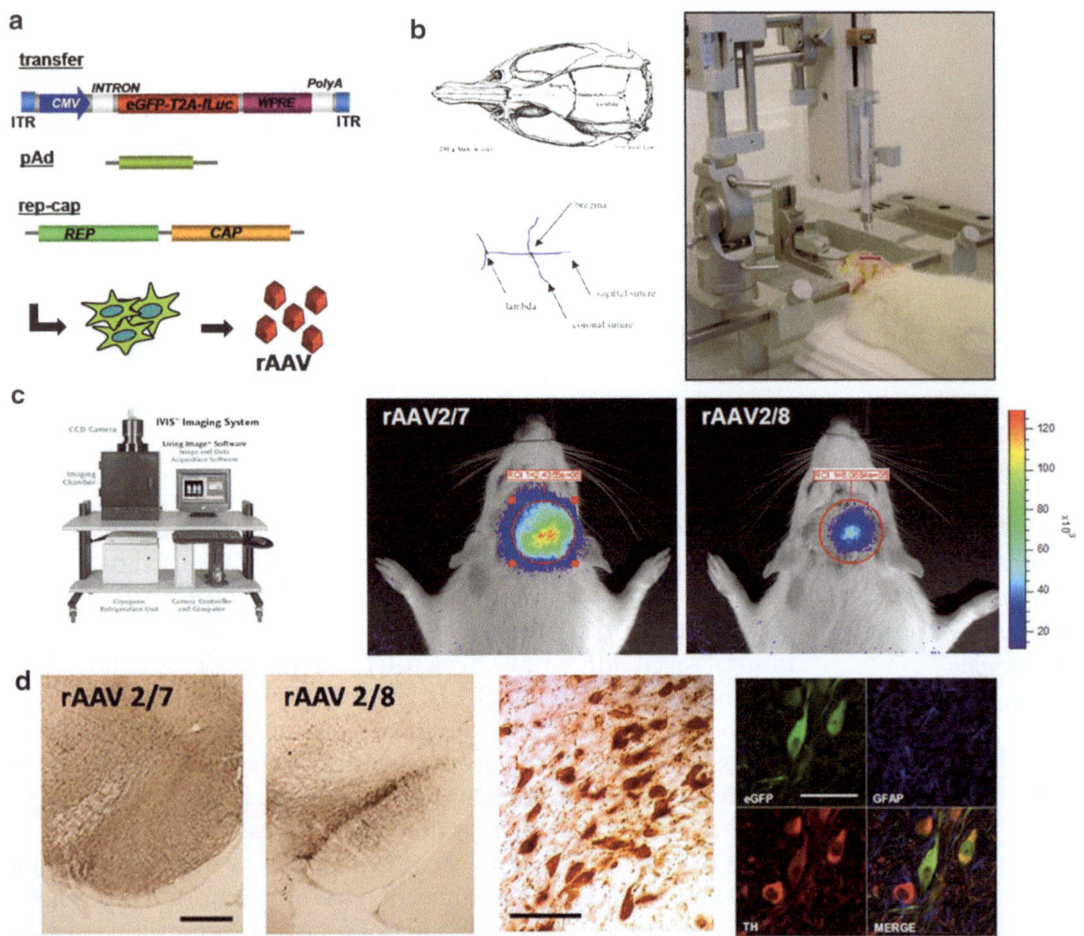

Fig. 4. In vivo validation of different rAAV serotypes using BLI and immunohistochemistry. (**a**) Production of rAAV vectors using eGFP-T2A-fLuc as transgenes. (**b**) Stereotactic injection of rAAV vectors in the rat brain. (**c**) In vivo bioluminescence imaging of fLuc expression after injection of rAAV 2/7 and 2/8 in the rat brain, *left*: IVIS imaging system, *middle*: BLI signal in overlay with a photographic image of the head region. Twelve days after injection, a light signal originating from the site of injection (*SN*) was detected. (**d**) Immunochemical staining for eGFP (*left*) 6 weeks after injection in SN. Scale bar 400 μm, *right*: confocal images of fluorescent double staining for eGFP and tyrosine hydroxylase (*TH*) shows extensive colocalization in dopaminergic neurons in the transduced region. *Scale bar* 50 μm.

major component of proteinaceous cellular inclusions typical of PD, termed Lewy bodies (113). For these reasons, it was hypothesized that overexpression of alpha-synuclein in the brain may lead to pathological features reminiscent of PD such as neuronal cell death and the accumulation of cellular inclusions. This hypothesis was initially tested using rAAV2 to overexpress alpha-synuclein (under control of a CBA promoter) in the main locus of neurodegeneration in PD, the substantia nigra. These experiments showed a very efficient transduction of nigral dopamine neurons with alpha-synuclein (90–95%) and led to a 30–80% reduction of nigral dopaminergic neurons at 8–16 weeks postinjection (28, 29).

By comparison, lentiviral vectors encoding alpha-synuclein were also capable of inducing neuronal cell loss, although this was comparatively delayed with 24–35% cell loss at 5 months (114) or 10–25% TH cell loss at 12 months postinjection. These findings that rAAV-mediated overexpression of alpha-synuclein can be used to model PD have also been observed using rAAV2/5 to overexpress alpha-synuclein in the ventral midbrain of marmoset (115).

One of the advantages of viral vectors for overexpression of genes is the possibility to test several gene variants relatively quickly compared to other techniques such as via the generation of transgenic animals. This is illustrated by studies which have explored the links between alpha-synuclein phosphorylation and PD pathology. Investigations of the phosphorylation of alpha-synuclein in PD and aged brains have shown that phosphorylation of alpha-synuclein at serine 129 (S129) is correlated with PD pathology (116–118). Thus far, three studies investigating the role of S129 phosphorylation have been performed using rAAV-mediated overexpression of alpha-synuclein mutants at S129 either mimicking (S129E/D) or blocking (S129A) phosphorylation (119–121). All three studies overexpressed wild-type alpha-synuclein as well as the S129A and S129D variants in the rat substantia nigra, using different rAAV subtypes and different promoters: rAAV2/5-CBA-alpha-synuclein (120), rAAV2/6-CMV-alpha-synuclein (119), or rAAV2/8-CBA-alpha-synuclein (122). All three studies showed expression of alpha-synuclein in a majority of nigral cells and confirmed toxicity of overexpressed alpha-synuclein. Furthermore, the studies with the rAAV2/5 and rAAV2/6 constructs showed enhanced toxicity for S129A compared to the wild type, while the S129D mutant showed reduced or no toxicity (119, 120). With rAAV2/8, the alpha-synuclein-mediated toxicity was equivalent for the wild type as well as for both S129 mutants (121), an observation potentially explained by increased expression levels obtained by rAAV2/8 compared to rAAV2/5 or rAAV2/6 in the substantia nigra (see Table 2 above and (121)).

7. Concluding Remarks

Since the first engineering of rAAV vectors in 1982, the rAAV vector field has confirmed that rAAV vectors are a powerful, safe, and versatile tool to deliver genes in vivo and particularly in the brain. Vector design and production processes have been optimized from which large quantities of high titer vectors can readily be obtained. As illustrated in this chapter, rAAV vectors have shown excellent gene transfer potential and good safety in the brain for applications ranging from cellular labeling and non-invasive imaging to disease modeling and clinical gene therapy. A large number of AAV serotypes have now been discovered and assessed for their gene transfer potential revealing serotype specific cellular tropisms, a finding that has spurred research to determine the relative performance of separate AAV serotypes in specific tissues. This continuing work offers researchers a broad rAAV toolbox from which to select the optimal vector for a given application and holds the promise of yet more effective development of rAAV-mediated disease models and gene therapy strategies in the brain.

References

1. Samulski, R. J., Berns, K. I., Tan, M., and Muzyczka, N. (1982) Cloning of adeno-associated virus into pBR322: rescue of intact virus from the recombinant plasmid in human cells, *Proc Natl Acad Sci USA 79*, 2077–2081.
2. Atchison, R. W., Casto, B. C., and Hammon, W. M. (1965) Adenovirus-Associated Defective Virus Particles, *Science 149*, 754–756.
3. Buller, R. M., Janik, J. E., Sebring, E. D., and Rose, J. A. (1981) Herpes simplex virus types 1 and 2 completely help adenovirus-associated virus replication, *J Virol 40*, 241–247.
4. Baekelandt, V., De Strooper, B., Nuttin, B., and Debyser, Z. (2000) Gene therapeutic strategies for neurodegenerative diseases, *Curr Opin Mol Ther. 2*, 540–554.
5. Grieger, J. C., and Samulski, R. J. (2005) Packaging capacity of adeno-associated virus serotypes: impact of larger genomes on infectivity and postentry steps, *J Virol 79*, 9933–9944.
6. Xiao, X., Li, J., and Samulski, R. J. (1996) Efficient long-term gene transfer into muscle tissue of immunocompetent mice by adeno-associated virus vector, *J Virol 70*, 8098–8108.
7. Miao, C. H., Snyder, R. O., Schowalter, D. B., Patijn, G. A., Donahue, B., Winther, B., and Kay, M. A. (1998) The kinetics of rAAV integration in the liver, *Nat Genet 19*, 13–15.
8. Chu, D., Thistlethwaite, P. A., Sullivan, C. C., Grifman, M. S., and Weitzman, M. D. (2004) Gene delivery to the mammalian heart using AAV vectors, *Methods Mol Biol 246*, 213–224.
9. Hellstrom, M., Ruitenberg, M. J., Pollett, M. A., Ehlert, E. M., Twisk, J., Verhaagen, J., and Harvey, A. R. (2009) Cellular tropism and transduction properties of seven adeno-associated viral vector serotypes in adult retina after intravitreal injection, *Gene Ther 16*, 521–532.
10. Tenenbaum, L., Chtarto, A., Lehtonen, E., Velu, T., Brotchi, J., and Levivier, M. (2004) Recombinant AAV-mediated gene delivery to the central nervous system, *J Gene Med 6*, S212–222.
11. Coura Rdos, S., and Nardi, N. B. (2007) The state of the art of adeno-associated virus-based vectors in gene therapy, *Virol J 4*, 99.
12. Mueller, C., and Flotte, T. R. (2008) Clinical gene therapy using recombinant adeno-associated virus vectors, *Gene Ther 15*, 858–863.
13. Zaiss, A. K., and Muruve, D. A. (2008) Immunity to adeno-associated virus vectors in animals and humans: a continued challenge, *Gene Ther 15*, 808–816.
14. Wu, Z., Asokan, A., and Samulski, R. J. (2006) Adeno-associated virus serotypes: vector toolkit for human gene therapy, *Mol Ther 14*, 316–327.
15. Gao, G., Vandenberghe, L. H., and Wilson, J. M. (2005) New recombinant serotypes of AAV vectors, *Curr Gene Ther 5*, 285–297.
16. Christine, C. W., Starr, P. A., Larson, P. S., Eberling, J. L., Jagust, W. J., Hawkins, R. A., VanBrocklin, H. F., Wright, J. F., Bankiewicz, K. S., and Aminoff, M. J. (2009) Safety and tolerability of putaminal AADC gene therapy for Parkinson disease, *Neurology 73*, 1662–1669.
17. Kaplitt, M. G., Feigin, A., Tang, C., Fitzsimons, H. L., Mattis, P., Lawlor, P. A., Bland, R. J., Young, D., Strybing, K., Eidelberg, D., and During, M. J. (2007) Safety and tolerability of gene therapy with an adeno-associated virus (AAV) borne GAD gene for Parkinson's disease: an open label, phase I trial, *Lancet 369*, 2097–2105.
18. Mandel, R. J. CERE-110, an adeno-associated virus-based gene delivery vector expressing human nerve growth factor for the treatment of Alzheimer's disease, *Curr Opin Mol Ther 12*, 240–247.
19. Samulski, R. J., Chang, L. S., and Shenk, T. (1987) A recombinant plasmid from which an

infectious adeno-associated virus genome can be excised in vitro and its use to study viral replication, *J Virol 61*, 3096–3101.

20. Hauck, B., Zhao, W., High, K., and Xiao, W. (2004) Intracellular viral processing, not single-stranded DNA accumulation, is crucial for recombinant adeno-associated virus transduction, *J Virol 78*, 13678–13686.
21. Thomas, C. E., Storm, T. A., Huang, Z., and Kay, M. A. (2004) Rapid uncoating of vector genomes is the key to efficient liver transduction with pseudotyped adeno-associated virus vectors, *J Virol 78*, 3110–3122.
22. Ferrari, F. K., Samulski, T., Shenk, T., and Samulski, R. J. (1996) Second-strand synthesis is a rate-limiting step for efficient transduction by recombinant adeno-associated virus vectors, *J Virol 70*, 3227–3234.
23. Fisher, K. J., Gao, G. P., Weitzman, M. D., DeMatteo, R., Burda, J. F., and Wilson, J. M. (1996) Transduction with recombinant adeno-associated virus for gene therapy is limited by leading-strand synthesis, *J Virol 70*, 520–532.
24. Nakai, H., Storm, T. A., and Kay, M. A. (2000) Recruitment of single-stranded recombinant adeno-associated virus vector genomes and intermolecular recombination are responsible for stable transduction of liver in vivo, *J Virol 74*, 9451–9463.
25. Rabinowitz, J. E., and Samulski, R. J. (2000) Building a better vector: the manipulation of AAV virions, *Virology 278*, 301–308.
26. Grimm, D., and Kay, M. A. (2003) From virus evolution to vector revolution: use of naturally occurring serotypes of adeno-associated virus (AAV) as novel vectors for human gene therapy, *Curr Gene Ther 3*, 281–304.
27. Deroose, C. M., Reumers, V., Debyser, Z., and Baekelandt, V. (2009) Seeing genes at work in the living brain with non-invasive molecular imaging, *Curr Gene Ther 9*, 212–238.
28. Kirik, D., Rosenblad, C., Burger, C., Lundberg, C., Johansen, T. E., Muzyczka, N., Mandel, R. J., and Bjorklund, A. (2002) Parkinson-like neurodegeneration induced by targeted overexpression of alpha-synuclein in the nigrostriatal system, *J Neurosci 22*, 2780–2791.
29. Kirik, D., Annett, L. E., Burger, C., Muzyczka, N., Mandel, R. J., and Bjorklund, A. (2003) Nigrostriatal alpha-synucleinopathy induced by viral vector-mediated overexpression of human alpha-synuclein: a new primate model of Parkinson's disease, *Proc Natl Acad Sci USA 100*, 2884–2889.
30. Lauwers, E., Debyser, Z., Van Dorpe, J., De Strooper, B., Nuttin, B., and Baekelandt, V. (2003) Neuropathology and neurodegeneration in rodent brain induced by lentiviral vector-mediated overexpression of alpha-synuclein, *Brain Pathol 13*, 364–372.
31. Klein, R. L., King, M. A., Hamby, M. E., and Meyer, E. M. (2002) Dopaminergic cell loss induced by human A30P alpha-synuclein gene transfer to the rat substantia nigra, *Hum Gene Ther 13*, 605–612.
32. Manfredsson, F. P., Burger, C., Sullivan, L. F., Muzyczka, N., Lewin, A. S., and Mandel, R. J. (2007) rAAV-mediated nigral human parkin over-expression partially ameliorates motor deficits via enhanced dopamine neurotransmission in a rat model of Parkinson's disease, *Exp Neurol 207*, 289–301.
33. Vercammen, L., Van der Perren, A., Vaudano, E., Gijsbers, R., Debyser, Z., Van den Haute, C., and Baekelandt, V. (2006) Parkin protects against neurotoxicity in the 6-hydroxydopamine rat model for Parkinson's disease, *Mol Ther 14*, 716–723.
34. Winklhofer, K. F. (2007) The parkin protein as a therapeutic target in Parkinson's disease, *Expert Opin Ther Targets 11*, 1543–1552.
35. Danos, O. (2008) AAV vectors for RNA-based modulation of gene expression, *Gene Ther 15*, 864–869.
36. Shevtsova, Z., Malik, J. M., Michel, U., Bahr, M., and Kugler, S. (2005) Promoters and serotypes: targeting of adeno-associated virus vectors for gene transfer in the rat central nervous system in vitro and in vivo, *Exp Physiol 90*, 53–59.
37. Gao, G., Qu, G., Burnham, M. S., Huang, J., Chirmule, N., Joshi, B., Yu, Q. C., Marsh, J. A., Conceicao, C. M., and Wilson, J. M. (2000) Purification of recombinant adeno-associated virus vectors by column chromatography and its performance in vivo, *Hum Gene Ther 11*, 2079–2091.
38. Clark, K. R., Voulgaropoulou, F., Fraley, D. M., and Johnson, P. R. (1995) Cell lines for the production of recombinant adeno-associated virus, *Hum Gene Ther 6*, 1329–1341.
39. Blouin, V., Brument, N., Toublanc, E., Raimbaud, I., Moullier, P., and Salvetti, A. (2004) Improving rAAV production and purification: towards the definition of a scaleable process, *J Gene Med 6 Suppl 1*, S223–228.
40. Wright, J. F. (2009) Transient transfection methods for clinical adeno-associated viral vector production, *Hum Gene Ther 20*, 698–706.
41. Grimm, D., Kay, M. A., and Kleinschmidt, J. A. (2003) Helper virus-free, optically controllable, and two-plasmid-based production of adeno-associated virus vectors of serotypes 1 to 6, *Mol Ther 7*, 839–850.

42. Gao, G. P., Qu, G., Faust, L. Z., Engdahl, R. K., Xiao, W., Hughes, J. V., Zoltick, P. W., and Wilson, J. M. (1998) High-titer adeno-associated viral vectors from a Rep/Cap cell line and hybrid shuttle virus, *Hum Gene Ther 9*, 2353–2362.
43. Liu, X. L., Clark, K. R., and Johnson, P. R. (1999) Production of recombinant adeno-associated virus vectors using a packaging cell line and a hybrid recombinant adenovirus, *Gene Ther 6*, 293–299.
44. Conway, J. E., Zolotukhin, S., Muzyczka, N., Hayward, G. S., and Byrne, B. J. (1997) Recombinant adeno-associated virus type 2 replication and packaging is entirely supported by a herpes simplex virus type 1 amplicon expressing Rep and Cap, *J Virol 71*, 8780–8789.
45. Urabe, M., Ding, C., and Kotin, R. M. (2002) Insect cells as a factory to produce adeno-associated virus type 2 vectors, *Hum Gene Ther 13*, 1935–1943.
46. Negrete, A., Yang, L. C., Mendez, A. F., Levy, J. R., and Kotin, R. M. (2007) Economized large-scale production of high yield of rAAV for gene therapy applications exploiting baculovirus expression system, *J Gene Med 9*, 938–948.
47. Lock, M., Alvira, M., Vandenberghe, L. H., Samanta, A., Toelen, J., Debyser, Z., and Wilson, J. M. (2010) Rapid, Simple and Versatile Manufacturing of Recombinant Adeno-Associated Virus Vectors at Scale, *Hum Gene Ther* 21 (10), 1259–1271.
48. Toelen J, V. d. P. A., Carlon M, Michiels M, Lock M, Vandenberghe L, Bannert N, Wilson JM, Gijsbers R, Debyser Z. (2010) Novel approach to research grade AAV vector manufacturing and separation of distinct AAV serotypes for in vivo applications. *Submitted.*
49. Eslamboli, A., Romero-Ramos, M., Burger, C., Bjorklund, T., Muzyczka, N., Mandel, R. J., Baker, H., Ridley, R. M., and Kirik, D. (2007) Long-term consequences of human alpha-synuclein overexpression in the primate ventral midbrain, *Brain 130*, 799–815.
50. Passini, M. A., Watson, D. J., Vite, C. H., Landsburg, D. J., Feigenbaum, A. L., and Wolfe, J. H. (2003) Intraventricular brain injection of adeno-associated virus type 1 (AAV1) in neonatal mice results in complementary patterns of neuronal transduction to AAV2 and total long-term correction of storage lesions in the brains of beta-glucuronidase-deficient mice, *J Virol 77*, 7034–7040.
51. Broekman, M. L., Comer, L. A., Hyman, B. T., and Sena-Esteves, M. (2006) Adeno-associated virus vectors serotyped with AAV8 capsid are more efficient than AAV-1 or -2 serotypes for widespread gene delivery to the neonatal mouse brain, *Neuroscience 138*, 501–510.
52. Foust, K. D., Nurre, E., Montgomery, C. L., Hernandez, A., Chan, C. M., and Kaspar, B. K. (2009) Intravascular AAV9 preferentially targets neonatal neurons and adult astrocytes, *Nat Biotechnol 27*, 59–65.
53. Towne, C., Raoul, C., Schneider, B. L., and Aebischer, P. (2008) Systemic AAV6 delivery mediating RNA interference against SOD1: neuromuscular transduction does not alter disease progression in fALS mice, *Mol Ther 16*, 1018–1025.
54. Foust, K. D., Poirier, A., Pacak, C. A., Mandel, R. J., and Flotte, T. R. (2008) Neonatal intraperitoneal or intravenous injections of recombinant adeno-associated virus type 8 transduce dorsal root ganglia and lower motor neurons, *Hum Gene Ther 19*, 61–70.
55. Schneider, B., Zufferey, R., and Aebischer, P. (2008) Viral vectors, animal models and new therapies for Parkinson's disease, *Parkinsonism Relat Disord 14 Suppl 2*, S169–171.
56. Paterna, J. C., Feldon, J., and Bueler, H. (2004) Transduction profiles of recombinant adeno-associated virus vectors derived from serotypes 2 and 5 in the nigrostriatal system of rats, *J Virol 78*, 6808–6817.
57. Burger, C., Gorbatyuk, O. S., Velardo, M. J., Peden, C. S., Williams, P., Zolotukhin, S., Reier, P. J., Mandel, R. J., and Muzyczka, N. (2004) Recombinant AAV viral vectors pseudotyped with viral capsids from serotypes 1, 2, and 5 display differential efficiency and cell tropism after delivery to different regions of the central nervous system, *Mol Ther 10*, 302–317.
58. Taymans, J. M., Vandenberghe, L. H., Haute, C. V., Thiry, I., Deroose, C. M., Mortelmans, L., Wilson, J. M., Debyser, Z., and Baekelandt, V. (2007) Comparative analysis of adeno-associated viral vector serotypes 1, 2, 5, 7, and 8 in mouse brain, *Hum Gene Ther 18*, 195–206.
59. McFarland, N. R., Lee, J. S., Hyman, B. T., and McLean, P. J. (2009) Comparison of transduction efficiency of recombinant AAV serotypes 1, 2, 5, and 8 in the rat nigrostriatal system, *J Neurochem 109*, 838–845.
60. Dodiya, H. B., Bjorklund, T., Stansell Iii, J., Mandel, R. J., Kirik, D., and Kordower, J. H. (2010) Differential Transduction Following Basal Ganglia Administration of Distinct Pseudotyped AAV Capsid Serotypes in Nonhuman Primates, *Mol Ther* 18 (3), 579–587.

61. Blits, B., Derks, S., Twisk, J., Ehlert, E., Prins, J., and Verhaagen, J. (2010) Adeno-associated viral vector (AAV)-mediated gene transfer in the red nucleus of the adult rat brain: Comparative analysis of the transduction properties of seven AAV serotypes and lentiviral vectors, *J Neurosci Methods* 185 (2), 257–263.
62. Passini, M. A., Watson, D. J., and Wolfe, J. H. (2004) Gene delivery to the mouse brain with adeno-associated virus, *Methods Mol Biol 246*, 225–236.
63. Bankiewicz, K. S., Eberling, J. L., Kohutnicka, M., Jagust, W., Pivirotto, P., Bringas, J., Cunningham, J., Budinger, T. F., and Harvey-White, J. (2000) Convection-enhanced delivery of AAV vector in parkinsonian monkeys; in vivo detection of gene expression and restoration of dopaminergic function using pro-drug approach, *Exp Neurol 164*, 2–14.
64. Cunningham, J., Oiwa, Y., Nagy, D., Podsakoff, G., Colosi, P., and Bankiewicz, K. S. (2000) Distribution of AAV-TK following intracranial convection-enhanced delivery into rats, *Cell Transplant 9*, 585–594.
65. Mastakov, M. Y., Baer, K., Xu, R., Fitzsimons, H., and During, M. J. (2001) Combined injection of rAAV with mannitol enhances gene expression in the rat brain, *Mol Ther 3*, 225–232.
66. Mastakov, M. Y., Baer, K., Kotin, R. M., and During, M. J. (2002) Recombinant adeno-associated virus serotypes 2- and 5-mediated gene transfer in the mammalian brain: quantitative analysis of heparin co-infusion, *Mol Ther 5*, 371–380.
67. Burger, C., Nguyen, F. N., Deng, J., and Mandel, R. J. (2005) Systemic mannitol-induced hyperosmolality amplifies rAAV2-mediated striatal transduction to a greater extent than local co-infusion, *Mol Ther 11*, 327–331.
68. Hadaczek, P., Mirek, H., Bringas, J., Cunningham, J., and Bankiewicz, K. (2004) Basic fibroblast growth factor enhances transduction, distribution, and axonal transport of adeno-associated virus type 2 vector in rat brain, *Hum Gene Ther 15*, 469–479.
69. Moskalenko, M., Chen, L., van Roey, M., Donahue, B. A., Snyder, R. O., McArthur, J. G., and Patel, S. D. (2000) Epitope mapping of human anti-adeno-associated virus type 2 neutralizing antibodies: implications for gene therapy and virus structure, *J Virol 74*, 1761–1766.
70. Rutledge, E. A., Halbert, C. L., and Russell, D. W. (1998) Infectious clones and vectors derived from adeno-associated virus (AAV) serotypes other than AAV type 2, *J Virol 72*, 309–319.
71. Gao, G. P., Alvira, M. R., Wang, L., Calcedo, R., Johnston, J., and Wilson, J. M. (2002) Novel adeno-associated viruses from rhesus monkeys as vectors for human gene therapy, *Proc Natl Acad Sci USA 99*, 11854–11859.
72. Gao, G., Vandenberghe, L. H., Alvira, M. R., Lu, Y., Calcedo, R., Zhou, X., and Wilson, J. M. (2004) Clades of Adeno-associated viruses are widely disseminated in human tissues, *J Virol 78*, 6381–6388.
73. Schmidt, M., Grot, E., Cervenka, P., Wainer, S., Buck, C., and Chiorini, J. A. (2006) Identification and characterization of novel adeno-associated virus isolates in ATCC virus stocks, *J Virol 80*, 5082–5085.
74. Gao, G., Alvira, M. R., Somanathan, S., Lu, Y., Vandenberghe, L. H., Rux, J. J., Calcedo, R., Sanmiguel, J., Abbas, Z., and Wilson, J. M. (2003) Adeno-associated viruses undergo substantial evolution in primates during natural infections, *Proc Natl Acad Sci USA 100*, 6081–6086.
75. Choi, V. W., McCarty, D. M., and Samulski, R. J. (2005) AAV hybrid serotypes: improved vectors for gene delivery, *Curr Gene Ther 5*, 299–310.
76. Ding, W., Zhang, L., Yan, Z., and Engelhardt, J. F. (2005) Intracellular trafficking of adeno-associated viral vectors, *Gene Ther 12*, 873–880.
77. Chen, S., Kapturczak, M., Loiler, S. A., Zolotukhin, S., Glushakova, O. Y., Madsen, K. M., Samulski, R. J., Hauswirth, W. W., Campbell-Thompson, M., Berns, K. I., Flotte, T. R., Atkinson, M. A., Tisher, C. C., and Agarwal, A. (2005) Efficient transduction of vascular endothelial cells with recombinant adeno-associated virus serotype 1 and 5 vectors, *Hum Gene Ther 16*, 235–247.
78. Seiler, M. P., Miller, A. D., Zabner, J., and Halbert, C. L. (2006) Adeno-associated virus types 5 and 6 use distinct receptors for cell entry, *Hum Gene Ther 17*, 10–19.
79. Negishi, A., Chen, J., McCarty, D. M., Samulski, R. J., Liu, J., and Superfine, R. (2004) Analysis of the interaction between adeno-associated virus and heparan sulfate using atomic force microscopy, *Glycobiology 14*, 969–977.
80. Wu, Z., Miller, E., Agbandje-McKenna, M., and Samulski, R. J. (2006) Alpha2,3 and alpha2,6 N-linked sialic acids facilitate efficient binding and transduction by adeno-associated virus types 1 and 6, *J Virol 80*, 9093–9103.
81. Summerford, C., and Samulski, R. J. (1998) Membrane-associated heparan sulfate proteoglycan is a receptor for adeno-associated virus type 2 virions, *J Virol 72*, 1438–1445.

82. Summerford, C., Bartlett, J. S., and Samulski, R. J. (1999) AlphaVbeta5 integrin: a co-receptor for adeno-associated virus type 2 infection, *Nat Med 5*, 78–82.
83. Qing, K., Mah, C., Hansen, J., Zhou, S., Dwarki, V., and Srivastava, A. (1999) Human fibroblast growth factor receptor 1 is a co-receptor for infection by adeno-associated virus 2, *Nat Med 5*, 71–77.
84. Kashiwakura, Y., Tamayose, K., Iwabuchi, K., Hirai, Y., Shimada, T., Matsumoto, K., Nakamura, T., Watanabe, M., Oshimi, K., and Daida, H. (2005) Hepatocyte growth factor receptor is a coreceptor for adeno-associated virus type 2 infection, *J Virol 79*, 609–614.
85. Akache, B., Grimm, D., Pandey, K., Yant, S. R., Xu, H., and Kay, M. A. (2006) The 37/67-kilodalton laminin receptor is a receptor for adeno-associated virus serotypes 8, 2, 3, and 9, *J Virol 80*, 9831–9836.
86. Rabinowitz, J. E., Bowles, D. E., Faust, S. M., Ledford, J. G., Cunningham, S. E., and Samulski, R. J. (2004) Cross-dressing the virion: the transcapsidation of adeno-associated virus serotypes functionally defines subgroups, *J Virol 78*, 4421–4432.
87. Blackburn, S. D., Steadman, R. A., and Johnson, F. B. (2006) Attachment of adeno-associated virus type 3 H to fibroblast growth factor receptor 1, *Arch Virol 151*, 617–623.
88. Walters, R. W., Yi, S. M., Keshavjee, S., Brown, K. E., Welsh, M. J., Chiorini, J. A., and Zabner, J. (2001) Binding of adeno-associated virus type 5 to 2,3-linked sialic acid is required for gene transfer, *J Biol Chem 276*, 20610–20616.
89. Kaludov, N., Brown, K. E., Walters, R. W., Zabner, J., and Chiorini, J. A. (2001) Adeno-associated virus serotype 4 (AAV4) and AAV5 both require sialic acid binding for hemagglutination and efficient transduction but differ in sialic acid linkage specificity, *J Virol 75*, 6884–6893.
90. Di Pasquale, G., Davidson, B. L., Stein, C. S., Martins, I., Scudiero, D., Monks, A., and Chiorini, J. A. (2003) Identification of PDGFR as a receptor for AAV-5 transduction, *Nat Med 9*, 1306–1312.
91. Fitzsimons, H. L., Bland, R. J., and During, M. J. (2002) Promoters and regulatory elements that improve adeno-associated virus transgene expression in the brain, *Methods 28*, 227–236.
92. Paterna, J. C., Moccetti, T., Mura, A., Feldon, J., and Bueler, H. (2000) Influence of promoter and WHV post-transcriptional regulatory element on AAV-mediated transgene expression in the rat brain, *Gene Ther 7*, 1304–1311.
93. Ayuso, E., Mingozzi, F., Montane, J., Leon, X., Anguela, X. M., Haurigot, V., Edmonson, S. A., Africa, L., Zhou, S., High, K. A., Bosch, F., and Wright, J. F. (2010) High AAV vector purity results in serotype- and tissue-independent enhancement of transduction efficiency, *Gene Ther* 17(4), 503–510.
94. Flotte, T. R., Fischer, A. C., Goetzmann, J., Mueller, C., Cebotaru, L., Yan, Z., Wang, L., Wilson, J. M., Guggino, W. B., and Engelhardt, J. F. Dual reporter comparative indexing of rAAV pseudotyped vectors in chimpanzee airway, *Mol Ther 18*, 594–600.
95. Cearley, C. N., and Wolfe, J. H. (2006) Transduction characteristics of adeno-associated virus vectors expressing cap serotypes 7, 8, 9, and Rh10 in the mouse brain, *Mol Ther 13*, 528–537.
96. Lawlor, P. A., Bland, R. J., Mouravlev, A., Young, D., and During, M. J. (2009) Efficient gene delivery and selective transduction of glial cells in the mammalian brain by AAV serotypes isolated from nonhuman primates, *Mol Ther 17*, 1692–1702.
97. Cearley, C. N., Vandenberghe, L. H., Parente, M. K., Carnish, E. R., Wilson, J. M., and Wolfe, J. H. (2008) Expanded repertoire of AAV vector serotypes mediate unique patterns of transduction in mouse brain, *Mol Ther 16*, 1710–1718.
98. Van Damme, P., Bogaert, E., Dewil, M., Hersmus, N., Kiraly, D., Scheveneels, W., Bockx, I., Braeken, D., Verpoorten, N., Verhoeven, K., Timmerman, V., Herijgers, P., Callewaert, G., Carmeliet, P., Van Den Bosch, L., and Robberecht, W. (2007) Astrocytes regulate GluR2 expression in motor neurons and their vulnerability to excitotoxicity, *Proc Natl Acad Sci USA 104*, 14825–14830.
99. Reimsnider, S., Manfredsson, F. P., Muzyczka, N., and Mandel, R. J. (2007) Time course of transgene expression after intrastriatal pseudotyped rAAV2/1, rAAV2/2, rAAV2/5, and rAAV2/8 transduction in the rat, *Mol Ther 15*, 1504–1511.
100. Van der Perren Anke, T. J., Carlon Marianne, Van den Haute Chris, Coun Frea, Heeman Bavo, Reumers Veerle, Vandenberghe Luk H, Wilson James M., Debyser Zeger, Baekelandt Veerle. (2011) Efficient and stable transduction of dopaminergic neurons in rat substantia nigra rAAV 2/1, 2/2, 2/5, 2/6.2, 2/7, 2/8 and 2/9, *Gene Ther* 18 (5), 517–527.
101. Klein, R. L., Dayton, R. D., Leidenheimer, N. J., Jansen, K., Golde, T. E., and Zweig, R. M. (2006) Efficient neuronal gene transfer with AAV8 leads to neurotoxic levels of tau or green fluorescent proteins, *Mol Ther 13*, 517–527.

102. Markakis, E. A., Vives, K. P., Bober, J., Leichtle, S., Leranth, C., Beecham, J., Elsworth, J. D., Roth, R. H., Samulski, R. J., and Redmond, D. E., Jr. (2010) Comparative Transduction Efficiency of AAV Vector Serotypes 1–6 in the Substantia Nigra and Striatum of the Primate Brain, *Mol Ther* 18 (3), 588–593.

103. Vite, C. H., Passini, M. A., Haskins, M. E., and Wolfe, J. H. (2003) Adeno-associated virus vector-mediated transduction in the cat brain, *Gene Ther 10*, 1874–1881.

104. Kornum, B. R., Stott, S. R., Mattsson, B., Wisman, L., Ettrup, A., Hermening, S., Knudsen, G. M., and Kirik, D. (2009) Adeno-associated viral vector serotypes 1 and 5 targeted to the neonatal rat and pig striatum induce widespread transgene expression in the forebrain, *Exp Neurol.*

105. Ibrahimi, A., Vande Velde, G., Reumers, V., Toelen, J., Thiry, I., Vandeputte, C., Vets, S., Deroose, C., Bormans, G., Baekelandt, V., Debyser, Z., and Gijsbers, R. (2009) Highly efficient multicistronic lentiviral vectors with peptide 2A sequences, *Hum Gene Ther 20*, 845–860.

106. Mandel, R. J., Rendahl, K. G., Snyder, R. O., and Leff, S. E. (1999) Progress in direct striatal delivery of L-dopa via gene therapy for treatment of Parkinson's disease using recombinant adeno-associated viral vectors, *Exp Neurol 159*, 47–64.

107. Shen, Y., Muramatsu, S. I., Ikeguchi, K., Fujimoto, K. I., Fan, D. S., Ogawa, M., Mizukami, H., Urabe, M., Kume, A., Nagatsu, I., Urano, F., Suzuki, T., Ichinose, H., Nagatsu, T., Monahan, J., Nakano, I., and Ozawa, K. (2000) Triple transduction with adeno-associated virus vectors expressing tyrosine hydroxylase, aromatic-L-amino-acid decarboxylase, and GTP cyclohydrolase I for gene therapy of Parkinson's disease, *Hum Gene Ther 11*, 1509–1519.

108. Kirik, D., Georgievska, B., Burger, C., Winkler, C., Muzyczka, N., Mandel, R. J., and Bjorklund, A. (2002) Reversal of motor impairments in parkinsonian rats by continuous intrastriatal delivery of L-dopa using rAAV-mediated gene transfer, *Proc Natl Acad Sci USA 99*, 4708–4713.

109. Hardy, J., and Singleton, A. (2009) Genomewide association studies and human disease, *N Engl J Med 360*, 1759–1768.

110. Gasser, T. (2009) Molecular pathogenesis of Parkinson disease: insights from genetic studies, *Expert Rev Mol Med 11*, e22.

111. Lees, A. J., Hardy, J., and Revesz, T. (2009) Parkinson's disease, *Lancet 373*, 2055–2066.

112. Hardy, J., Lewis, P., Revesz, T., Lees, A., and Paisan-Ruiz, C. (2009) The genetics of Parkinson's syndromes: a critical review, *Curr Opin Genet Dev 19*, 254–265.

113. Savitt, J. M., Dawson, V. L., and Dawson, T. M. (2006) Diagnosis and treatment of Parkinson disease: molecules to medicine, *J Clin Invest 116*, 1744–1754.

114. Lo Bianco, C., Ridet, J. L., Schneider, B. L., Deglon, N., and Aebischer, P. (2002) alpha -Synucleinopathy and selective dopaminergic neuron loss in a rat lentiviral-based model of Parkinson's disease, *Proc Natl Acad Sci USA 99*, 10813–10818.

115. Eslamboli, A., Romero-Ramos, M., Burger, C., Bjorklund, T., Muzyczka, N., Mandel, R. J., Baker, H., Ridley, R. M., and Kirik, D. (2007) Long-term consequences of human alpha-synuclein overexpression in the primate ventral midbrain, *Brain: a journal of neurology 130*, 799–815.

116. Anderson, J. P., Walker, D. E., Goldstein, J. M., de Laat, R., Banducci, K., Caccavello, R. J., Barbour, R., Huang, J., Kling, K., Lee, M., Diep, L., Keim, P. S., Shen, X., Chataway, T., Schlossmacher, M. G., Seubert, P., Schenk, D., Sinha, S., Gai, W. P., and Chilcote, T. J. (2006) Phosphorylation of Ser-129 is the dominant pathological modification of alpha-synuclein in familial and sporadic Lewy body disease, *The Journal of biological chemistry 281*, 29739–29752.

117. Fujiwara, H., Hasegawa, M., Dohmae, N., Kawashima, A., Masliah, E., Goldberg, M. S., Shen, J., Takio, K., and Iwatsubo, T. (2002) alpha-Synuclein is phosphorylated in synucleinopathy lesions, *Nature cell biology 4*, 160–164.

118. Hasegawa, M., Fujiwara, H., Nonaka, T., Wakabayashi, K., Takahashi, H., Lee, V. M.-Y., Trojanowski, J. Q., Mann, D., and Iwatsubo, T. (2002) Phosphorylated alpha-synuclein is ubiquitinated in alpha-synucleinopathy lesions, *The Journal of biological chemistry 277*, 49071–49076.

119. Azeredo da Silveira, S., Schneider, B. L., Cifuentes-Diaz, C., Sage, D., Abbas-Terki, T., Iwatsubo, T., Unser, M. l., and Aebischer, P. (2009) Phosphorylation does not prompt, nor prevent, the formation of alpha-synuclein toxic species in a rat model of Parkinson's disease, *Human molecular genetics 18*, 872–887.

120. Gorbatyuk, O. S., Li, S., Sullivan, L. F., Chen, W., Kondrikova, G., Manfredsson, F. P.,

Mandel, R. J., and Muzyczka, N. (2008) The phosphorylation state of Ser-129 in human alpha-synuclein determines neurodegeneration in a rat model of Parkinson disease, *Proceedings of the National Academy of Sciences of the United States of America 105*, 763–768.

121. McFarland, N. R., Fan, Z., Xu, K., Schwarzschild, M. A., Feany, M. B., Hyman, B. T., and McLean, P. J. (2009) Alpha-synuclein S129 phosphorylation mutants do not alter nigrostriatal toxicity in a rat model of Parkinson disease, *J Neuropathol Exp Neurol 68*, 515–524.

Chapter 4

Using Lentiviral Vectors as Delivery Vehicles for Gene Therapy

Gregory A. Dissen, Jodi McBride, Alejandro Lomniczi, Valerie Matagne, Mauricio Dorfman, Tanaya L. Neff, Francesco Galimi, and Sergio R. Ojeda

Abstract

Viral-vector-based gene therapy is a powerful tool that allows experimental studies of species that previously were not amenable to genetic manipulation. Nonhuman primates (NHPs) are an invaluable resource for the study of genetic regulation of disease mechanisms. The main disadvantage of using NHPs as a preclinical model of human disease is the difficulty of manipulating the monkey genome using conventional gene modifying strategies. Lentiviruses offer the possibility of circumventing this difficulty by infecting and transducing dividing or nondividing cells, without eliciting an immune response. In addition, lentiviruses can permanently integrate into the genome of host cells and are able to maintain long-term expression. In this chapter, we describe the lentiviral vectors that we use to both express transgenes and suppress expression of endogenous genes via RNA interference (RNAi) in NHPs. We also discuss the safety features of currently available vectors that are especially important when lentiviral vectors are used in a species as closely related to humans as NHPs. In addition, we describe in detail the lentiviral vector production protocol we use and provide specific examples of how this vector can be employed to target peripheral tissues and the brain. Finally, we conclude by comparing and contrasting the use of lentiviral vectors with adeno-associated viral vectors with a particular emphasis on the use of these vectors in preclinical and clinical studies.

Key words: Lentiviral vectors, Adeno-associated viruses, Gene therapy, Gene regulation, Nonhuman primates, RNA interference, microRNA

1. Introduction

Nonhuman primates (NHPs) are close phylogenetic relatives of humans and thus represent a valuable animal model for the study of the relationship that exists between genetics and disease in higher primates. The difficulty with using NHPs is that conventional genetic tools, such as the production of transgenic animals

Alexei Morozov (ed.), *Controlled Genetic Manipulations*, Neuromethods, vol. 65,
DOI 10.1007/978-1-61779-533-6_4,

or the use of gene-targeting techniques, are impractical because of the extended time required for these gene manipulations to generate expected phenotypes and the large number of animals required to achieve the desired gene modification. Given these major obstacles, somatic cell gene transfer emerges as a viable approach to study the role that genetic modifications have in the genesis of various pathologies, as well as a tool to investigate potential therapeutic intervention in NHPs. An ideal gene therapy tool would allow investigators to enhance or suppress the expression of specific genes in a cell- or tissue-specific, and temporally defined, manner. Lentiviruses, or slow retroviruses, naturally perform some of these activities as they infect cells as directed by the glycoprotein specificity of their envelope, permanently integrate into the host cell genome, and upon integration express viral proteins (1, 2). An important characteristic that sets lentiviruses apart from other viral vectors and enhances their value as agents for gene therapy is their ability to transduce nondividing cells, as well as dividing cells in contrast to nonlenti retroviruses that transduce only dividing cells (1–4). Although the adeno-associated virus (AAV) system also shares this capability, lentiviral vectors hold two key advantages over AAV vectors: (a) lentiviruses allow for a larger packaging capacity (8–10 kb) (3, 5) compared to less than 5 kb for AAVs (3), and (b) the majority of self-inactivating (SIN) HIV-based infective lentiviral particles become integrated into the genome 3 days after infection (6) as compared to less than 10% for AAV (3). Because of these advantages, lentiviruses appear uniquely poised for studies aimed at manipulating gene expression in NHPs.

2. Materials

2.1. Lentiviral-Derived Vector Systems

The first lentiviral vectors described were based on HIV-1 (2) and although other vector systems have been developed using lentiviruses that are specific for other species (7), the HIV-1 vector system remains the vector of choice. The life cycle of HIV-1 lends itself well to the purpose of viral-mediated gene transfer, such as host cell attachment, receptor-mediated entry into host cells, viral mediated reverse transcription, and integration of the viral genome into the host cell chromatin (2, 8). Another feature of HIV-1 that makes it an ideal gene therapy vector is its ability to escape from cellular immune responses (9). Following HIV-1 infection, neutralizing antibodies are rarely generated in vivo (10), and lentiviral vectors similarly integrate their genome into that of target cells without an inflammatory response (2). This is not to imply that the current versions of lentiviral vectors are ready for all uses and methods of administration. For example, lentiviral vectors that have been pseudotyped with the most commonly used envelope protein are inactivated by human serum complement, preventing the use of these vectors in protocols involving systemic administration (11). While lentiviral vectors might not be ready for use in all situations, HIV-1-based lentiviral vectors are promising gene therapy tools (1–4, 7, 12).

2.2. Vector Components

An HIV-1 viral vector system consists of a replication-incompetent, nonpathogenic version of the virus where the *cis*- and *trans*-acting components required to generate a viral particle are separated on different plasmids. The viral sequences required in *cis* are included in the vector genome plasmid (see below; Fig. 1a, b), which contains the transgene (TG) or RNA interference (RNAi) (TG-RNAi) cassette. Once transcribed into RNA, these *cis* viral and transgene sequences are packaged into the vector particles and ultimately transferred into the target cells. The *trans*-acting sequences, typically encoding proteins required to assemble functional viral particles, are located on separate plasmids (Fig. 1c). These plasmids are used during the vector production phase (called "packaging"), but the genes for the *trans* elements are not included in the viral particle and therefore are not transduced into target cells. This results in the production of a viral particle that, upon infecting a cell, does not propagate the infection further.

2.3. cis Elements

The vector genome plasmid is the structure that will provide the *cis* elements required for efficient packaging, reverse transcription, and integration (2). Importantly, the vector carrying the TG-RNAi cassette also contains the regulatory components necessary for expression of the cassette. The vector is a stripped version of the HIV genome; it contains less than 5% of the parental genome (3). Although the 5′ end of the HIV-1 Gag (including the ATG) is included in the vector, the coding sequence for Gag has been interrupted with stop codons (the first of which occurs 21 codons after the ATG). The rest of the vector contains no other HIV-1 open reading frames. It is bounded by long terminal repeats (LTRs; Fig. 1a) and includes the packaging signal (ψ) which is required for incorporating the RNA in the virion (13). It also contains the primer binding site, the HIV-1 splice donor site, the splice acceptor site (all located in the same vicinity as the ψ signal), the *rev* response element (RRE), the central polypurine tract (cPPT), and a foreign *cis* element known as the woodchuck hepatitis posttranscriptional regulatory element (wPRE; Fig. 1a, b; (2, 14)). Each LTR is divided into the U3, retroviral promoter; R, repeat region/transcription start site; and U5, polyadenylation region; the LTRs are responsible for controlling viral replication and integration of the viral genome into the host genome (13, 14).

The mechanism by which the virus genome is replicated allows the production of the SIN viral vector (15, 16), resulting from a modification made to the 3′ LTR. Self-inactivation is achieved by deleting a 400-bp segment of the U3 retroviral promoter region in the 3′ LTR of the DNA used to produce the vector RNA (15, 17).

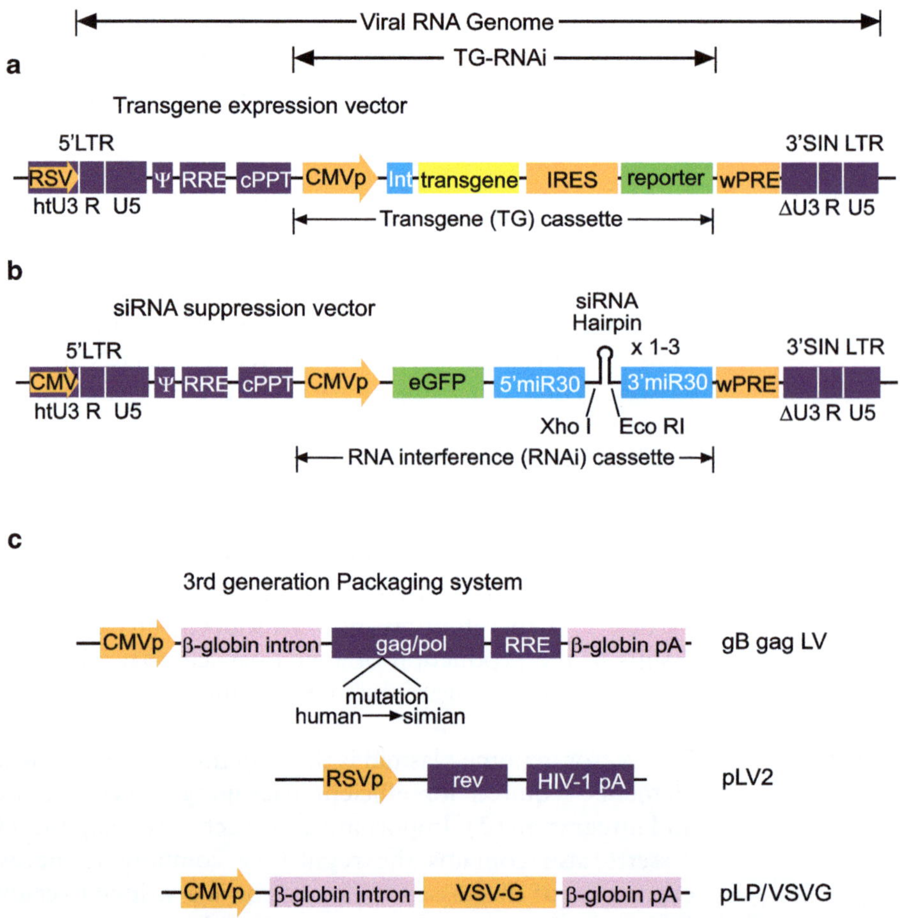

Fig. 1. Maps of the third-generation lentiviral vector system containing either a TG or RNAi (*TG-RNAi*) cassette. The 5′ LTR contains the heterologous U3 promoter (*htU3*), the repeat region/transcription initiation site (*R*), and the polyadenylation region (*U5*). The R and U5 regions are also present in the 3′ LTR, but importantly there is a 400-bp deletion in the 3′ LTR U3 region (*ΔU3*) that results in the self-inactivation (*SIN*) of these vectors. The heterologous promoters are either the Rous sarcoma virus promoter (*RSVp*) or the cytomegalovirus promoter (*CMVp*); the CMVp also serves as the TG-RNAi cassette promoter. The other components include the packaging signal (ψ), the Rev response element binding site (*RRE*), the central polypurine tract (*cPPT*), and the woodchuck hepatitis virus posttranslational regulatory element (*wPRE*). See Sect. 2.3 for descriptions of the vector components. (**a**) Map of a lentivirus delivery vector (33) containing a bicistronic TG cassette composed of a human nerve growth factor (*hNGF*) cDNA linked to an enhanced green fluorescent protein (*eGFP*) cDNA via an internal ribosome entry site (*IRES*) and placed downstream of the rat insulin II intron A (*Int*). (**b**) Map of the pPRIME lentivirus delivery vector (42) containing an artificial miR30 and replaceable hairpin structure able to accommodate 1–3 hairpins (47); Xho I and Eco RI restriction enzyme sites are used as cloning sites to insert the hairpin(s). (**c**) Maps of the packaging system containing the plasmids gB gag LV (32), pLP2, and pLP/VSVG (Invitrogen). The mutation noted in the *gag* gene increases the ability of the HIV-1-based lentiviral vector to transduce simian cells (32). The other components in the packaging plasmids are the β-globin polyadenylation signal sequence (*β-globin pA*), HIV-1 polyadenylation signal sequence (*HIV-1 pA*), and the vesicular stomatitis virus glycoprotein (*VSV-G*) (See Sect. 2.4 for more information regarding the packaging plasmid components).

During reverse transcription of either the wild-type HIV-1 or the SIN attenuated viral vector, the 3′ LTR is duplicated and replaces the 5′ LTR. As a consequence of this duplication, the deletion in the U3 region of the 3′ LTR results in a vector that loses the ability to replicate once integrated into the host genome (17). Although the native HIV-1 LTR promoter is not active in the absence of Tat, the developers of these vectors have used the viruses own mechanism of replication to completely eliminate the LTR promoter in the SIN vectors. Advantages of using a SIN vector include the reduced probability that target cell coding sequences adjacent to the vector integration site will be aberrantly expressed either due to the promoter activity of the 3′ LTR or through an enhancer effect (17), and the prevention of any potential transcriptional interference between the LTR and the internal TG-RNAi promoter (17).

The RRE is the binding site for Rev which allows for the nuclear export of full-length RNA vector genomes (8, 14). The remaining two *cis* elements known to enhance the effectiveness of the vector are the cPPT (18) and the wPRE (19); together, they synergize to provide enhanced transduction and transgene expression (20).

2.4. trans Elements

HIV-1 contains nine genes that play a role in the life cycle of the virus and determine its pathogenic properties (1, 2). Six of the genes (*vif, vpr, vpu, nef, env, tat*), known as accessory genes, are either directly related to pathogenesis or not necessary for vector production and functionality. Consequently, these genes are usually deleted from the vector system or replaced with another gene as in the case of *env* (2, 5, 8). Because the remaining three genes, *gag*, *pol*, and *rev* (Fig. 1c), are necessary for viral packaging, they are retained in the vector system (2, 5, 21, 22). The *gag* and *pol* genes code for the p55 Gag and p160 Gag-Pol polypeptide precursors which are cleaved into mature products by viral proteases and are required for packaging of the viral vector. The p55 Gag precursor is cleaved into p17 matrix, p24 capsid, p9 nucleocapsid, p6 proteins, and two spacer peptides (14). The p160 Gag-Pol protein is the precursor of mature Gag protein, p12 viral protease, p51/66 reverse transcriptase, and p31 integrase (14). Some of the proteins serve a structural function in the virus capsule. More importantly, the Gag-Pol cleavage products have several functions in the viral life cycle including: (1) reverse transcriptase which is responsible for reverse transcription of viral DNA from the RNA genome; and (2) integrase, required for nuclear import and integration of the viral DNA into the host genome (14, 23). The *rev* (Fig. 1c) is a single 21-kDa protein that serves a single but important role. It binds mRNAs, removing them from the spliceosome complex, and links the rescued mRNA to a nuclear pore such that full-length and partially spliced mRNAs are transported out of the nucleus and into the cytoplasm (8, 14). This transfer is required for some of the transcripts from *gag* and *pol*, but more importantly, it is absolutely critical for the production of the full-length viral RNA genome (8).

There is an additional *trans* element required to generate active packaged particles surrounded by a protein envelope. The *env* gene from HIV-1 provides host cell specificity such that HIV-1 only infects CD4-producing cells (1). By replacing the HIV-1 *env* gene with the vesicular stomatitis virus glycoprotein (VSV-G; Fig. 1c; (24, 25)), a process known as pseudotyping, three important objectives are achieved: (1) the tropism of the vector is greatly broadened, (2) the vector particles are stabilized such that they can withstand ultracentrifugation to allow for vector concentration, and (3) the VSV-G directs the vector to an endocytic pathway, reducing the requirement for HIV-1 accessory proteins for infectivity (26). The tropism is determined by the target molecule on the surface of the host cell that the VSV-G protein attaches to. In, this case, the target molecules are phosphatidylserine, phosphatidylinositol, and GM3 ganglioside (25). While the VSV-G envelope is the most commonly used envelope, there are potential advantages to pseudotyping with other envelope proteins. Modifying the envelope protein is one possible strategy for targeting the vector specifically to cells of a given phenotype. Among several examples (4, 27), Ross River virus pseudotyped viruses stand out because of their selectivity for glial cells (see below for additional examples).

The proteins described above all act in *trans* to achieve packaging of the lentiviral vector; the protein products are coded by genes present on the HIV genome but can be separated from the RNA genome backbone (see below) creating room for the insertion of additional genetic material, such as the TG-RNAi cassette we describe below.

2.5. Packaging Cells

Packaging cells are used as the viral factory where the *trans* and *cis* elements are brought together to produce the vector particles. The packaging cells used to produce lentiviral vectors are 293T/17 cells (ATCC#CRL-11268), a highly transfectable derivative of the 293 human fetal kidney cell line, into which the temperature-sensitive gene for simian virus 40 (SV40) large T antigen has been inserted (28). There are several versions of the HIV-1-based lentiviral packaging systems; they are referred to as first (29), second (30), and third (31) generation. The differences between these versions reside in the number of plasmids used for packaging, allowing the *cis* and *trans* elements to be split on different plasmids and thereby improving the biosafety profile of the vector preparations. The first and second generations consist of three plasmids: a vector plasmid, a packaging plasmid, and an envelope plasmid. The number of accessory genes present on the packaging plasmid varies depending on the system (29, 30). The third generation system includes four expression cassettes: a vector plasmid, a *gag/pol* plasmid, a *rev* plasmid, and an *env* plasmid (31). There is an additional system that comprises four or more plasmids; the key feature to this system is that the *gag/pol* cassette has been split into two plasmids (21, 22, 26). This system also includes *tat* and a *vpr* fusion protein, and it is being marketed as a fourth generation system (Clontech, La Jolla, CA). We have primarily used the third generation system, which provides the *gag* and *pol* genes on one plasmid and the *rev* on a separate plasmid (Fig. 1c). In our work with rhesus monkeys, we have utilized a packaging plasmid in which the *gag* gene has been altered such that the HIV-1 Gag protein has been mutated in one site to be more similar to the Gag of simian immunodeficiency virus (SIV) (32). In simian cells, HIV-1 fails to replicate because of an early postentry block, so alteration of the *gag* gene significantly increases the ability of HIV-based lentiviral vectors to transduce simian cell lines (32).

2.6. Constructs

The TG-RNAi vector plasmid contains all the *cis* elements necessary for production of the viral RNA genome, infection of target cells, integration into the host genome, and expression of TG-RNAi cassettes. The transcription of the RNA genome is directed by the 5′ LTR (Fig. 1a, b). In the third generation system, the promoter sequences of the 5′ LTR have been replaced by a constitutively active promoter, such as the cytomegalovirus (CMV) or the Rous sarcoma virus (RSV) promoter, creating a heterologous U3 (htU3) promoter (31, 33). This replacement eliminates the need for *tat* (31). Because the second-generation and fourth-generation vector plasmids have retained the HIV 5′ LTR U3 promoter, both systems require a packaging plasmid that supplies *tat* (21, 22, 30).

The TG-RNAi cassette in the transfer vector is the region in which the gene or RNAi molecule of interest can be inserted (Fig. 1a). Lentiviruses are limited to an overall TG insert size of 8–10 kb (3, 5). Primarily two types of constructs are used; they either express a transgene of interest or produce RNAi molecules to suppress expression of a target gene. In either case, a promoter must be provided, as the upstream LTR promoter activity has been eliminated once the viral genome is reverse transcribed and integrated into the host genome. The transgene promoter sequence can be a constitutively active promoter, a tissue-specific promoter, or a regulated promoter. For our work, we have used the constitutively active CMV promoter (Fig. 1a). Because of size limitations, we have found it is best to use just the coding region, including the Kozak sequence (34) of the cDNA encoding the transgene of interest. In one of our studies, we used the coding region of the human nerve growth factor (hNGF) gene. By inserting an internal ribosome entry site (IRES) after the transgene, we produced a lentiviral construct that is capable of producing a bicistronic mRNA (Fig. 1a). The IRES is from the encephalomyocarditis virus and was generated by PCR from the pERV3 plasmid (Stratagene, La Jolla, CA). The PCR primers used to generate the IRES were as follows: upstream 5′-ACGCGTCCCCCCTCTCCCT-3′ and downstream 5′-ACG CGTGATCGTGTTTTTCAAAGG-3′. An Mlu I site was inserted at the 5′ end of both primers to facilitate cloning. Downstream of the IRES, a cell marker can be used to identify infected cells. We commonly use enhanced green fluorescent protein (eGFP), but other cell markers such as different color fluorescent proteins can replace the eGFP. Alternatively, an antibiotic resistance gene can be inserted in place of the eGFP (Fig. 1a).

The placement of cDNAs in relation to the IRES greatly influences relative expression. For example, while studying the involvement of phospholemman (FXYD1), a modulator of the sodium-potassium ATPase (35), in the neuropathology of Rett syndrome (36), the cDNA of *Fxyd1* was inserted either upstream or downstream of the eGFP reporter (Fig. 2a). As shown in Fig. 2, the cDNA upstream of the IRES

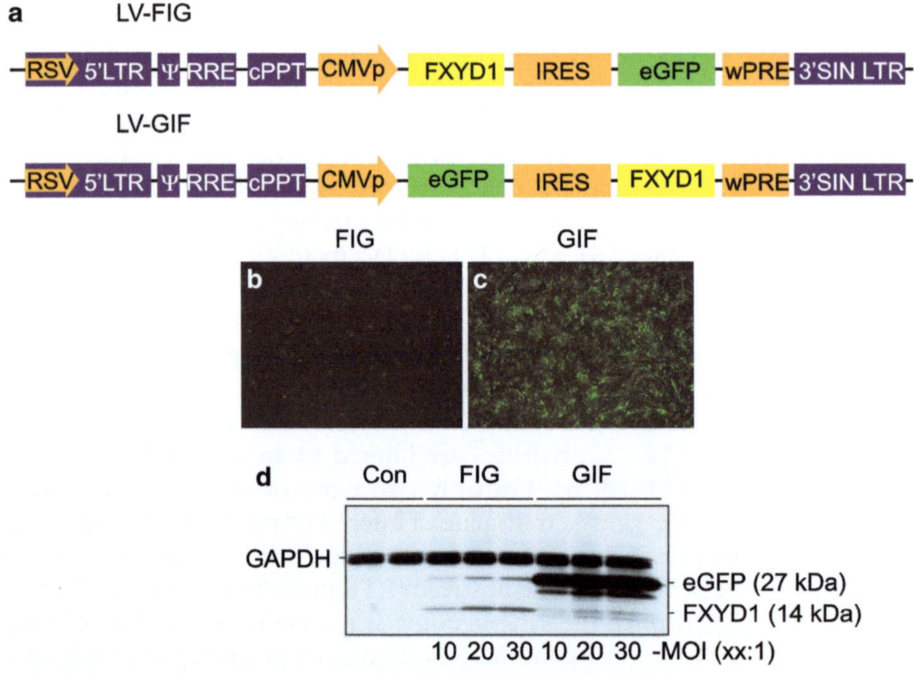

Fig. 2. Position of the IRES element greatly influences transgene expression. (**a**) Maps of lentiviral vectors containing the same cDNAs in alternate order with respect to the IRES. The cDNAs used in this case code for phospholemman (*FXYD1*) and enhanced green fluorescent protein (*eGFP*). (**b–d**) Cells derived from rat hippocampal neurons (*HiB5*) and devoid of FXYD1 expression were infected with both lentiviral constructs in the presence of polybrene (10 μg/ml). (**b**) Photomicrograph of cells infected with the LV-FIG (**F**xyd1-**I**RES-e**G**FP) construct (at the highest multiplicity of infection, MOI, 30:1) under fluorescent light. (**c**) Photomicrograph of cells infected with the LV-GIF (e**G**FP-**I**RES-**F**xyd1) construct (MOI 20:1) showing *green* fluorescent cells. (**d**) Western blot of HiB5 cells (plated at 200,000 cells/wells in 12-well culture plates, 3 wells/condition) infected with constructs in **a** at 3 MOIs (10:1, 20:1, 30:1, as determined by flow cytometry of green fluorescence). Cells were pooled and homogenized in lysis buffer (25 mM Tris, pH 7.4, 50 mM glycerophosphate, 1% Triton X-100, 1.5 mM EGTA, 0.5 mM EDTA, 1 mM sodium pyrophosphate, 1 mM sodium orthovanadate, 10 μg/ml leupeptin and pepstatin A, 10 μg/ml aprotinin, and 1 mM PMSF). Samples were loaded as follows: control HiB5 cells (Con, 10 μg/ml polybrene only), cells infected with LV-FIG, cells infected with LV-GIF. Twenty micrograms of protein was loaded on a Tris-glycine protein gel (denaturing conditions, NOVEX system, Invitrogen) and transferred on an Immobilon-P membrane (Millipore, Billerica, MA). The membrane was blocked in 5% milk-tris-buffered saline plus 0.5% tween 20 (TBST) and incubated overnight at 4°C with antibodies against glyceraldehyde 3-phosphate dehydrogenase (GAPDH) (0.1 μg/ml, Abcam Inc., Cambridge, MA) and FXYD1 (1/10,000, PLM-C2, kindly provided by R. Moorman). Membranes were washed (four changes of TBST every 15 min), incubated with a goat anti-mouse (0.1 μg/ml, Zymed Laboratories, San Francisco, CA), then washed and incubated with a donkey anti-goat (8 ng/ml, Santa Cruz Biotechnology, Inc., Santa Cruz, CA) secondary antibody. After another wash, the membrane was incubated overnight at 4°C with an anti-GFP antibody (1 μg/ml, Abcam). The membrane was washed and incubated with a goat anti-rabbit secondary antibody (0.1 μg/ml, Zymed Laboratories) then washed again. The immunoreactions were visualized using ECL reagents (Thermo Fisher Scientific, Rockford, IL). All secondary antibodies were conjugated to horseradish peroxidase.

is expressed at higher levels than the cDNA downstream of the IRES. This was demonstrated both by observation of epifluorescence in infected cultured cells (Fig. 2b, c) and by protein expression determined in Western blots (Fig. 2d).

A heterologous intron inserted into the transgene construct (Figs. 1a and 3a) has been shown in some circumstances to enhance gene expression of the transgene. It is well known that heterologous introns enhance expression of transgenes in mice (37). In order to determine if the heterologous intron has the same effect in the context of a lentiviral vector, we inserted the rat insulin II intron A sequence (38), used in transgenic mice (37, 39), between the CMV promoter and the cDNA for hNGF (Figs. 1a and 3a) and compared NGF production from infected 293T/17 cells. NGF released into the culture medium was determined with the NGF Emax immunoassay system (Promega Corp. Madison, WI). It was undetectable in culture medium from 293T/17 cells infected with virus lacking the NGF cDNA, clearly measurable in the medium of cells transduced with a viral construct lacking the heterologous intron, and greatly enhanced (>3-fold) when the intron was present (Fig. 3). Extending these studies to another cDNA, the same intron

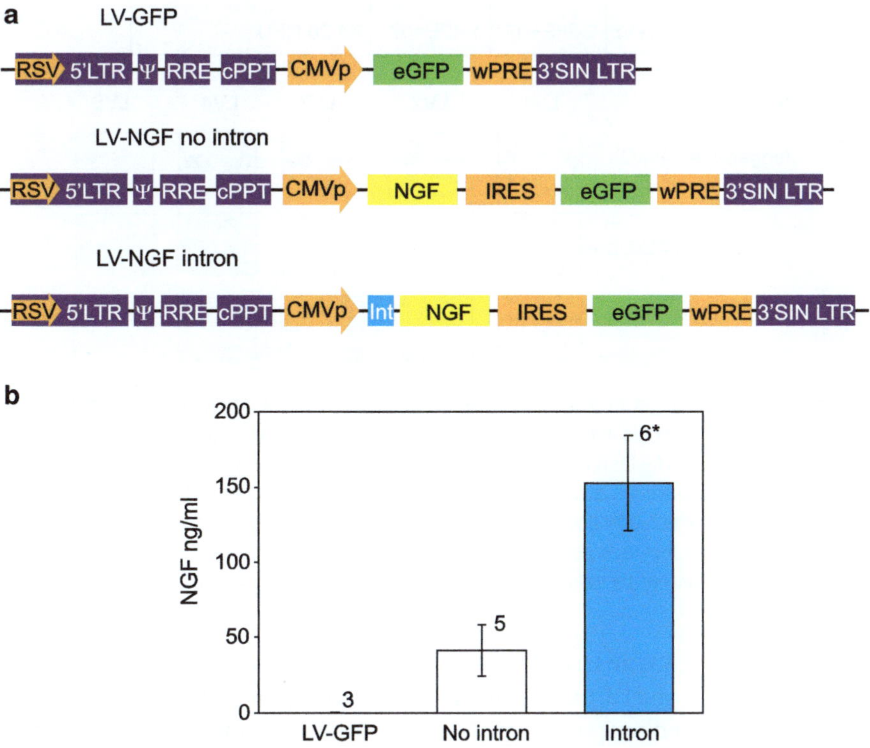

Fig. 3. A heterologous intron incorporated into a TG cassette of a lentiviral vector results in increased transgene expression, measured as an increase in NGF output. (**a**) Maps of the lentiviruses carrying TG expression vector constructs. (**b**) NGF release into the culture medium of 293T cells transduced with the viral vectors shown in **a**. Bars = mean ± SEM; numbers above bars are numbers of wells per group; $^{*}P < 0.05$ (*t* test) vs. no intron group. NGF was measured after 24 h of culture (See Fig. 1 and text for descriptions and definitions of vector components and abbreviations).

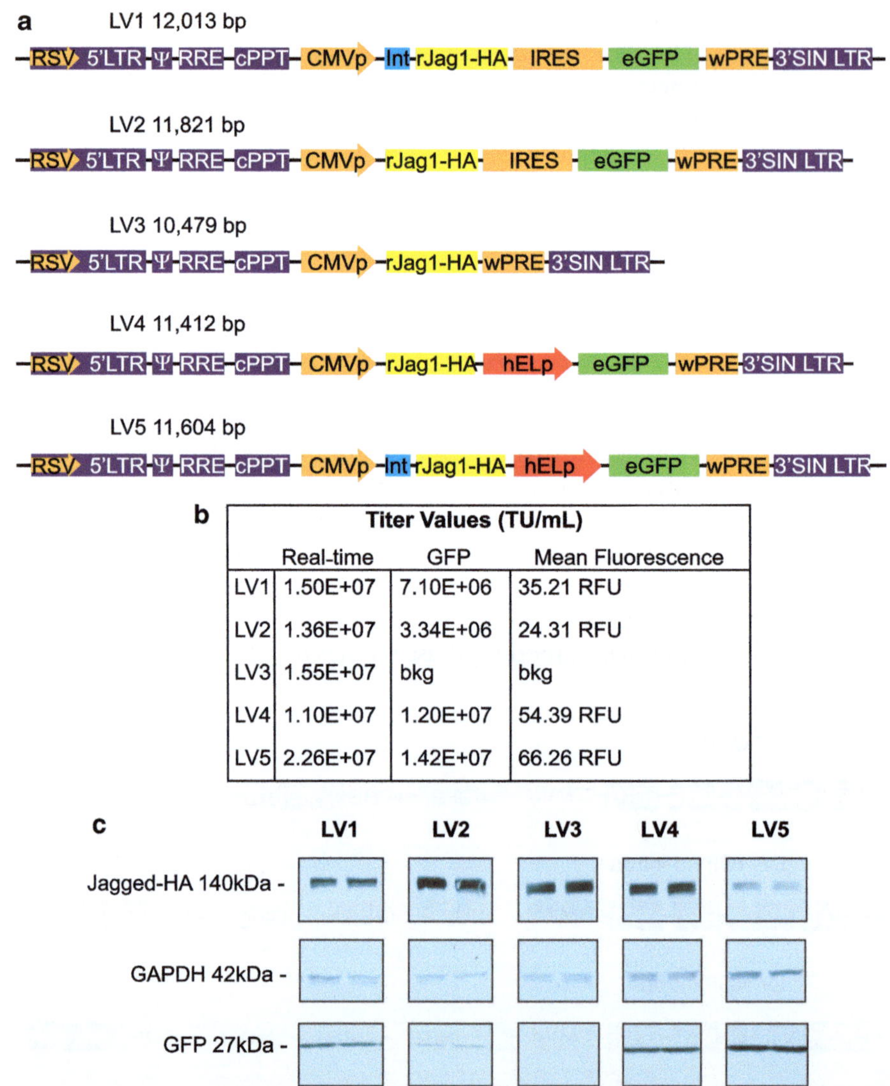

Titer Values (TU/mL)			
	Real-time	GFP	Mean Fluorescence
LV1	1.50E+07	7.10E+06	35.21 RFU
LV2	1.36E+07	3.34E+06	24.31 RFU
LV3	1.55E+07	bkg	bkg
LV4	1.10E+07	1.20E+07	54.39 RFU
LV5	2.26E+07	1.42E+07	66.26 RFU

Fig. 4. Five constructs were produced to test if the presence of a heterologous intron influenced transgene expression, if deletion of downstream elements would improve transgene expression, and if expression of a downstream reporter would be enhanced by the human elongation factor promoter (hELp) compared to the IRES element. (**a**) Maps of the five constructs used in this experiment. (**b**) Table of titer values as determined by real-time PCR and GFP-dependent flow cytometry. The mean fluorescence values of green cells detected by flow cytometry are also presented. (**c**) The protein expression of Jagged1, GFP, and GAPDH of the five constructs are shown as determined by Western blot analysis. Western blots were performed as described in Fig. 2; Jagged-HA was detected with a mouse monoclonal antibody anti-HA (HA.11, Covance).

was inserted upstream of the much larger cDNA for rat Jagged1, one of the ligands for notch receptors (40). In contrast to the clear enhancement in NGF expression from its relatively small (0.8 kb) cDNA observed previously (Fig. 3), the opposite result was observed with the much larger Jagged1 cDNA (3.8 kb; Fig. 4). The Jagged1 protein was tagged with the human influenza hemagglutinin (HA) epitope tag. Five different constructs were compared to examine the relationship between enhancer position and transgene expression (Fig. 4a). The viral preparations of the constructs shown in Fig. 4a were produced and titered using both flow cytometry, via detection of GFP (GFP; Fig. 4b), and infection rate detected by real-time PCR (Real-time; Fig. 4b). The presence of the heterologous intron (Fig. 4a; LV1 and LV5) resulted in reduced Jagged1 expression, measured in Western blot (Fig. 4c) using a monoclonal antibody (HA.11; Covance, Berkeley, CA), in comparison to constructs lacking this intron (Fig. 4a; LV2–LV4). Interestingly, the expression of the GFP fluorescence (Fig. 4b) and protein (Fig. 4c) appeared to be enhanced by the presence of the upstream intron. In order to determine if the downstream presence of the IRES and eGFP interfere with the expression of Jagged1, construct LV3 was produced in which these components were eliminated (Fig. 4a). There was no enhancement of Jagged1 expression by deletion of the downstream elements (Fig. 4c; LV3 vs. LV2). In an attempt to enhance GFP expression, the IRES was removed, and in its place, the promoter for human elongation factor (41) was inserted. The second promoter greatly enhanced the expression of GFP fluorescence (Fig. 4b) and protein (Fig. 4c).

The second type of construct we use is intended to suppress expression of target genes. The system called pPRIME (potent RNAi using miR expression) was developed in the Elledge lab (42) and has been made available through Addgene (www.addgene.org). Artificial microRNAs (miRNAs) have improved host safety over short hairpin construct's (43) potentially harmful oversaturation of cellular pathways (43, 44). The pPRIME vectors use an RNA polymerase II to direct the expression of a marker gene (eGFP, RFP, or an antibiotic resistance gene such as neomycin) followed by an artificial miRNA precursor (Fig. 1b). The artificial miRNA is derived from miR-30 (45) and is inserted downstream from the marker coding sequence; in our studies, the marker sequence is eGFP. The miRNA sequence is engineered to allow replacement of the hairpin sequence so that any mRNA sequence can be targeted with this system (42, 46). We improved the pPRIME system by placing multiple hairpins in the same miRNA, a modification shown to increase the suppressive effectiveness of the RNAi cassette (47).

3. Methods

3.1. Screening of NHPs for Studies Involving Lentivirus Administration

The major safety concern with using HIV-1 lentivirus is that a recombination event could occur resulting in the production of replication-competent lentivirus (RCL). While no such event has been reported to occur, vigilance in this area during production is warranted. A brief discussion of a screening method to detect RCLs is presented below. A heightened concern regarding recombination events arises when the host animals are NHPs, because of the presence of retroviruses native to NHPs. Prescreening of candidate animals for previous presence of retroviruses that have the potential for recombination is a prudent preventative measure. All animals used for our studies are screened for antibodies specific for SIV, simian retrovirus (SRV; types 1, 2, 3, and 5), and simian T-cell leukemia virus (STLV; types 1 and 2). Animals positive for any of these viruses are not included in any study involving administration of lentiviruses.

3.2. Packaging

The lentivirus is prepared by transient transfection of the 293T/17 packaging cells. The cells are grown, maintained, and transfected in antibiotic-free, high-glucose Dulbecco's modified eagle medium (D-MEM) (Invitrogen, Carlsbad, CA) plus 10% defined fetal bovine serum (Hyclone-ThermoFisher, Waltham, MA). The day before transfection, the cells are plated at 70–80% confluency (six million cells/dish) in 10-cm tissue culture dishes (BD-Falcon, Franklin Lakes, NJ) that have been precoated with 20 μg/ml poly-L-lysine (Sigma-Aldrich, St Louis, MO) in phosphate-buffered saline. The next day, each dish is transfected with a mixture of the TG-RNAi expression plasmid (10 μg) plus the packaging plasmids (6.5 μg gB gag LV; 2.5 μg pLP2; 3.5 μg pLP/VSVG; Fig. 1a–c). The gB gag LV was kindly provided by Dr. I. Verma (32); the pLP2 and pLP/VSVG were acquired as part of the ViraPower Lentiviral Expression System (Invitrogen). The transfection is carried out using a calcium-phosphate solution consisting of a 1:1 mixture of 0.25 M $CaCl_2$ (J.T. Baker, Phillipsburg, NJ): 2XBBS (0.28 M NaCl (Sigma-Aldrich); 0.05 M N,N-*bis*-(2-hydroxyethyl)-2-aminoethanesulfonic acid (BES) (Calbiochem, La Jolla, CA); 1.5 mM Na_2HPO_4 (Sigma-Aldrich)). The calcium-phosphate-DNA mixture is added dropwise to each dish, and the cells are incubated at 3% CO_2/37°C. After 18 h, the transfection medium is replaced with fresh D-MEM, and the cells are incubated overnight in an atmosphere of 10% CO_2/37°C. The virus is then harvested by collection and filtration of the medium through 0.22 μM steriflip filters (Millipore, Billerica, MA) followed by ultracentrifugation at 20,000 rpm using a swinging bucket rotor (Beckman SW28, Beckman Coulter Inc., Fullerton, CA). The viral pellets are redissolved in Hank's balanced salt solution (HBSS; Invitrogen), shaken at room temperature for 45 min, and stored overnight at 4°C. Fresh D-MEM is added to the dishes, and cells are again incubated overnight at 10% CO_2/37°C. The virus is harvested as mentioned above, and viral pellets are combined with the resuspended virus from the previous day and shaken again to mix. The virus is concentrated by an additional ultracentrifugation spin at 21,000 rpm, in an SW55 rotor (Beckman Coulter Inc.) over 20% sucrose (Sigma-Aldrich) in phosphate-buffered saline. The viral pellet is resuspended in HBSS, shaken for 45 min at room temperature, aliquoted, and stored at −85°C.

Titration of the virus is performed using flow cytometry to analyze infected cells for GFP expression. 293T/17 cells are plated at 400,000 cells per well in six-well plates (Costar, Corning, NY), and various volumes of viral supernatant or dilutions of concentrated virus are added to each well, along with 0.16 μg/ml polybrene (hexadimethrine bromide, Sigma-Aldrich) to enhance infection efficiency. The plates are incubated at 10% CO_2/37°C for approximately 72 h, and the cells are lifted with stable trypsin replacement enzyme (TrypLE, Invitrogen), fixed with 0.5% paraformaldehyde in phosphate-buffered saline (to inactivate remaining virus), transferred to FACS cell strainer cap tubes (BD-Falcon, Franklin Lakes, NJ), and analyzed by flow cytometry (FACScalibur, BD Biosciences, San Jose CA) for GFP fluorescence. Calculations are made using the percentage of GFP-positive cells to yield titer values that represent the number of transforming units (TU) per ml. For example, if a 2 μl aliquot of conditioned medium resulted in 10% of the cells being counted as positive for GFP, the calculation would be as follows: 400,000 cells × 0.1 (percent GFP-positive cells) × 1,000/2 μl (dilution factor correction) = 2×10^6 TU/ml. Concentrated viral preparations routinely have titer values of at least one million viral particles (TU) per μl. If it is not possible to include the reporter GFP protein to facilitate titration of the viral preps, real-time PCR or p24 ELISA (see below) can be used to titer viral preparations (48, 49).

To ensure that replication-competent lentivirus (RCL) is not produced by recombination events in the host cell, a suspension of SUP-T1 (ATCC#CRL-1942) lymphoblast cells, seeded at one million cells per 25 cm^2 flask (Corning, Lowell, MA), is infected with ten million virus particles, and cells are propagated for 2 weeks in antibiotic-free RPMI-1640 medium (Sigma-Aldrich) plus 10% defined fetal bovine serum (Hyclone-ThermoFisher). The medium is collected from the flasks containing the infected cells and assayed for the HIV-1 p24 antigen (29, 50) using a commercial ELISA kit (Zeptometrix Corp., Buffalo, N.Y.). The sensitivity of the ELISA kit is approximately 7.8 pg/ml. However, because the cells are propagated for 2 weeks prior to assay, an RCL would be amplified during this propagation step, making it unlikely that the lack of RCL detection is due to low sensitivity of the assay. It is important to note that should an RCL develop, the most likely envelope protein present would be VSV-G, which would result in the virus having a very broad tropism.

We have used the above-described lentiviral vector system for both gene expression and gene suppression in the ovary and brain of NHPs. We will use these studies as examples of the two major applications, one involving a peripheral tissue and the other involving the brain.

3.3. Delivery of Lentiviruses to a Peripheral Tissue

Our work with a peripheral tissue has been focused on the ovary. The ovary is a suitable example of a peripheral tissue amenable to lentiviral gene transfer by localized injection, because it has clearly defined boundaries and is of a size that allows good penetration by injection. The goal of this study was to overexpress the gene encoding hNGF (Fig. 1a). The lentivirus was delivered to the ovary via laparotomy and direct injection using a 30-ga. needle attached to a 100 μl syringe, as described (51). The needle was inserted into the ovary adjacent to the utero-ovarian ligament, and 12 μl of lentivirus (200,000 TU/μl) was injected into each of four ovarian quadrants, for a total of 48 μl/ovary. After the monkeys were injected with the lentivirus, they were separated from other animals of the rhesus colony in an animal biological safety level II facility. The monkeys were then observed for changes in the menstrual cycle, both by daily observation of menses and regular blood sampling. We found that collecting blood samples three times a week is sufficient to characterize the length of the monkey's cycle and assess its normalcy.

3.4. Delivery of Lentiviruses to the Brain

For our studies, we were interested in the effects of knockdown of a gene known as enhanced at puberty 1 (EAP1). EAP1 mRNA expression was initially discovered to be upregulated in NHPs at puberty (52). Further, EAP1 was shown to be required for normal cyclicity and to be an important regulator of luteinizing-hormone-releasing hormone (LHRH) release in rodents (52). Thus, it was important to test the hypothesis that EAP1 is critical for cyclicity in NHPs. We targeted cells expressing EAP1 mRNA, which are located in a region of the hypothalamus known as the arcuate nucleus. The pPRIME lentivirus was constructed as described above with three copies of the hairpin specific for the target mRNA embedded into the body of miR30 (Fig. 1b; (42, 47)). In adult rhesus monkeys, the arcuate nucleus is an oval-shaped nucleus at the base of the brain and lateral to the third ventricle, and it is 3–4.5 × 1–2 × 2–3 mm (anterior-posterior × lateral × dorsal-ventral) in size (53). An important challenge posed by this location is delivering the lentivirus particles with accuracy to cells included in this area; the surgery has been described in detail ((51); Fig. 5a–c).

The injections were made 1–2 weeks after imaging of the brain and securing fiducial markers to the skull of the anesthetized monkeys (51). Using the MRI images, the distance from the midline of the brain to the posterior fiducial marker (Fig. 5c, d) was measured using the software Slicer3D (www.slicer.org). This measurement was used as a correction to establish the midline of the monkey once it was mounted in the stereotaxic apparatus on the day of injections. A lateral X-ray was then obtained to establish the target (Fig. 5a, e); the MRI image in Fig. 5d shows the arcuate nucleus (arrow). The target is 1 mm dorsal of the line drawn between the points of the anterior and posterior clinoids of the sella turcica bone structure on the X-ray image (Fig. 5e). To access the target

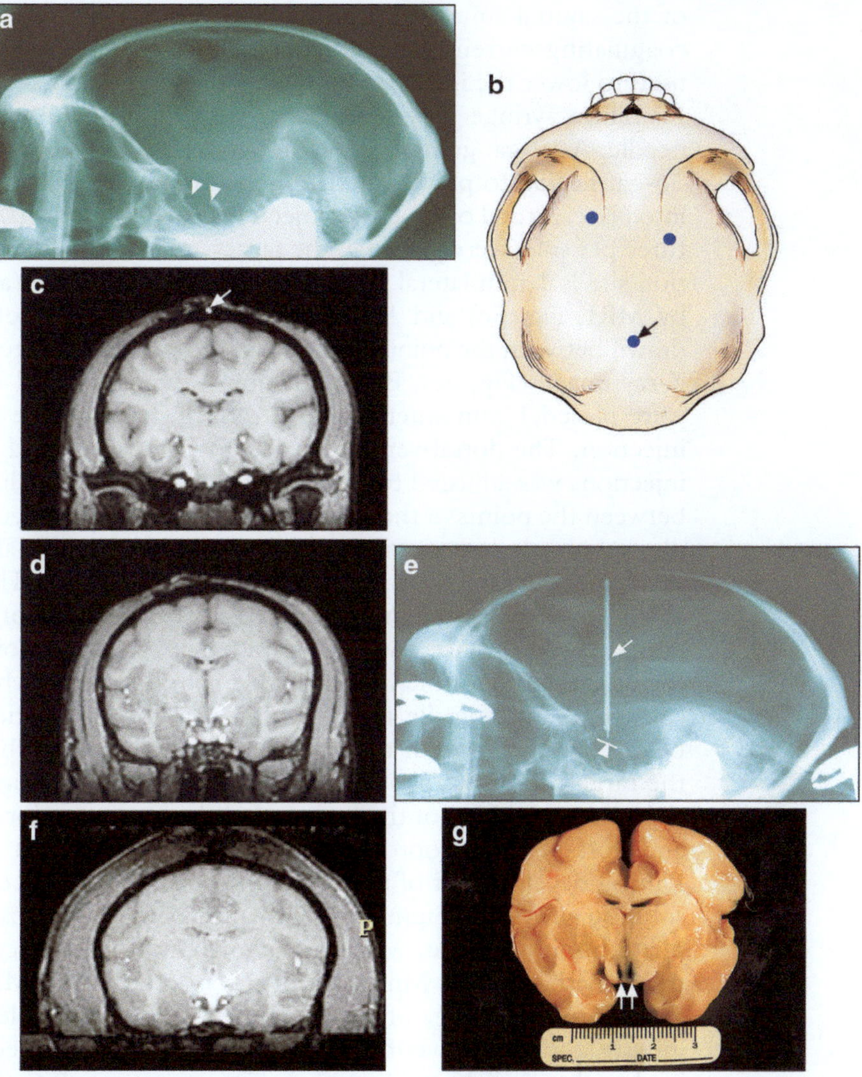

Fig. 5. X-ray- and MRI-guided stereotaxic injection of lentiviral particles into the medial basal hypothalamus-arcuate nucleus of the hypothalamus in rhesus macaques. (**a**) Initial X-ray showing the clinoids of the sella turcica (*arrowheads*). (**b**) Drawing of a monkey skull showing the placement of fiducial markers. The most posterior marker (*arrow*) was placed on the midline and used to establish the true midline of the brain. (**c**) MRI in which the posterior fiducial marker can be seen (*arrow*). (**d**) MRI showing the arcuate nucleus of the hypothalamus (*arrow*). (**e**) Final X-ray showing the injection needle (*arrow*); the tip of the needle shows the injection site, which is 1 mm dorsal to the line drawn between the clinoids (*arrowhead*). (**f**) MRI following bilateral injection of 1.5 μl of a 1 μM solution of MnCl in 0.4% trypan blue (in phosphate-buffered saline) into the arcuate nucleus of the medial basal hypothalamus (*arrow*). (**g**) Cross section of a monkey brain following bilateral injection of MnCl-trypan blue showing the sites of the injections (*arrows*).

site, the animal was anesthetized and positioned in the stereotaxic apparatus as previously described. A linear sagittal incision was then made over the dorsal scalp to expose the skull. A circular 1.5-cm-diameter craniotomy was created using a Hall air drill (Conmed Corp., Largo, FL) and medium round bur. The dura mater was incised on either side of the sagittal sinus, and the lateral branches of the sagittal sinus were cauterized as necessary using a bipolar coagulating current. The stereotaxic micromanipulator was then used to lower the injection syringe to the targets. We used a 10-μl Hamilton syringe (Hamilton Co.) with a blunt 26-ga., 52-mm needle. A 19-ga. guide tube set in a nylon ring was fitted over the 26-ga. needle to prevent deviation of the needle as it was lowered into place. A total of six 1.5-μl injections containing 1×10^6 TU/μl, three per side, were placed in the target area. The first target injection site is 1 mm lateral from the midline of the brain (as defined by MRI; Fig. 5c) and 1 mm dorsal of the midpoint of the line drawn between the points of the sella turcica bone structure on the X-ray image (Fig. 5e). From this point, two additional injections were placed 1 mm anterior and 1 mm posterior of the midpoint injection. The dorsal-ventral depth for the anterior and posterior injections was adjusted to maintain 1 mm dorsal of the line drawn between the points of the sella turcica (Fig. 5e). The injection needle was slowly lowered into place, for each injection, and placement was confirmed on the left side by lateral X-ray. The needle placement can be seen in Fig. 5e, which is an example of the midpoint injection. Once it was confirmed that the needle point was in the target, 1.5 μl of the viral construct solution was slowly injected (30–45 s). The needle remained in place for an additional 30 s before it was removed, to allow the solution time to absorb into the target. Because the injections on the right side (1 mm to the right of the midline of the brain) used the same anterior-posterior and dorsal-ventral coordinates, lateral X-rays were not repeated. During development of the injection procedure, 1.5 μl of a 1 μM solution of MnCl (Sigma-Aldrich) in trypan blue (0.4% in phosphate-buffered saline, MP Biomedicals Inc.) was injected using these methods. The MnCl solution was detected by MRI in the arcuate nucleus (Fig. 5f), and the brain was then collected and grossly sectioned to verify that the trypan blue was correctly localized to the medial basal hypothalamus- arcuate nucleus (Fig. 5g).

After the injections were complete, the craniotomy site was filled with Gelfoam (Pharmacia & Upjohn Co., Kalamazoo, MI), and the incision was closed using 4–0 Monocryl in a simple interrupted pattern to appose the subcutaneous tissue followed by 4–0 Monocryl intradermal pattern to appose the skin layer. Following treatment, the monkeys were housed as described above, in standard housing segregated from the rest of the colony. The monkeys were then observed, as described above, for changes in menstrual cycles resulting from siRNA-induced EAP1 knockdown.[1]

[1]The study described under section 3.4 Delivery of Lentiviruses to the Brain, has been accepted for publication and is in press at the time of final revisions of this Chapter; Dissen GA, Lomniczi A, Heger S, Neff TL, Ojeda SR, Hypothalamic EAP1 (Enhanced at Puberty 1) is Required for Menstrual Cyclicity in Nonhuman Primates, Endocrinology 2012.

4. Notes

4.1. Comparing and Contrasting Lentiviral Vectors with Adeno-Associated Viral Vectors

Recombinant LVs and AAVs have garnered a significant amount of interest in the past decade for use in both preclinical and clinical gene therapy studies. A detailed description of AAV biology and production can be found in the chapter by Veerle Baekelandt et al. in this book. Here, we provide only a brief description of the AAV genome and compare and contrast LV to AAV with respect to the benefits and limitations of both vector systems, including cell trafficking and transgene expression, packaging capacity, persistence and integration, tropism and pseudotyping, and immune response (Table 1). These considerations are especially paramount when designing preclinical studies that have the potential to translate into the clinic as viable gene therapies for human patients.

4.2. AAV Genome

AAVs are small, nonenveloped viruses that are members of the Parvoviridae family, belonging to the genus *Dependovirus.* AAVs harbor a single-stranded DNA genome approximately 4.7 kb long. The wild-type AAV genome contains inverted terminal repeats (ITRs) at both ends of the DNA strand (145 bases each) and two

Table 1
Comparison and contrast of vector properties

Vector characteristic	Lentiviral vectors	Adeno-associated viral vectors
Family	Retroviridae	Parvoviridae
Coat	Enveloped	Nonenveloped
Genome	ssRNA	ssDNA
Packaging capacity	8–10 kb	~4.7 kb
Host genome interaction	Integrates into host DNA	Nonintegrating, episomal concatemer
Tropism	Dividing and nondividing cells, ability to pseudotype	Dividing and nondividing cells, ability to pseudotype
Transgene expression	Long-term, persists with cell division	Long-term, does not persist with cell division
Immune profile	Low immunogenicity	Low immunogenicity, prevalence of NAbs in the human population
Biosafety level containment	Biosafety level 2	Biosafety level 1

open reading frames for the genes *rep* and *cap*. Rep proteins are required for site-specific integration, nicking, helicase activity, and promoter function (54). The Cap proteins (VP1, VP2, and VP3) interact together to form the outer viral capsid, which is approximately 20 nm in diameter. Similar to LVs, genes crucial to viral replication and capsid production are removed during the generation of AAV used in gene therapy studies. The *rep* and *cap* from the DNA of the vector are removed, and the transgene and promoter of choice are inserted between the ITRs.

4.3. Cellular Trafficking and Expression of Transgenes

Both LVs and AAVs bind to specific receptors on the cell surface to initiate infection, depending on the viral pseudotype or serotype. Both viruses require endocytosis to achieve cell entry (26, 55–57). With LV infection, fusion between the virus envelope and the endosome occurs, and the particles move toward the perinuclear area of the cell (Fig. 6) where the contents of the particle are

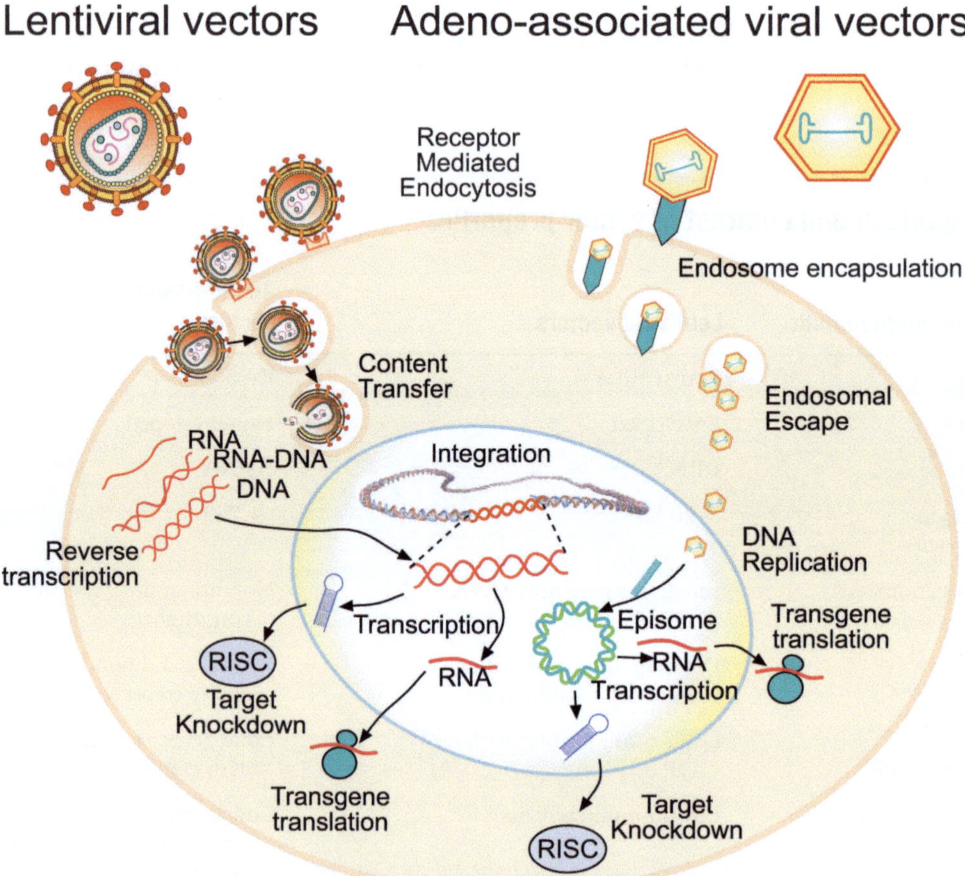

Fig. 6. Diagram comparing lentiviral and AAV vector infection, and distinctive methods of expression of transgene products. Both vector systems are capable of transgene expression and endogenous gene knockdown using siRNA technology.

transferred to the cytoplasm. In the cytoplasm, reverse transcriptase (RT) synthesizes double-stranded DNA (dsDNA) from the single-stranded RNA viral genome (Fig. 6). The dsDNA is incorporated into a preintegration complex, which facilitates nuclear entry and integration into the host genome. Viral integrase (IN) is the enzyme that catalyzes the random and stable integration of the vector transgene into the genome (13). AAVs enter the host cell following binding to the cell surface receptor heparin sulfate proteoglycan (HSPG); thereafter, they are rapidly internalized through clathrin-coated pits through a process involving $\alpha_v\beta_5$ integrin (57). The capsids remain intact through endosomal trafficking from the early endosome into the late endosome and finally through endosomal escape (Fig. 6). Following endosome processing and release, AAVs accumulate perinuclearly and slowly transverse the nuclear pore complex into the nucleus. Following capsid protein degradation, which occurs inside the nucleus, second-strand synthesis occurs to generate a dsDNA that is capable of expressing the transgene of interest (57). Because the virus does not encode a polymerase, it relies on host cell polymerases for genome replication. Unlike LV and wild-type AAV, which integrate into the host genome (wild-type AAV integrates in a specific site on Chromosome 19; (58)), recombinant AAV vector DNA remains as circular, duplex episomal concatemers in the host cell nucleus. In nondividing cells, these concatemers remain for the life of the host cell. However, in dividing cells, AAV DNA is reduced and eventually lost through cell division, since the episomal DNA is not replicated along with the host cell DNA. Random integration of recombinant AAV DNA into the host genome has been reported to varying degrees but at a very low (<0.01%) frequency (59).

While LV vector integration into the host genome can lead to stable and long-lasting expression of a desired transgene, a salient feature of its integration is the potentially dangerous lack of specificity. While gamma retroviruses preferentially integrate in the 5′ flanking regions of genes near transcription start sites, LVs tend to integrate across the entire gene, making them a safer and lessgenotoxic choice of vector for gene-therapy-based studies (3). Promiscuous integration of retroviruses may result in insertional mutagenesis, which has been shown to be associated with oncogenesis in both animal and human gene therapy studies (60, 61). In a clinical trial at the Hôpital Necker Enfants Malades (Paris, France) aimed at correcting X-linked severe combined immunodeficiency (X-SCID) with retroviral delivery of γc interleukin receptor subunit, five of the treated children developed T-cell leukemia (60). A follow-up report confirmed an inappropriate insertion of the retroviral vector near the proto-oncogene LIM domain only 2 (*LMO2)* promoter, which encodes a transcription factor required for hematopoiesis. The insertion led to uncontrolled proliferation of mature T cells (60). Alternatives to avoid random integration include fusing the integrase enzyme to a sequence-specific DNA-binding protein, which can direct integration into specific safe sites in the gene (62). Another strategy that has emerged over the past few years consists of introducing different point mutations that affect the integration process, creating integration-deficient LVs (IDLVs) that are identical to AAV in that they lead to sustained transgene expression without integration into the host genome (63, 64).

4.4. Transgene Persistence

LVs and AAVs both express transgenes for long periods of time in multiple animal species and in multiple tissues types. Robust transgene expression from LVs can be measured 4–7 days postinjection, whereas maximal transgene expression from AAVs is seen after 3–4 weeks. The longer latency to maximal gene expression with AAVs is due to the prolonged time to undergo endosomal processing, trafficking, and second-strand DNA synthesis (57). Because LVs integrate into the genome, they have a high transgene persistence rate, usually for the lifetime of the animal, with expression levels varying depending on the tissue infected, transgene promoter, and the specific transgene being evaluated (65–67). Although AAVs remain episomal, they also drive transgene expression for long periods of time. AAVs have been shown to express transgenes in different tissues throughout the lifetime of rodents (68–70) and have been shown to express transgenes for as long as 10 years postinjection (last time-point studied) into the NHP brain (Piotr Hadaczek, UCSF, personal communication). Both LVs and AAVs have high transduction efficiencies targeting many human and experimental animal cell types, including stem cells, neurons and glia, muscle, skin, liver, pancreas, lung, primary T lymphocytes, and fetal tissues (68, 71–80). Because LVs integrate into the host genome, transgene expression persists through cell division of nonquiescent cells and is observed in daughter cells (81). This becomes an important consideration when targeting cells that undergo frequent mitosis such as epithelial cells of the lungs and gastrointestinal tract, splenocytes, epidermal cells, monocytes, and stem cells. Because of their potent ability to integrate into the genome and confer transgene expression in progeny cells, LVs have recently been used to create transgenic animals in species where conventional transgenic methodologies are difficult, such as in rats (82) and larger species including nonhuman primates (83). Chan and colleagues developed the first transgenic monkey model of Huntington's disease in 2008 by injecting mature oocytes with LV expressing the mutant *Htt* gene with expanded polyglutamine repeats, which is the primary cause of HD. Oocytes were fertilized by intracytoplasmic sperm injection, and the resulting embryos were transferred into surrogate mothers. The resulting pregnancies lead to the birth of transgenic monkeys bearing neuropathological hallmarks and behavioral manifestations of the disease (83).

4.5. Vector Tropism and Pseudotyping

The range of host tissues for both LVs and AAVs can be greatly increased by genetically altering either the envelope (in the case of LV) or the capsid (in the case of AAV) proteins of the respective viral vectors. Many gene therapy vectors have now been developed in which the endogenous viral envelope proteins have been replaced by proteins from other viruses. The majority of LVs used in gene therapy studies are pseudotyped with VSV-G, affording the vector a very broad tropism that mediates gene transfer to a wide variety of cell types (24, 25). However, when transduction of a specific cell population is desired, LV can be pseudotyped with viral proteins to tightly restrict the population of host cells transduced. The list of LV pseudotypes used successfully in gene therapy studies has grown substantially over the past decade (84). The most prominent include lyssavirus pseudotypes (including those from Mokola virus (85) and rabies virus (85), both of which exhibit strong neurotropism), alphavirus pseudotypes (such as Ross River virus (4, 27), Semliki Forest virus (86), and Sindbis virus (87) which show increased expression in muscle and liver), filovirus pseudotypes (including Marburg virus (88) and Ebola virus (89) which show high tropism in liver and lung), and gamma-retrovirus pseudotypes ((90) which show tropism for hematopoietic progenitor cells, T lymphocytes, and mesenchymal stem cells).

To date, several different serotypes of AAV have been characterized (as well as over 100 variants), which are defined as isolates of AAV that do not exhibit cross-reaction with neutralizing sera from any other known serotype (54). Human AAV1–6 and AAV7–9 are defined as true serotypes, while AAVs 6, 10, and 11 are considered variants, although they are commonly listed as serotypes. AAV serotypes exhibit unique tropism for different tissues throughout the body and use differing cell surface receptors to gain entry into the host cell (54).

Historically, AAV2 has been the most extensively studied serotype, and hundreds of publications have detailed both preclinical and clinical use of AAV2 to deliver transgenes in varying tissues throughout the body (91). Similar to the numerous pseudotyped LVs used in clinical and preclinical gene therapy studies, many different AAV pseudotypes have been created to direct expression of a particular transgene in restricted populations of cells. The majority of AAV pseudotypes contain the ITRs of AAV2 flanking the transgene of interest and capsid proteins from alternate serotypes creating hybrid vectors such as AAV2/1, AAV2/4, AAV2/5, AAV2/6, AAV2/8, and AAV2/9 with different patterns of transduction in diverse tissue types (92–95). While AAV2 has been the historical gold standard for gene-therapy-based studies, new publications have emerged directly comparing AAV2 with pseudotyped AAV2/1, AAV2/5, and AAV2/8 and demonstrated that the pseudotyped vectors lead to higher transgene expression levels, particularly in the CNS. Burger and coworkers showed significantly increased expression of GFP (under control of a chicken beta-actin promoter) expressed from AAV2/1 and AAV2/5 compared to AAV2/2 in the rat striatum, globus pallidus, substantia nigra, spinal cord, and hippocampus (92). Increases in expression level were noted in terms of both spread (volume of tissue transduced) and intensity (fluorescence) of the GFP transgene. A follow-up report by Dodiya et al. extended these results into the nonhuman primate (cynomolgus macaques), showing significantly more GFP-positive neurons in the caudate and putamen when expressed from AAV2/1 and AAV2/5 compared to AAV2/8 (95).

4.6. Immune Response

Mitigating host immune responses to viral vectors remains a critical challenge in gene therapy studies. Successful studies require suitable levels of transgene expression in desired cell populations without provoking a host immune response that will impede gene expression, or worse, lead to toxicity in the target tissues. Compared to other vector systems used to confer gene transfer, such as first-generation adenoviral vectors, both LV and AAV have a relatively low immunogenic profile, which is in large part why their use has become prevalent in recent gene therapy clinical trials. Notwithstanding, both LV and AAV elicit some degree of innate and adaptive immune response upon administration, the repercussions of which are dependent on the injection site and target tissues being transduced, the titer/dose of vector injected, prior exposure of the host animal to envelope and capsid proteins, the level of preexisting neutralizing antibodies to the virus being used, and immunogenicity of the transgene being expressed. First, the site of injection plays a large role in determining the effect of the host immune response to the vector. The preeminent example of this is LV and AAV delivery to the CNS. Due to the presence of the blood–brain barrier and the immune compartmentalization of the brain, injection of both vectors into brain parenchyma elicits a very low immune response (96). Acute inflammation following local parenchymal injection into rodents can include microglial activation, recruitment of macrophages, and T-cell infiltration, the latter two likely owing to temporary disruption of the dura or puncture of vasculature during the injection. Injection of LV and AAV into the ventricular system, however, has been associated with a more profound immune response, as the meninges and choroid plexus contain all of the cellular, vascular, and lymphatic components of the immune system, including dendritic cells (96). Moreover, cerebrospinal fluid carrying unbound vector is ultimately drained from the subarachnoid space into the deep cervical lymph nodes and venous sinuses where vectors may be susceptible to immune surveillance. In contrast, injection of both LV and AAV into the periphery has shown variable degrees of immune response, depending on the tissue type transduced. Following both LV and AAV administration, the rodent host immune system responds with a fast innate response that can include transient (usually less than 24 h) and local production of proinflammatory cytokines and chemokines, and infiltration of monocytes and macrophages into target tissues (97, 98). In addition, an adaptive immune response mounts against LV envelope and AAV capsid proteins, resulting in the production of neutralizing antibodies (Nabs) that remain in the host animal's circulation, usually precluding a second administration using the same vector unless the envelope or capsid proteins are altered.

One major consideration when choosing LV or AAV for use in preclinical (especially in larger animals that are captured from the wild vs. bred in the laboratory) and clinical studies is that a significant portion of NHPs and humans have been naturally exposed to different serotypes of AAV and, therefore, already have developed circulating Nabs. Over 90% of humans have antibodies that cross-react to one or more AAV serotypes (99), and Halbert and coworkers demonstrated data from a collection of 37 normal adults (recruited in the Seattle, WA area) showing that 30% of normal adults were seropositive for AAV2, 20–30% were seropositive for AAV6, and 10–20% were seropositive for AAV5 (100). In a separate study, Bouin and colleagues showed that in the normal human patient population in France, prevalences of anti-AAV1 and anti-AAV2 total IgGs were higher (67% and 72%) than those of anti-AAV5 (40%), anti-AAV6 (46%), anti-AAV8 (38%), and anti-AAV9 (47%) (101). The highest levels of Nabs were observed for AAV2 (59%) and AAV1 (50.5%), and the lowest were observed for AAV8 (19%) and AAV5 (3.2%). The short-term administration of immunosuppressive drugs, while the viral capsid epitopes (VP1, VP2, and VP3) are presented on MHC-1 during vector uncoating, may be one way to circumvent this issue (96). Additionally, newer generation mosaic and chimeric AAVs to which humans have not yet been exposed should provide better alternatives as clinical vectors in the future. In contrast, because LVs are modified HIV-1-based vectors, the general human population has had little to no natural exposure to these vectors, and the levels of circulating Nabs are negligible (10, 102).

Acknowledgments

This work was supported by NIH grants HD-24870, HD-25123, the Eunice Kennedy Shriver NICHD/NIH through cooperative agreement HD-18185 as part of the Specialized Cooperative Centers Program in Reproduction and Infertility Research, RR-000163 for the operation of the Oregon National Primate Research Center, the International Rett Syndrome Association, and the Northwest Rett Syndrome Foundation.

References

1. Verma IM, Somia N (1997) Gene therapy - promises, problems and prospects. Nature 389:239–242.
2. Kafri T (2004) Gene delivery by lentivirus vectors an overview. Methods Mol Biol 246:367–390.
3. Thomas CE, Ehrhardt A, Kay MA (2003) Progress and problems with the use of viral vectors for gene therapy. Nat Rev Genet 4:346–358.
4. Wong LF, Goodhead L, Prat C, Mitrophanous KA, Kingsman SM, Mazarakis ND (2006) Lentivirus-mediated gene transfer to the central nervous system: therapeutic and research applications. Hum Gene Ther 17:1–9.
5. Salmon P, Trono D (2007) Production and titration of lentiviral vectors. Curr Protoc Hum Genet Chapter 12:Unit 12.10.1-Unit 12.10.24.
6. Butler SL, Johnson EP, Bushman FD (2002) Human immunodeficiency virus cDNA metabolism: notable stability of two-long terminal repeat circles. J Virol 76:3739–3747.
7. Watson DJ, Karolewski BA, Wolfe JH (2004) Stable gene delivery to CNS cells using lentiviral vectors. Methods Mol Biol 246:413–428.
8. Debyser Z (2003) Biosafety of lentiviral vectors. Curr Gene Ther 3:517–525.
9. Wodarz D, Nowak MA (1999) Evolutionary dynamics of HIV-induced subversion of the immune response. Immunol Rev 168:75–89.
10. Stebbing J, Patterson S, Gotch F (2003) New insights into the immunology and evolution of HIV. Cell Res 13:1–7.
11. DePolo NJ, Reed JD, Sheridan PL, Townsend K, Sauter SL, Jolly DJ, Dubensky TW, Jr. (2000) VSV-G pseudotyped lentiviral vector particles produced in human cells are inactivated by human serum. Mol Ther 2:218–222.
12. Zhou HS, Liu DP, Liang CC (2004) Challenges and strategies: the immune responses in gene therapy. Med Res Rev 24:748–761.
13. Debyser Z (2003) A short course on virology/vectorology/gene therapy. Curr Gene Ther 3:495–499.
14. Federico M (2003) From lentiviruses to lentivirus vectors In: Federico M (ed) Methods in Molecular Biology, Vol 229: Lentivirus Gene Engineering Protocols edn. Humana Press Inc, Totowa, New Jersey.
15. Yu S-F, von Rüden T, Kantoff PW, Garber C, Seiberg M, Rüther U, Anderson WF, Wagner EF, Gilboa E (1986) Self-inactivating retroviral vectors designed for transfer of whole genes into mammalian cells. Proc Natl Acad Sci USA 83:3194–3198.
16. Miyoshi H, Blomer U, Takahashi M, Gage FH, Verma IM (1998) Development of a self-inactivating lentivirus vector. J Virol 72:8150–8157.
17. Zufferey R, Dull T, Mandel RJ, Bukovsky A, Quiroz D, Naldini L, Trono D (1998) Self-inactivating lentivirus vector for safe and efficient in vivo gene delivery. J Virol 72:9873–9880.
18. Zennou V, Petit C, Guetard D, Nerhbass U, Montagnier L, Charneau P (2000) HIV-1 genome nuclear import is mediated by a central DNA flap. Cell 101:173–185.
19. Zufferey R, Donello JE, Trono D, Hope TJ (1999) Woodchuck hepatitis virus posttranscriptional regulatory element enhances expression of transgenes delivered by retroviral vectors. J Virol 73:2886–2892.
20. Barry SC, harder B, Brzezinski M, Flinit LY, Seppen J, Osborne WRA (2001) Lentivirus vectors encoding both central polypurine tract and posttranscriptional regulatory element provide enhanced transduction and transgene expression. Hum Gene Ther 12:1103–1108.
21. Wu X, Wakefield JK, Liu H, Xiao H, Kralovics R, Prchal JT, Kappes JC (2000) Development of a novel trans-lentiviral vector that affords predictable safety. Mol Ther 2:47–55.
22. Wu X, Liu H, Xiao H, Conway JA, Hehl E, Kalpana GV, Prasad V, Kappes JC (1999) Human immunodeficiency virus type 1 integrase protein promotes reverse transcription through specific interactions with the nucleoprotein reverse transcription complex. J Virol 73:2126–2135.
23. Frankel AD, Young JA (1998) HIV-1: fifteen proteins and an RNA. Annu Rev Biochem 67:1–25.
24. Burns JC, Friedmann T, Driever W, Burrascano M, Yee JK (1993) Vesicular stomatitis virus G glycoprotein pseudotyped retroviral vectors: concentration to very high titer and efficient gene transfer into mammalian and nonmammalian cells. Proc Natl Acad Sci USA 90: 8033–8037.
25. Hanazono Y, Terao K, Ozawa K (2001) Gene transfer into nonhuman primate hematopoietic stem cells: implications for gene therapy. Stem Cells 19:12–23.
26. Cockrell AS, Kafri T (2007) Gene delivery by lentivirus vectors. Mol Biotechnol 36:184–204.
27. Kahl CA, Marsh J, Fyffe J, Sanders DA, Cornetta K (2004) Human immunodeficiency virus type 1-derived lentivirus vectors pseudotyped with envelope glycoproteins derived

from Ross River virus and Semliki Forest virus. J Virol 78:1421–1430.

28. Pear WS, Nolan GP, Scott ML, Baltimore D (1993) Production of high-titer helper-free retroviruses by transient transfection. Proc Natl Acad Sci USA 90:8392–8396.
29. Naldini L, Blomer U, Gallay P, Ory D, Mulligan R, Gage FH, Verma IM, Trono D (1996) In vivo gene delivery and stable transduction of nondividing cells by a lentiviral vector. Science 272:263–267.
30. Zufferey R, Nagy D, Mandel RJ, Naldini L, Trono D (1997) Multiply attenuated lentiviral vector achieves efficient gene delivery in vivo. Nat Biotechnol 15:871–875.
31. Dull T, Zufferey R, Kelly M, Mandel RJ, Nguyen M, Trono D, Naldini L (1998) A third-generation lentivirus vector with a conditional packaging system. J Virol 72:8463–8471.
32. Kootstra NA, Munk C, Tonnu N, Landau NR, Verma IM (2003) Abrogation of postentry restriction of HIV-1-based lentiviral vector transduction in simian cells. Proc Natl Acad Sci USA 100:1298–1303.
33. Follenzi A, Ailles LE, Bakovic S, Geuna M, Naldini L (2000) Gene transfer by lentiviral vectors is limited by nuclear translocation and rescued by HIV-1 pol sequences. Nat Genet 25:217–222.
34. Kozak M (1987) An analysis of 5′-noncoding sequences from 699 vertebrate messenger RNAs. Nucleic Acids Res 15:8125–8148.
35. Crambert G, Füzesi M, Garty H, Karlish S, Geering K (2002) Phospholemman (FXYD1) associates with Na,K-ATPase and regulates its transport properties. Proc Natl Acad Sci USA 99:11476–11481.
36. Deng V, Matagne V, Banine F, Frerking M, Ohliger P, Budden S, Pevsner J, Dissen GA, Sherman LS, Ojeda SR (2007) FXYD1 is an MeCP2 target gene overexpressed in the brains of Rett syndrome patients and Mecp2-null mice. Hum Mol Genet 16:640–650.
37. Palmiter RD, Sandgren EP, Avarbock MR, Allen DD, Brinster RL (1991) Heterologous introns can enhance expression of transgenes in mice. Proc Natl Acad Sci USA 88: 478–482.
38. Lomedico P, Rosenthal N, Efstratidadis A, Gilbert W, Kolodner R, Tizard R (1979) The structure and evolution of the two nonallelic rat preproinsulin genes. Cell 18:545–558.
39. Hoyle GW, Graham RM, Finkelstein JB, Nguyen K-PT, Gozal D, Friedman M (1998) Hyperinnervation of the airways in transgenic mice overexpressing nerve growth factor. Am J Respir Cell Mol Biol 18:149–157.
40. Lindsell CE, Shawber CJ, Boulter J, Weinmaster G (1995) Jagged: a mammalian ligand that activates Notch1. Cell 80:909–917.
41. Hobbs S, Jitrapakdee S, Wallace JC (1998) Development of a bicistronic vector driven by the human polypeptide chain elongation factor 1α promoter for creation of stable mammalian cell lines that express very high levels of recombinant proteins. Biochem Biophys Res Commun 252:368–372.
42. Stegmeier F, Hu G, Rickles RJ, Hannon GJ, Elledge SJ (2005) A lentiviral microRNA-based system for single-copy polymerase II-regulated RNA interference in mammalian cells. Proc Natl Acad Sci USA 102:13212–13217.
43. Boudreau RL, Martins I, Davidson BL (2009) Artificial microRNAs as siRNA shuttles: improved safety as compared to shRNAs in vitro and in vivo. Mol Ther 17:169–175.
44. Grimm D, Streetz KL, Jopling CL, Storm TA, Pandey K, Davis CR, Marion P, Salazar F, Kay MA (2006) Fatality in mice due to oversaturation of cellular microRNA/short hairpin RNA pathways. Nature 441:537–541.
45. Lagos-Quintana M, Rauhut R, Lendeckel W, Tuschl T (2001) Identification of novel genes coding for small expressed RNAs. Science 294:853–858.
46. Silva JM, Li MZ, Chang K, Ge W, Golding MC, Rickles RJ, Siolas D, Hu G, Paddison PJ, Schlabach MR, Sheth N, Bradshaw J, Burchard J, Kulkarni A, Cavet G, Sachidanandam R, McCombie WR, Cleary MA, Elledge SJ, Hannon GJ (2005) Second-generation shRNA libraries covering the mouse and human genomes. Nat Genet 37:1281–1288.
47. Sun D, Melegari M, Sridhar S, Rogler CE, Zhu L (2006) Multi-miRNA hairpin method that improves gene knockdown efficiency and provides linked multi-gene knockdown. BioTechniques 41:59–63.
48. Tiscornia G, Tergaonkar V, Galimi F, Verma IM (2004) CRE recombinase-inducible RNA interference mediated by lentiviral vectors. Proc Natl Acad Sci USA 101:7347–7351.
49. Galimi F, Saez E, Gall J, Hoong N, Cho G, Evans RM, Verma IM (2005) Development of ecdysone-regulated lentiviral vectors. Mol Ther 11:142–148.
50. Sinn PL, Sauter SL, McCray PB, Jr. (2005) Gene therapy progress and prospects: development of improved lentiviral and retroviral vectors--design, biosafety, and production. Gene Ther 12:1089–1098.
51. Dissen GA, Lomniczi A, Neff TL, Hobbs TR, Kohama SG, Kroenke CD, Galimi F, Ojeda SR (2009) In vivo manipulation of gene

expression in non-human primates using lentiviral vectors as delivery vehicles. Methods 49:70–77.

52. Heger S, Mastronardi C, Dissen GA, Lomniczi A, Cabrera R, Roth CL, Jung H, Galimi F, Sippell W, Ojeda SR (2007) Enhanced at puberty 1 (EAP1) is a new transcriptional regulator of the female neuroendocrine reproductive axis. J Clin Invest 117:2145–2154.
53. Paxinos G, Huang X-F, Toga AW (2000) The Rhesus Monkey Brain in Stereotaxic Coordinates. Academic Press, San Diego, CA.
54. Wu Z, Asokan A, Samulski RJ (2006) Adeno-associated virus serotypes: vector toolkit for human gene therapy. Mol Ther 14:316–327.
55. Miyauchi K, Kim Y, Latinovic O, Morozov V, Melikyan GB (2009) HIV enters cells via endocytosis and dynamin-dependent fusion with endosomes. Cell 137:433–444.
56. Uchil PD, Mothes W (2009) HIV Entry Revisited. Cell 137:402–404.
57. Bartlett JS, Wilcher R, Samulski RJ (2000) Infectious entry pathway of adeno-associated virus and adeno-associated virus vectors. J Virol 74:2777–2785.
58. Kotin RM, Siniscalco M, Samulski RJ, Zhu XD, Hunter L, Laughlin CA, McLaughlin S, Muzyczka N, Rocchi M, Berns KI (1990) Site-specific integration by adeno-associated virus. Proc Natl Acad Sci USA 87:2211–2215.
59. Towne C, Pertin M, Beggah AT, Aebischer P, Decosterd I (2009) Recombinant adeno-associated virus serotype 6 (rAAV2/6)-mediated gene transfer to nociceptive neurons through different routes of delivery. Mol Pain 5:52.
60. Hacein-Bey-Abina S, Von KC, Schmidt M, McCormack MP, Wulffraat N, Leboulch P, Lim A, Osborne CS, Pawliuk R, Morillon E, Sorensen R, Forster A, Fraser P, Cohen JI, de Saint BG, Alexander I, Wintergerst U, Frebourg T, Aurias A, Stoppa-Lyonnet D, Romana S, Radford-Weiss I, Gross F, Valensi F, Delabesse E, Macintyre E, Sigaux F, Soulier J, Leiva LE, Wissler M, Prinz C, Rabbitts TH, Le DF, Fischer A, Cavazzana-Calvo M (2003) LMO2-associated clonal T cell proliferation in two patients after gene therapy for SCID-X1. Science 302:415–419.
61. Shou Y, Ma Z, Lu T, Sorrentino BP (2006) Unique risk factors for insertional mutagenesis in a mouse model of XSCID gene therapy. Proc Natl Acad Sci USA 103:11730–11735.
62. Su K, Wang D, Ye J, Kim YC, Chow SA (2009) Site-specific integration of retroviral DNA in human cells using fusion proteins consisting of human immunodeficiency virus type 1 integrase and the designed polydactyl zinc-finger protein E2C. Methods 47:269–276.
63. Philippe S, Sarkis C, Barkats M, Mammeri H, Ladroue C, Petit C, Mallet J, Serguera C (2006) Lentiviral vectors with a defective integrase allow efficient and sustained transgene expression in vitro and in vivo. Proc Natl Acad Sci USA 103:17684–17689.
64. Banasik MB, McCray PB, Jr. (2010) Integrase-defective lentiviral vectors: progress and applications. Gene Ther 17:150–157.
65. Naldini L, Blömer U, Gage FH, Trono D, Verma IM (1996) Efficient transfer, integration, and sustained long-term expression of the transgene in adult rat brains injected with a lentiviral vector. Proc Natl Acad Sci USA 93:11382–11388.
66. Imren S, Payen E, Westerman KA, Pawliuk R, Fabry ME, Eaves CJ, Cavilla B, Wadsworth LD, Beuzard Y, Bouhassira EE, Russell R, London IM, Nagel RL, Leboulch P, Humphries RK (2002) Permanent and pan-erythroid correction of murine beta thalassemia by multiple lentiviral integration in hematopoietic stem cells. Proc Natl Acad Sci USA 99:14380–14385.
67. Kim YJ, Kim YS, Larochelle A, Renaud G, Wolfsberg TG, Adler R, Donahue RE, Hematti P, Hong BK, Roayaei J, Akagi K, Riberdy JM, Nienhuis AW, Dunbar CE, Persons DA (2009) Sustained high-level polyclonal hematopoietic marking and transgene expression 4 years after autologous transplantation of rhesus macaques with SIV lentiviral vector-transduced CD34+ cells. Blood 113:5434–5443.
68. Kaplitt MG, Leone P, Samulski RJ, Xiao X, Pfaff DW, O'Malley KL, During MJ (1994) Long-term gene expression and phenotypic correction using adeno-associated virus vectors in the mammalian brain. Nat Genet 8:148–154.
69. McCown TJ, Xiao X, Li J, Breese GR, Samulski RJ (1996) Differential and persistent expression patterns of CNS gene transfer by an adeno-associated virus (AAV) vector. Brain Res 713:99–107.
70. Xiao X, Li J, Samulski RJ (1996) Efficient long-term gene transfer into muscle tissue of immunocompetent mice by adeno-associated virus vector. J Virol 70:8098–8108.
71. Khan IF, Hirata RK, Wang PR, Li Y, Kho J, Nelson A, Huo Y, Zavaljevski M, Ware C, Russell DW (2010) Engineering of Human Pluripotent Stem Cells by AAV-mediated Gene Targeting. Mol Ther 18:1192–1199.
72. Grieger JC, Samulski RJ (2005) Adeno-associated virus as a gene therapy vector: vector development, production and clinical applications. Adv Biochem Eng Biotechnol 99:119–145.

73. Cui Y, Golob J, Kelleher E, Ye Z, Pardoll D, Cheng L (2002) Targeting transgene expression to antigen-presenting cells derived from lentivirus-transduced engrafting human hematopoietic stem/progenitor cells. Blood 99:399–408.
74. Hengge UR, Mirmohammadsadegh A (2000) Adeno-associated virus expresses transgenes in hair follicles and epidermis. Mol Ther 2:188–194.
75. Chirmule N, Xiao W, Truneh A, Schnell MA, Hughes JV, Zoltick P, Wilson JM (2000) Humoral immunity to adeno-associated virus type 2 vectors following administration to murine and nonhuman primate muscle. J Virol 74:2420–2425.
76. Koeberl DD, Alexander IE, Halbert CL, Russell DW, Miller AD (1997) Persistent expression of human clotting factor IX from mouse liver after intravenous injection of adeno-associated virus vectors. Proc Natl Acad Sci USA 94:1426–1431.
77. Prasad KM, Yang Z, Bleich D, Nadler JL (2000) Adeno-associated virus vector mediated gene transfer to pancreatic beta cells. Gene Ther 7:1553–1561.
78. Conrad CK, Allen SS, Afione SA, Reynolds TC, Beck SE, Fee-Maki M, Barrazza-Ortiz X, Adams R, Askin FB, Carter BJ, Guggino WB, Flotte TR (1996) Safety of single-dose administration of an adeno-associated virus (AAV)-CFTR vector in the primate lung. Gene Ther 3:658–668.
79. Zhang PX, Fuleihan RL (1999) Transfer of activation-dependent gene expression into T cell lines by recombinant adeno-associated virus. Gene Ther 6:182–189.
80. Mitchell M, Jerebtsova M, Batshaw ML, Newman K, Ye X (2000) Long-term gene transfer to mouse fetuses with recombinant adenovirus and adeno-associated virus (AAV) vectors. Gene Ther 7:1986–1992.
81. Mikkola H, Woods NB, Sjogren M, Helgadottir H, Hamaguchi I, Jacobsen SE, Trono D, Karlsson S (2000) Lentivirus gene transfer in murine hematopoietic progenitor cells is compromised by a delay in proviral integration and results in transduction mosaicism and heterogeneous gene expression in progeny cells. J Virol 74:11911–11918.
82. Remy S, Nguyen TH, Menoret S, Tesson L, Usal C, Anegon I (2010) The use of lentiviral vectors to obtain transgenic rats. Methods Mol Biol 597:109–125.
83. Yang SH, Cheng PH, Banta H, Piotrowska-Nitsche K, Yang JJ, Cheng EC, Snyder B, Larkin K, Liu J, Orkin J, Fang ZH, Smith Y, Bachevalier J, Zola SM, Li SH, Li XJ, Chan AW (2008) Towards a transgenic model of Huntington's disease in a non-human primate. Nature 453:921–924.
84. Cronin J, Zhang XY, Reiser J (2005) Altering the tropism of lentiviral vectors through pseudotyping. Curr Gene Ther 5:387–398.
85. Mochizuki H, Schwartz JP, Tanaka K, Brady RO, Reiser J (1998) High-titer human immunodeficiency virus type 1-based vector systems for gene delivery into nondividing cells. J Virol 72:8873–8883.
86. Strang BL, Takeuchi Y, Relander T, Richter J, Bailey R, Sanders DA, Collins MK, Ikeda Y (2005) Human immunodeficiency virus type 1 vectors with alphavirus envelope glycoproteins produced from stable packaging cells. J Virol 79:1765–1771.
87. Morizono K, Bristol G, Xie YM, Kung SK, Chen IS (2001) Antibody-directed targeting of retroviral vectors via cell surface antigens. J Virol 75:8016–8020.
88. Sinn PL, Hickey MA, Staber PD, Dylla DE, Jeffers SA, Davidson BL, Sanders DA, McCray PB, Jr. (2003) Lentivirus vectors pseudotyped with filoviral envelope glycoproteins transduce airway epithelia from the apical surface independently of folate receptor alpha. J Virol 77:5902–5910.
89. Kobinger GP, Weiner DJ, Yu QC, Wilson JM (2001) Filovirus-pseudotyped lentiviral vector can efficiently and stably transduce airway epithelia in vivo. Nat Biotechnol 19:225–230.
90. Reiser J, Harmison G, Kluepfel-Stahl S, Brady RO, Karlsson S, Schubert M (1996) Transduction of nondividing cells using pseudotyped defective high-titer HIV type 1 particles. Proc Natl Acad Sci USA 93: 15266–15271.
91. Mueller C, Flotte TR (2008) Clinical gene therapy using recombinant adeno-associated virus vectors. Gene Ther 15:858–863.
92. Burger C, Gorbatyuk OS, Velardo MJ, Peden CS, Williams P, Zolotukhin S, Reier PJ, Mandel RJ, Muzyczka N (2004) Recombinant AAV viral vectors pseudotyped with viral capsids from serotypes 1, 2, and 5 display differential efficiency and cell tropism after delivery to different regions of the central nervous system. Mol Ther 10:302–317.
93. Taymans JM, Vandenberghe LH, Haute CV, Thiry I, Deroose CM, Mortelmans L, Wilson JM, Debyser Z, Baekelandt V (2007) Comparative analysis of adeno-associated viral vector serotypes 1, 2, 5, 7, and 8 in mouse brain. Hum Gene Ther 18:195–206.
94. Zincarelli C, Soltys S, Rengo G, Rabinowitz JE (2008) Analysis of AAV serotypes 1–9 mediated gene expression and tropism in mice

after systemic injection. Mol Ther 16: 1073–1080.

95. Dodiya HB, Bjorklund T, Stansell J, III, Mandel RJ, Kirik D, Kordower JH (2010) Differential transduction following basal ganglia administration of distinct pseudotyped AAV capsid serotypes in nonhuman primates. Mol Ther 18:579–587.
96. Lowenstein PR, Mandel RJ, Xiong WD, Kroeger K, Castro MG (2007) Immune responses to adenovirus and adeno-associated vectors used for gene therapy of brain diseases: the role of immunological synapses in understanding the cell biology of neuroimmune interactions. Curr Gene Ther 7:347–360.
97. Zaiss AK, Liu Q, Bowen GP, Wong NC, Bartlett JS, Muruve DA (2002) Differential activation of innate immune responses by adenovirus and adeno-associated virus vectors. J Virol 76:4580–4590.
98. Follenzi A, Santambrogio L, Annoni A (2007) Immune responses to lentiviral vectors. Curr Gene Ther 7:306–315.
99. Kotin RM (1994) Prospects for the use of adeno-associated virus as a vector for human gene therapy. Hum Gene Ther 5:793–801.
100. Halbert CL, Miller AD, McNamara S, Emerson J, Gibson RL, Ramsey B, Aitken ML (2006) Prevalence of neutralizing antibodies against adeno-associated virus (AAV) types 2, 5, and 6 in cystic fibrosis and normal populations: Implications for gene therapy using AAV vectors. Hum Gene Ther 17:440–447.
101. Boutin S, Monteilhet V, Veron P, Leborgne C, Benveniste O, Montus MF, Masurier C (2010) Prevalence of Serum IgG and Neutralizing Factors Against Adeno-Associated Virus (AAV) Types 1, 2, 5, 6, 8, and 9 in the Healthy Population: Implications for Gene Therapy Using AAV Vectors. Hum Gene Ther 21: 1–9.
102. Karlsson Hedestam GB, Fouchier RA, Phogat S, Burton DR, Sodroski J, Wyatt RT (2008) The challenges of eliciting neutralizing antibodies to HIV-1 and to influenza virus. Nat Rev Microbiol 6:143–155.

Chapter 5

Targeting Neurological Disease with siRNA

Jan Christoph Koch, Mathias Bähr, and Paul Lingor

Abstract

RNA interference (RNAi) is a mechanism that specifically inhibits gene expression via small RNAs such as small interfering RNA (siRNA) and microRNA (miRNA). It not only plays a pivotal role under physiological conditions as cell differentiation and development but it also permits a systematic research on gene function and the identification of potential drug targets. In neuroscientific research, RNAi technologies are broadly used to unravel gene function under physiological and pathological conditions. Moreover, RNAi has already been successfully employed as a therapeutic approach in different animal models of neurological disease in vivo. Here, we present an overview on RNAi-mediated gene silencing focusing on issues of design, delivery, and detrimental effects of siRNAs. We review the current use of RNAi in different models of neurological disease and give an outlook on the possibilities of implementation of this technology in clinical therapy.

Key words: RNA interference, Neurological disease, Transfection methods, Neuron culture, Off-target effects

1. Introduction

The phenomenon of sequence-specific posttranslational gene silencing was initially observed in plants. In order to deepen the pigmentation of petunia, plant biologists introduced a chimeric chalcone synthase gene into the plant genome, but surprisingly this genetic modification resulted in a more or less pronounced depigmentation (1). The underlying mechanism, however, remained unresolved until in 1998 Fire and Mello described the mechanism of RNA-mediated gene silencing, which they observed in the nematode *Caenorhabditis elegans* (2). For this groundbreaking work, Fire and Mello were awarded the 2006 Nobel Prize in Physiology or Medicine. These initial findings were quickly followed by the description of functional RNAi in *Drosophila melanogaster* (3) and, interestingly, also in mammals (4). It thus became obvious that most components of the RNAi machinery are greatly

Alexei Morozov (ed.), *Controlled Genetic Manipulations*, Neuromethods, vol. 65,
DOI 10.1007/978-1-61779-533-6_5, © Springer Science+Business Media, LLC 2012

conserved throughout different species underscoring its essential physiological role. One of the biological functions that could be attributed to the enzymatic machinery involved in RNAi is providing a defense mechanism for cells against viral infections (5) and transposable genetic elements (6). Another crucial function of the RNAi machinery is the control of cell differentiation and tissue specificity via translational gene regulation by (miRNAs) (7). In fact, analyses have suggested that 1% of all human genes code for miRNAs which regulate the expression of at least 10% of all human coding genes (8).

Beside these novel insights into cellular regulation of gene expression, RNAi has opened completely new avenues for research in neuroscience and for putative therapeutic applications in the treatment of neurological disease. The possibility to specifically downregulate the expression of virtually any target protein by introduction of synthetically designed siRNAs into cells has greatly advanced research on gene function and intracellular signaling cascades, for example, in the study of cell death and apoptosis. Furthermore, RNAi has been successfully employed to improve disease phenotypes in numerous animal models of neurological disease (9).

In this chapter, we will give an overview on the mechanisms of RNAi, delivery methods, and off-target effects of siRNAs focusing on neurological disease. We will discuss examples of RNAi application in models of neurological disease and give an outlook on putative therapeutic approaches in humans.

2. The RNAi Machinery

The mechanism of RNAi-mediated gene silencing is initiated by small regulatory RNAs which have a length of 20–30 nucleotides. These small regulatory RNAs can be grouped into three main classes: short interfering RNAs (siRNAs), microRNAs (miRNAs), and PIWI-interacting RNAs (piRNAs) (10). Longer double-stranded RNAs (dsRNA) serve as precursors and are either produced in the nucleus (as for miRNAs) or can be artificially introduced into a cell (as in the case of siRNAs).

In the nucleus, primary miRNA transcripts (so-called pri-miRNAs), that form intramolecular stem-loop structures (hairpins), are transcribed from the genome. These pri-miRNAs are processed by the Drosha–DGCR8 complex to form pre-miRNAs with a length of 60–70 nucleotides. The pre-miRNAs are then transported from the nucleus to the cytoplasm by exportin-5. In the cytoplasm, these pre-miRNAs are further processed by the Dicer enzyme, an RNAse III family member, which removes the loop region to form the mature miRNA.

The Dicer enzyme also cleaves other long dsRNAs that result from RNA virus replication, mobile genetic elements, self-annealing transcripts, or experimental transfection. This cleavage leads to the generation of dsRNA fragments of 21 or 22 nucleotides, the mature siRNAs (11).

In the cytoplasm, mature siRNAs and miRNAs associate with members of the Argonaute family of proteins, which function as the core components of a diverse set of protein–RNA complexes called RNA-induced silencing complexes (RISC) (12). Upon integration into RISC, the siRNA strand which is antisense to the targeted RNA (guide strand) is incorporated, while its complementary strand (passenger strand) is cleaved and ejected. Subsequently, the guide strand permits the recognition of the complementary mRNA. SiRNAs contain perfectly complementary sequences to their target mRNAs. This leads to cleavage of the target mRNA catalyzed by the Argonaute protein (extensively reviewed in (13, 14)).

miRNAs, on the other hand, are only partly complementary to sequences in the 3′ untranslated regions of their target mRNAs. Due to this imperfect complementarity, the target mRNA is usually not cleaved by the Argonaute protein, but its translation is clearly diminished. The exact mechanism of this miRNA-mediated silencing remains unresolved but is thought to occur by repression of target mRNA translation and mRNA deadenylation, which in the end leads to mRNA degradation (15, 16).

The knowledge about the RNAi machinery allows us to specifically design siRNAs or miRNAs that can be transfected into cells and subsequently lead to a silencing of a target gene of interest.

3. siRNA Design

The intelligent design of effective siRNA sequences has been one of the major challenges for the successful implementation of RNAi technology because not all siRNA sequences result in equal silencing of the mRNA target. Large-scale screening programs have been performed to find the most potent inhibitors for a given target. These studies provided some basic rules for intelligent siRNA design (17).

Conventional siRNAs consist of 21 nucleotides and have a 3′-dinucleotide overhang to mimic Dicer cleavage products (11). For effective gene silencing, the proportion of the nucleotides guanosine and cytidine should be lower than 50% (18) and inverted repeats should be avoided (19). The 5′ end of the antisense (guide) strand should be designed to have a lower thermodynamic stability than the 5′ end of the sense strand as this facilitates incorporation into the RISC (20). The siRNA should not target protein-binding sites in mRNA regulatory regions because binding of regulatory proteins might block siRNA–target pairing. Intramolecular structures in the target should be avoided. Nowadays, intelligent RNAi design is often conducted with the help of software that is able to identify the optimal silencing sequence for a certain target (21).

Next to the appropriate siRNA sequence RNAi-mediated gene silencing is also dependent on the target gene: the intracellular protein synthesis and turnover may limit silencing effects for proteins with a long half-life, such as cytoskeletal structure proteins (22).

Another method for optimizing RNAi-mediated gene silencing, especially for the in vivo usage, is chemical modification of the siRNA. The most common chemical modification is siRNA methylation. A single 2′-O-methyl group on the passenger strand of the siRNA duplex can block the activation of Toll-like receptors and thereby prevent toxicity caused by the activation of type I interferon pathway gene expression (23). Addition of phosphorothioate linkages or substitution of 2′-fluoropyrimidines or a 2′-O-methyl for the 2′-ribose at certain positions increases resistance to ribonuclease without altering the siRNA activity (24). Important for systemic application of siRNA, it could be shown that fluoro-β-D-arabinonucleic acid (25) or arabinonucleic acid modifications (26) can increase both the serum stability and the potency of siRNAs in vivo. Other chemical modifications aim at decreasing the activity of the siRNA sense (passenger) strand in order to reduce specific off-target effects.

4. Transfection Methods

siRNAs cannot easily cross cell membranes due to their size and negative charge. Application of naked siRNA only leads to moderate effects on gene regulation. Even though naked siRNA is taken up by endosomes in primary neurons, only insufficient amounts reach the cytoplasm to induce functional RNAi (27).

Therefore, the delivery of siRNA to cells has been one of the major challenges for RNAi technology. In particular, the delivery of siRNA into neuronal cells is very challenging. For reasons that are yet not completely understood, neurons seem more difficult to transfect than many other cell types, and primary neurons again represent the biggest challenge. Numerous delivery methods have been experimentally assessed and offer different possibilities in respect to transfection efficiency, cytotoxicity, and cell type specificity. The two main strategies are delivery of chemically synthesized siRNAs (nonviral delivery) and delivery of shRNA-encoding genes by engineered viruses that will ultimately generate siRNAs by transcription in the target cells (viral delivery).

The chemical reagents used most commonly to improve siRNA transfection of neuronal cells in vitro are liposomic agents (28). For neuronal cell lines, transfection rates up to 70% are possible under optimized conditions, but a strong cytotoxicity usually limits their application (29). Besides, cells often display morphological changes after transfection (e.g., dendrites and axons often detach from the culture surface) (30). Artificial viruslike particles (AVP) and peptide-based transfection reagents, for example stearylated octaarginine, may represent an alternative to conventional liposomic agents, although they are less commercially available (29).

Conventional electroporation usually also has the drawback of a reduced cell viability due to the long-lasting current pulses that lead to the generation of cytotoxic anions (30). Recently, an alternative electroporation method, so-called nucleofection, has proven to be one of the most effective methods for siRNA transfection of primary neurons in regard to transfection efficiency and cell viability (31). In nucleofection, high-voltage pulses in combination with a cell type–specific buffer are used. The short time of the pulses together with the high buffer capacity reduces the problem of cytotoxic anion generation. The major drawback of the nucleofection technique next to the relatively high costs is that the procedure is until now only established for neuronal cells in suspension. This means that primary neurons have to be transfected immediately after being isolated from the animal. However, different approaches have recently been published to transfect adherent cells in a 96-well culture-plate format (e.g., (32)). It remains to be evaluated whether this system can be applied for neurons.

For viral delivery, short hairpin RNAs (shRNAs) are expressed from the viral vector genome and are intracellularly processed by Dicer for functional RNAi (33). Viral vectors permit long-term shRNA expression for several months (e.g., stable expression for over 18 months with AAV in the rat retina, personal observation), while single-time siRNA transfections only result in silencing effects of up to 7 days in nondifferentiated cells or up to a few weeks in differentiated cells (34). Importantly, viral vectors offer the possibility to limit the expression to a specific cell type using cell type specific promoters and capsid serotypes (35).

Two viral systems are most commonly used for transduction of nervous system tissue—lentivirus (LV) and adenoassociated virus (AAV). LV belongs to the retrovirus class and has been shown to be suitable for the transduction of nondividing cells, such as neurons. One of the biggest advantages of LV is their large packaging capacity of up to ~9 kb. This allows for the expression of shRNA plasmids in addition to larger reporter genes or other genes of interest (36, 37). Besides, the immune response to LV is very low or even completely missing. However, LV randomly integrates into the host genome, which might cause mutations and even activation of oncogenes (38). Another safety concern is the possible emergence of replication-competent viruses. Both problems have been effectively addressed by different modifications of the viral vector genome including the development of self-inactivating oncoretroviral vectors. These modifications should minimize the risk of mutagenesis and production of self-replicative virus for the LV currently used for in vivo applications. Major obstacles for the usage of LV next to the safety concerns are the low titers of the virus stocks produced by currently available production systems (~10^5–10^7 infectious particles per ml) (39) and the inherent instability of retroviral vector particles.

AAV vectors transduce both dividing and nondividing cells making them very attractive for neuroscientific research. After receptor-mediated entering of the cell, AAVs predominantly stay in the episomal compartment where they allow for long-term expression of the transgene for time periods over many months. Integration into the host genome occurs only at a very low rate and with a preference for a well-defined region on human chromosome 19q13.42, where it is not known to cause any harmful mutations. The predominant episomal state of AAV vectors in quiescent cells ensures an extraordinarily high safety profile for clinical applications (40). So far, there is no human disease known to be caused by AAV. The disadvantage of AAV is their small packaging capacity of ~3.5 kb. However, this size permits the expression of a shRNA together with a fluorescent protein, for example, EGFP or dsRed. Up to now, 12 different serotypes (AAV1–12) have been described, each having a defined cell type specificity. AAV2 is the serotype used most commonly in neuroscience because of its tropism for neurons and muscle cells. Using different purification methods including virus gradient centrifugation and protein liquid chromatography (FPLC), high virus titers can be produced (up to ~10^{12} infectious particles per ml). However, careful optimization of the best virus titer for a certain experimental setting must be performed as high virus titers were shown to be toxic in vitro as well as in vivo (41).

Taken together, the choice of the suitable viral vector system has to be custom-tailored for each experimental paradigm depending on transfected cell type and brain region, construct size, and expected expression kinetics.

The introduction of siRNA to the brain parenchyma in vivo represents an even greater obstacle than in vitro transfection. In contrast to other organs, for example, the liver, where injection of naked siRNA has been shown to elicit functional RNAi (42), direct injection of siRNA to the adult rat brain parenchyma did not result in detectable gene silencing (43). Intraventricular infusion of siRNA was shown to reduce the expression of the reporter gene EGFP and was able to downregulate the endogenous dopamine transporter. This approach, however, required high amounts of siRNA, and the silencing was less pronounced in regions not directly adjacent to the injection site (44). Thus, both methods do not represent a real alternative in terms of practical use.

A systemic intravenous application of siRNA can also hardly target the brain because the blood–brain barrier prevents the passage of naked siRNA from blood serum to the brain parenchyma (45). In addition, the stability of naked siRNA in the serum remains limited, even though RNA double strands are more resistant towards RNAse digestion compared to single strands. Chemical modifications (see above) may increase the serum stability and improve pharmacokinetic properties of siRNA for in vivo applications (46). Pegylated immunoliposomes conjugated to antibodies binding to receptors of the brain vascular endothelium have been shown to overcome the blood–brain barrier and lead to a better cell type specificity depending on the targeted receptor (47). In another successful approach for transvascular delivery of siRNA to the brain, a short peptide derived from rabies virus glycoprotein (RVG) was used that specifically binds to the acetylcholine receptor expressed on neuronal cells. Following intravenous injection in mice, the synthesized chimeric RVG peptide was able to transduce siRNA to neuronal cells and resulted in specific gene silencing within the brain (48). Recently, intravenously applied AAV9 was shown to bypass the blood–brain barrier and efficiently target neonatal neurons and adult astrocytes (49) and thus could be useful for siRNA delivery in the future. Apart from these systemic approaches, siRNA-expressing viruses are effectively injected into defined brain regions in different animal models to locally modulate protein translation (e.g. (50)).

5. Off-Target Effects

Although siRNA-mediated gene silencing acts in a highly specific manner, it may also induce serious side effects.

A well-described side effect is the so-called interferon response. Unlike in *C. elegans* or *Drosophila* in mammalian cells, long dsRNAs induce an interferon response via activation of the protein kinase PKR which causes a shutdown of global protein synthesis and finally leads to cell death (51, 52). Until recently, it was believed that only dsRNAs greater than 30 base pairs in length could elicit this response. It could be shown though that in vivo and in cultured mammalian cells, shorter siRNAs are also capable of inducing an interferon response that is mediated by the binding of single- or double-stranded RNAs to Toll-like receptors (TLRs) localized in the endosomes. These TLRs trigger the translocation of the transcription factors nuclear factor-κB (NFκB), interferon regulatory factor (IRF) and activating transcription factor 2 (ATF2) into the nucleus and thereby can elicit an interferon response (53). The TLRs serve as pattern-recognition sensors of specific sequence motifs that act immunostimulatory and that should be avoided in siRNA design (e.g., 5′-GUCCUUCAA-3′ and 5′-UGUGU-3′) (54, 55). Interferon responses can also be attenuated by using shRNAs expressed from lentiviral vectors because these shRNAs enter the intracellular RNAi machinery from the nucleus and avoid endosomal incorporation (56). Besides, blunt-ended siRNAs lacking the two-nucleotide 3′ overhangs characteristic of Dicer processing are immunostimulatory being recognized by the retinoic acid gene 1 encoded RIG-I helicase (57).

siRNAs can affect the expression levels of many nontargeted transcripts (58). This effect is mainly mediated by a miRNA-like function of the siRNA that can inhibit protein translation and may lead to degradation of different mRNAs which are partly complementary to the siRNA sequence (59). This off-target effect can be reduced by a 2′-OMe modification at the second ribose from the 5′ end of the siRNA (60).

Several recent studies suggest that high levels of siRNA expression can be toxic. Delivery of siRNA may dose-dependently result in metabolic impairment and unspecific gene silencing in primary neuronal cultures (27). In vivo, it could be shown that AAV-mediated high-level expression of shRNAs may cause saturation of exportin-5 and subsequent inhibition of endogenous pre-miRNA nuclear export, which leads to cell death (61). Therefore, expression levels of shRNA should be carefully optimized.

Several other strategies have been developed to attenuate off-target effects. One approach is to transfect a mixture of different siRNAs against one target and thereby lower the concentration of each single siRNA in the cell. The mixture is generated by Dicer digestion of long double-stranded RNA resulting in so-called endoribonuclease-derived siRNAs (esiRNA). Not all of the siRNAs in the mixture equally downregulate the target gene, but due to the low concentration of each single siRNA, off-target effects are minimized. Comparably, a mixture of different commercially synthesized siRNAs with each siRNA in a relatively low contraction may reduce off-target effects (62).

It is therefore very important to assure good quality controls for RNAi experiments, including an appropriate control siRNA and possibly the additional use of siRNA mixtures or pools. Favorably, the control siRNA should be functional in a cellular context, for example, silencing a nonexpressed reporter gene. The downregulation of the target protein should be confirmed by Western blot. Additional PCR for demonstration of reduced mRNA levels (RT-PCR) can be performed, but usually plays a less important role in the experimental setting. If appropriate, immunostainings should confirm the RNAi effect in comparison to control. For final verification, physiological testing of the lacking protein should be performed.

6. RNAi in Neurological Diseases

In the last years, various studies have been published that employ RNAi methods for the evaluation and treatment of neurological disease models. In 2002, it was shown for the first time that RNAi functions in neurons in vitro, using cationic lipid-mediated transfection of primary mammalian cortical neurons (63). However, further progress in this field only became possible with the development of more elaborated delivery methods, such as viral vectors or the coupling of siRNA to immunoliposomes to specifically transfect target cells. Here, we will discuss some of the most relevant approaches and studies targeting different neurological diseases with RNAi.

Among the first examples for successful implementation of RNAi in the experimental treatment of neurological disorders were models of polyglutamine repeat diseases such as spinocerebellar ataxias (SCA) and Huntington's disease (HD). All of these diseases are caused by a toxic gain-of-function mutation due to a CAG trinucleotide repeat expansion. Expression of the polyglutamine-expanded protein leads to transcriptional dysregulation, disturbed protein homeostasis, formation of aggregates, and finally cell death (64). Due to this explicit pathomechanism, it is intriguing to positively influence the course of these diseases by RNAi-mediated downregulation of the polyglutamine-expanded proteins. The efficacy of this concept was first demonstrated in a mouse model of SCA-1 that expresses a pathogenic form of full-length human ataxin-1 (82 CAG repeats) in cerebellar Purkinje cells. An AAV1-vector expressing shRNAs against the CAG repeat region in the mutant human ataxin-1 was injected into the cerebellum and resulted in a downregulation of ataxin-1 in Purkinje cells. The treated animals not only showed an impressive attenuation of the cerebellar pathology but also an improvement of the motor performance on the rotarod. However, several injections to separate sites were required to achieve sufficient transduction of the cerebellum (65). Later, the same group performed a similar approach in a mouse model for HD (HD-N171-82Q). These mice express an N-terminal fragment of human huntingtin with 82 CAG repeats in neurons. An AAV1-vector expressing shRNA targeting human huntingtin was injected into the striatum of the mice. This shRNA treatment reduced mutant huntingtin expression in vivo and prevented the formation of neuronal aggregates. Moreover, the treated mice showed improved gait and rotarod performance compared to their untreated littermates (66). These results were confirmed in the HD-R6/1 transgenic mouse line that shows an even more aggressive phenotype compared to the HD-N171-82Q mice due to a longer polyglutamine repeat of 144 CAG repeats. Accordingly, in this model, AAV5-expressed shRNAs against huntingtin led to improved performance of the treated animals in different motor tests (67). In the same mouse model, a nonviral delivery approach using siRNAs against mutant human huntingtin complexed with lipofection reagents could also improve neuropathological findings as well as the performance in motor and behavioral testing although this transfection method only allowed for transient gene silencing for up to 14 days (68).

Parkinson's disease (PD) is one of the most common neurodegenerative diseases and characterized by a prominent, but not exclusive loss of dopaminergic neurons in the substantia nigra pars compacta. Several gene loci have been described to be mutated in familial PD. The exact pathophysiological role of these genes, however, still remains elusive. Alpha-synuclein is among the most intensively studied gene targets in PD because different single point mutations as well as a triplication in the alpha-synuclein gene have been found to cause familial PD in humans. Effective silencing of human wild-type and mutant A53T alpha-synuclein has been demonstrated in vitro using shRNA expressed by lentiviral vectors in SH-SY5Y cells and in vivo in a rat model overexpressing human alpha-synuclein (69). The therapeutic effect of this treatment, however, still needs to be evaluated. It is likely that silencing of alpha-synuclein may not represent a therapeutical option for PD, but it may help in resolving the function of the protein in greater detail. Another gene associated with familial PD is the PTEN-induced kinase 1 (Pink1). siRNA-mediated downregulation of wild-type Pink1 in SH-SY5Y cells resulted in mitochondrial fragmentation, induction of autophagy, and increased cell death (70). These results contribute to a better understanding of the pathophysiology of Pink1 in PD. One of the most common mutations in familial PD affects the leucine-rich repeat kinase 2 (LRRK2). The use of siRNAs against different kinases of the autophagy pathway could rescue the degenerative phenotype induced by overexpression of the G2019S LRRK2 mutant protein in SH-SY5Y cells and therefore link induction of autophagy to the pathological changes caused by LRRK2 mutations (71). This could point to putative therapeutic targets in PD associated with mutant LRRK2 in the future. However, in vivo experiments targeting these mutant proteins of familial PD are still lacking. Downregulation of the rate-limiting enzyme in dopamine synthesis, tyrosine hydroxylase (TH), via AAV-mediated expression of shRNA against TH has been shown to induce behavioral deficits similar to neurotoxin-induced models of the disease (72). This approach might thus serve as another disease model for PD. A proof that gene therapy might act as an effective treatment in animal models of PD was recently published, where a lentiviral vector was employed to express all three genes necessary for dopamine synthesis—tyrosine hydroxylase (TH), aromatic L-amino acid decarboxylase (AADC), and guanosine 5′triphosphate cyclohydrolase 1 (GCH1). The viral vector was injected locally into the putamen of macaques in which parkinsonism had been induced by prior systemic administration of the neurotoxin 1-methyl-4-phenyl-1,2,3,6-tetrahydropyridine (MPTP). Animals that received the vector showed a marked improvement of parkinsonian symptoms, including tremor and slow movements. These effects were reported to be stable for 1 year (73).

Amyotrophic lateral sclerosis (ALS) is a fatal disease that causes selective degeneration of motoneurons. Up to now, there is no effective pharmacologic treatment available; the disease leads to death within a few years after onset. The most commonly used mouse model of familial ALS is the superoxide dismutase 1 (SOD1) G93A mutant mouse that expresses a mutant human SOD1 transgene. Intraspinal injection of a lentiviral vector expressing shRNAs against SOD1 retarded both the onset and the progression rate of the disease (74). Moreover, multiple intramuscular injections of these lentiviral vectors were able to delay the onset of the disease markedly and to prolong the survival of the animals (75). The viral vectors were shown to have been transported retrogradely in axons from the injection sites in the muscle and to lead to a significant expression of the siRNAs in motoneurons. To prove that allele-specific RNAi is possible and might yield therapeutic benefits, transgenic mice that expressed a shRNA against mutant SOD1 (G93A) were crossed with mutant and wild-type SOD1 mice. Specific silencing was shown for the mutant SOD1 protein and significantly delayed disease onset and extended survival (76). However, SOD1 mutations only account for the minority of ALS cases; therefore other therapeutical targets will have to be considered in the future, for example, TDP-43.

The most common form of dementia is Alzheimer's disease (AD). As at least parts of the pathophysiology of AD are understood by now, several putative molecular targets for RNAi therapies have emerged. The beta-secretase BACE1 (β-site of APP cleaving enzyme) is involved in the cleavage of amyloid precursor protein (APP) generating amyloid-β peptides that form extracellular aggregates in brains of AD patients. Although no mutations of the BACE1 gene have been described yet, levels of this enzyme have been shown to be elevated in late-onset sporadic AD. Accordingly, downregulation of BACE1 by lentiviral vector-mediated RNAi reduced the formation of amyloid plaques in the injected brain regions in an AD mouse model. Moreover, the animals showed improvements in spatial learning and memory tests (77). RNAi against mutant APP itself was explored in a recent study where AAV expressing allele-specific shRNA against mutant APP (Swedish variant of APP (APPsw)) were injected into the hippocampi of AD transgenic mice (APP/PS1). This led to a long-term reduction of APPsw while the levels of wild-type APP remained unaltered. Moreover, the long-term expression of APPsw-shRNAs in the hippocampus resulted in a reduction of soluble amyloid beta levels and a significant improvement of the AD symptoms in the mutant mice compared to untreated animals (78). Mutations in the tau protein not only play a role in AD but also in other forms of dementia, like frontotemporal dementia, progressive supranuclear palsy, and corticobasal degeneration. siRNA-mediated downregulation of mutated tau protein (TauV337M) was shown to work effectively in vitro, but the effects in vivo still need to be evaluated (79).

Other neurological diseases where RNAi has shown promising results in animal model systems include prion diseases (80), neuropathic pain (81), ischemic stroke (82), and glioma (83).

7. Outlook

Although the RNAi machinery was discovered only a little more than a decade ago, clinical testing of siRNA-based therapeutics is already performed in a variety of diseases and has become a multimillion dollar business (one billion US Dollar in 2010 (20)). It is still too soon to evaluate whether or not RNAi-based therapeutics will live up to their expectations. The list of diseases for which RNAi is tested in human clinical trials includes age-related macular degeneration (AMD) (phases I, II, and III for different siRNA constructs and target genes, respectively), diabetic macular edema (phase II), respiratory syncytial virus (RSV) infection (phase I), pachyonychia congenita (phase I), hepatitis C (phase I), acute renal failure (phase I), metastatic melanoma (phase I), and AIDS lymphoma (phase I) (84). Neurological diseases are not yet on the list for human clinical trials, but several animal models have been successfully employed for RNAi testing as described above. Good candidates for human RNAi trials are surely the polyglutamine expansion diseases (SCA and HD), ALS, and PD. Clinical implementation of RNAi for the treatment of neurological disease will crucially depend on the ability to deliver the siRNA across the blood–brain barrier, to effectively downregulate the putative target gene over a long period of time, and to minimize the side effects. Preferably, the gene silencing should be regulable. Although many technical questions remain to be answered, the promising in vivo animal studies as well as the first clinical trials in humans with RNAi therapeutics give hope that these technical problems can be overcome soon. After achieving these goals, RNAi technology could be transferred from the bench to the clinic for the treatment of several severe disorders of the nervous system giving new hope for many patients.

References

1. Napoli C, Lemieux C, Jorgensen R (1990) Introduction of a Chimeric Chalcone Synthase Gene into Petunia Results in Reversible Co-Suppression of Homologous Genes in trans. Plant Cell 2: 279–289
2. Fire A, Xu S, Montgomery MK et al. (1998) Potent and specific genetic interference by double-stranded RNA in *Caenorhabditis elegans*. Nature 391: 806–811
3. Caplen NJ, Fleenor J, Fire A et al. (2000) dsRNA-mediated gene silencing in cultured *Drosophila* cells: a tissue culture model for the analysis of RNA interference. Gene 252: 95–105
4. Wianny F, Zernicka-Goetz M (2000) Specific interference with gene function by double-stranded RNA in early mouse development. Nat Cell Biol 2: 70–75
5. Gitlin L, Karelsky S, Andino R (2002) Short interfering RNA confers intracellular antiviral immunity in human cells. Nature 418: 430–434
6. Ketting RF, Haverkamp TH, van Luenen HG et al. (1999) Mut-7 of *C. elegans*, required for transposon silencing and RNA interference, is a homolog of Werner syndrome helicase and RNaseD. Cell 99: 133–141
7. Lau NC, Lim LP, Weinstein EG et al. (2001) An abundant class of tiny RNAs with probable regulatory roles in *Caenorhabditis elegans*. Science 294: 858–862
8. John B, Enright AJ, Aravin A et al. (2004) Human MicroRNA targets. PLoS Biol 2: e363
9. Boudreau RL, Davidson BL (2010) RNAi therapeutics for CNS disorders. Brain Res
10. Moazed D (2009) Small RNAs in transcriptional gene silencing and genome defence. Nature 457: 413–420
11. Elbashir SM, Lendeckel W, Tuschl T (2001) RNA interference is mediated by 21- and 22-nucleotide RNAs. Genes Dev 15: 188–200.
12. Hutvagner G, Simard MJ (2008) Argonaute proteins: key players in RNA silencing. Nat Rev Mol Cell Biol 9: 22–32
13. Tomari Y, Zamore PD (2005) Perspective: machines for RNAi. Genes Dev 19: 517–529
14. Sontheimer EJ (2005) Assembly and function of RNA silencing complexes. Nat Rev Mol Cell Biol 6: 127–138
15. Jinek M, Doudna JA (2009) A three-dimensional view of the molecular machinery of RNA interference. Nature 457: 405–412
16. Pillai RS, Bhattacharyya SN, Filipowicz W (2006) Repression of protein synthesis by miRNAs: how many mechanisms? Trends Cell Biol
17. Huesken D, Lange J, Mickanin C et al. (2005) Design of a genome-wide siRNA library using an artificial neural network. Nat Biotechnol 23: 995–1001
18. Elbashir SM, Harborth J, Weber K et al. (2002) Analysis of gene function in somatic mammalian cells using small interfering RNAs. Methods 26: 199–213
19. Reynolds A, Leake D, Boese Q et al. (2004) Rational siRNA design for RNA interference. Nat Biotechnol 22: 326–330
20. Castanotto D, Rossi JJ (2009) The promises and pitfalls of RNA-interference-based therapeutics. Nature 457: 426–433
21. Li W, Cha L (2007) Predicting siRNA efficiency. Cell Mol Life Sci 64: 1785–1792
22. Harborth J, Elbashir SM, Bechert K et al. (2001) Identification of essential genes in cultured mammalian cells using small interfering RNAs. J Cell Sci 114: 4557–4565
23. Robbins M, Judge A, Liang L et al. (2007) 2′-O-methyl-modified RNAs act as TLR7 antagonists. Mol Ther 15: 1663–1669
24. Morrissey DV, Lockridge JA, Shaw L et al. (2005) Potent and persistent in vivo anti-HBV activity of chemically modified siRNAs. Nat Biotechnol 23: 1002–1007
25. Dowler T, Bergeron D, Tedeschi AL et al. (2006) Improvements in siRNA properties mediated by 2′-deoxy-2′-fluoro-beta-D-arabinonucleic acid (FANA). Nucleic Acids Res 34: 1669–1675
26. Fisher M, Abramov M, Van Aerschot A et al. (2007) Inhibition of MDR1 expression with altritol-modified siRNAs. Nucleic Acids Res 35: 1064–1074
27. Lingor P, Michel U, Scholl U et al. (2004) Transfection of "naked" siRNA results in endosomal uptake and metabolic impairment in cultured neurons. Biochem Biophys Res Commun 315: 1126–1133
28. Dalby B, Cates S, Harris A et al. (2004) Advanced transfection with Lipofectamine 2000 reagent: primary neurons, siRNA, and high-throughput applications. Methods 33: 95–103
29. Tonges L, Lingor P, Egle R et al. (2006) Stearylated octaarginine and artificial virus-like particles for transfection of siRNA into primary rat neurons. Rna

30. Zeitelhofer M, Vessey JP, Xie Y et al. (2007) High-efficiency transfection of mammalian neurons via nucleofection. Nat Protoc 2: 1692–1704
31. Gresch O, Engel FB, Nesic D et al. (2004) New non-viral method for gene transfer into primary cells. Methods 33: 151–163
32. Bradburne C, Robertson K, Thach D (2009) Assessment of methods and analysis of outcomes for comprehensive optimization of nucleofection. Genet Vaccines Ther 7: 6
33. Brummelkamp TR, Bernards R, Agami R (2002) A system for stable expression of short interfering RNAs in mammalian cells. Science 296: 550–553
34. Omi K, Tokunaga K, Hohjoh H (2004) Long-lasting RNAi activity in mammalian neurons. FEBS Lett 558: 89–95
35. Michel U, Malik I, Ebert S et al. (2005) Long-term in vivo and in vitro AAV-2-mediated RNA interference in rat retinal ganglion cells and cultured primary neurons. Biochem Biophys Res Commun 326: 307–312
36. Trono D (2000) Lentiviral vectors: turning a deadly foe into a therapeutic agent. Gene Ther 7: 20–23
37. Wong LF, Goodhead L, Prat C et al. (2006) Lentivirus-mediated gene transfer to the central nervous system: therapeutic and research applications. Hum Gene Ther 17: 1–9
38. Themis M, Waddington SN, Schmidt M et al. (2005) Oncogenesis following delivery of a nonprimate lentiviral gene therapy vector to fetal and neonatal mice. Mol Ther 12: 763–771
39. Rodrigues T, Carrondo MJ, Alves PM et al. (2007) Purification of retroviral vectors for clinical application: biological implications and technological challenges. J Biotechnol 127: 520–541
40. Peel AL, Klein RL (2000) Adeno-associated virus vectors: activity and applications in the CNS. J Neurosci Methods 98: 95–104
41. Ehlert EM, Eggers R, Niclou SP et al. (2010) Cellular toxicity following application of adeno-associated viral vector-mediated RNA interference in the nervous system. BMC Neurosci 11: 20
42. Song E, Lee SK, Wang J et al. (2003) RNA interference targeting Fas protects mice from fulminant hepatitis. Nat Med 9: 347–351.
43. Isacson R, Kull B, Salmi P et al. (2003) Lack of efficacy of 'naked' small interfering RNA applied directly to rat brain. Acta Physiol Scand 179: 173–177
44. Thakker DR, Natt F, Husken D et al. (2004) Neurochemical and behavioral consequences of widespread gene knockdown in the adult mouse brain by using nonviral RNA interference. Proc Natl Acad Sci USA 101: 17270–17275
45. Mayhan WG (2001) Regulation of blood-brain barrier permeability. Microcirculation 8: 89–104
46. Soutschek J, Akinc A, Bramlage B et al. (2004) Therapeutic silencing of an endogenous gene by systemic administration of modified siRNAs. Nature 432: 173–178
47. Boado RJ (2005) RNA interference and nonviral targeted gene therapy of experimental brain cancer. NeuroRx 2: 139–150
48. Kumar P, Wu H, McBride JL et al. (2007) Transvascular delivery of small interfering RNA to the central nervous system. Nature 448: 39–43
49. Foust KD, Nurre E, Montgomery CL et al. (2009) Intravascular AAV9 preferentially targets neonatal neurons and adult astrocytes. Nat Biotechnol 27: 59–65
50. Sanftner LM, Sommer JM, Suzuki BM et al. (2005) AAV2-mediated gene delivery to monkey putamen: evaluation of an infusion device and delivery parameters. Exp Neurol 194: 476–483
51. Clemens MJ, Elia A (1997) The double-stranded RNA-dependent protein kinase PKR: structure and function. J Interferon Cytokine Res 17: 503–524
52. Gil J, Esteban M (2000) Induction of apoptosis by the dsRNA-dependent protein kinase (PKR): mechanism of action. Apoptosis 5: 107–114
53. Seth RB, Sun L, Chen ZJ (2006) Antiviral innate immunity pathways. Cell Res 16: 141–147
54. Judge AD, Sood V, Shaw JR et al. (2005) Sequence-dependent stimulation of the mammalian innate immune response by synthetic siRNA. Nat Biotechnol 23: 457–462
55. Hornung V, Guenthner-Biller M, Bourquin C et al. (2005) Sequence-specific potent induction of IFN-alpha by short interfering RNA in plasmacytoid dendritic cells through TLR7. Nat Med 11: 263–270
56. Robbins MA, Li M, Leung I et al. (2006) Stable expression of shRNAs in human CD34+ progenitor cells can avoid induction of interferon responses to siRNAs in vitro. Nat Biotechnol 24: 566–571
57. Marques JT, Devosse T, Wang D et al. (2006) A structural basis for discriminating between self and nonself double-stranded RNAs in mammalian cells. Nat Biotechnol 24: 559–565
58. Jackson AL, Bartz SR, Schelter J et al. (2003) Expression profiling reveals off-target gene

regulation by RNAi. Nat Biotechnol 21: 635–637

59. Jackson AL, Burchard J, Schelter J et al. (2006) Widespread siRNA "off-target" transcript silencing mediated by seed region sequence complementarity. RNA 12: 1179–1187
60. Jackson AL, Burchard J, Leake D et al. (2006) Position-specific chemical modification of siRNAs reduces "off-target" transcript silencing. RNA 12: 1197–1205
61. Grimm D, Streetz KL, Jopling CL et al. (2006) Fatality in mice due to oversaturation of cellular microRNA/short hairpin RNA pathways. Nature 441: 537–541
62. Jackson AL, Linsley PS (2004) Noise amidst the silence: off-target effects of siRNAs? Trends Genet 20: 521–524
63. Krichevsky AM, Kosik KS (2002) RNAi functions in cultured mammalian neurons. Proc Natl Acad Sci USA 99: 11926–11929.
64. Saudou F, Finkbeiner S, Devys D et al. (1998) Huntingtin acts in the nucleus to induce apoptosis but death does not correlate with the formation of intranuclear inclusions. Cell 95: 55–66
65. Xia H, Mao Q, Eliason SL et al. (2004) RNAi suppresses polyglutamine-induced neurodegeneration in a model of spinocerebellar ataxia. Nat Med 10: 816–820
66. Harper SQ, Staber PD, He X et al. (2005) RNA interference improves motor and neuropathological abnormalities in a Huntington's disease mouse model. Proc Natl Acad Sci USA 102: 5820–5825
67. Rodriguez-Lebron E, Denovan-Wright EM, Nash K et al. (2005) Intrastriatal rAAV-mediated delivery of anti-huntingtin shRNAs induces partial reversal of disease progression in R6/1 Huntington's disease transgenic mice. Mol Ther 12: 618–633
68. Wang YL, Liu W, Wada E et al. (2005) Clinicopathological rescue of a model mouse of Huntington's disease by siRNA. Neurosci Res 53: 241–249
69. Sapru MK, Yates JW, Hogan S et al. (2006) Silencing of human alpha-synuclein in vitro and in rat brain using lentiviral-mediated RNAi. Exp Neurol 198: 382–390
70. Dagda RK, Cherra SJ, 3rd, Kulich SM et al. (2009) Loss of PINK1 function promotes mitophagy through effects on oxidative stress and mitochondrial fission. J Biol Chem 284: 13843–13855
71. Plowey ED, Cherra SJ, 3rd, Liu YJ et al. (2008) Role of autophagy in G2019S-LRRK2-associated neurite shortening in differentiated SH-SY5Y cells. J Neurochem 105: 1048–1056
72. Hommel JD, Sears RM, Georgescu D et al. (2003) Local gene knockdown in the brain using viral-mediated RNA interference. Nat Med 9: 1539–1544
73. Jarraya B, Boulet S, Ralph GS et al. (2009) Dopamine gene therapy for Parkinson's disease in a nonhuman primate without associated dyskinesia. Sci Transl Med 1: 2ra4
74. Raoul C, Abbas-Terki T, Bensadoun JC et al. (2005) Lentiviral-mediated silencing of SOD1 through RNA interference retards disease onset and progression in a mouse model of ALS. Nat Med 11: 423–428
75. Ralph GS, Radcliffe PA, Day DM et al. (2005) Silencing mutant SOD1 using RNAi protects against neurodegeneration and extends survival in an ALS model. Nat Med 11: 429–433
76. Xia X, Zhou H, Huang Y et al. (2006) Allele-specific RNAi selectively silences mutant SOD1 and achieves significant therapeutic benefit in vivo. Neurobiol Dis 23: 578–586
77. Singer O, Marr RA, Rockenstein E et al. (2005) Targeting BACE1 with siRNAs ameliorates Alzheimer disease neuropathology in a transgenic model. Nat Neurosci 8: 1343–1349
78. Rodriguez-Lebron E, Gouvion CM, Moore SA et al. (2009) Allele-specific RNAi mitigates phenotypic progression in a transgenic model of Alzheimer's disease. Mol Ther 17: 1563–1573
79. Miller VM, Gouvion CM, Davidson BL et al. (2004) Targeting Alzheimer's disease genes with RNA interference: an efficient strategy for silencing mutant alleles. Nucleic Acids Res 32: 661–668
80. Pfeifer A, Eigenbrod S, Al-Khadra S et al. (2006) Lentivector-mediated RNAi efficiently suppresses prion protein and prolongs survival of scrapie-infected mice. J Clin Invest 116: 3204–3210
81. Dore-Savard L, Roussy G, Dansereau MA et al. (2008) Central delivery of Dicer-substrate siRNA: a direct application for pain research. Mol Ther 16: 1331–1339
82. Sun HS, Jackson MF, Martin LJ et al. (2009) Suppression of hippocampal TRPM7 protein prevents delayed neuronal death in brain ischemia. Nat Neurosci 12: 1300–1307
83. Gondi CS, Lakka SS, Dinh DH et al. (2004) RNAi-mediated inhibition of cathepsin B and uPAR leads to decreased cell invasion, angiogenesis and tumor growth in gliomas. Oncogene 23: 8486–8496
84. Tiemann K, Rossi JJ (2009) RNAi-based therapeutics-current status, challenges and prospects. EMBO Mol Med 1: 142–151

Chapter 6

In Utero Electroporation for Cellular Transgenesis in the Developing Mammalian Forebrain

Brady J. Maher and Joseph J. LoTurco

Abstract

The ability to introduce gain- or loss-of-function constructs into developing organisms is responsible for the majority of advances made in the field of developmental biology. Here we provide step-by-step methods on how to achieve cellular transgenesis in the developing mammalian neocortex. In utero electroporation (IUE) is a surgical preparation that allows for rapid manipulation of gene expression and is becoming an experimental standard in the field of neocortical development. We provide detailed description of the equipment required and the procedures necessary for successful embryonic brain transfection. In addition, we will discuss several different experimental approaches that demonstrate how IUE can be adapted to spatially and temporally control gene expression.

Key words: In utero electroporation, Cellular transgenesis, Cortical development, Forebrain development, Transfection, Neocortex

1. Introduction

Methods for cellular transgenesis are central to all experimental approaches aimed at determining the functional importance of genetic programs that guide the development of neural cells. These genetic programs contain spatially and temporally defined components, and therefore, it is essential to use methods that provide gain and loss of function in defined populations of developing neural cells during defined times in development. In utero electroporation (IUE) is a method that provides a versatile platform to rapidly execute experiments that require spatial and temporal manipulation of gene expression in the developing mammalian forebrain. Because this method uses standard plasmid DNA as genetic vectors, and can deliver up to four plasmids with a high degree of efficiency to developing neurons and glia in vivo, it offers

Alexei Morozov (ed.), *Controlled Genetic Manipulations*, Neuromethods, vol. 65,
DOI 10.1007/978-1-61779-533-6_6,

many advantages over viral approaches and is less time-consuming and less expensive that the production of transgenic animals.

Successful IUE in the mammalian forebrain was first demonstrated in 2001 by three groups that used electroporation to transfect populations of neural progenitors at the ventricular surface of developing neocortex in both mouse and rat embryos (1–3). Early application of this technique demonstrated that local manipulation of FGF8 expression respecifies regional patterning in neocortex (1) and revealed mechanisms necessary for radial neuronal migration in neocortex (4, 5). Subsequent technical developments of IUE have resulted in approaches for cell-type and temporal control using cell-specific promoters and tamoxifen-gated cre recombinase (6–8).

In this chapter, we first describe the specific technical requirements to perform the IUE surgical procedure. This includes a detailed description of the equipment and procedures that are necessary for successful embryonic brain transfection. In the second part of the chapter, we discuss several plasmid systems currently in use for experiments that require cell-type-specific expression, conditional expression, and lineage mapping. Because IUE can be generalized to virtually any combination of plasmids (up to 4–5 in each cell), and can be performed in transgenic mice and rats, there are an increasingly large number of experiment types possible with IUE.

2. Materials

2.1. Electroporator Systems

The electroporator provides a source of current and voltage that drives the DNA into progenitor cells that line the ventricle wall. For IUE, the current density required to generate a uniform voltage across a mammalian embryo in utero is more than what most standard voltage or current-stimulating devices used in electrophysiology can generate. The BTX ECM 8300 (Fig. 1a, c; Harvard Apparatus, Holliston, MA) is a good system that generates sufficient current and allows the user to adjust several parameters of the voltage pulse, including amount of voltage, pulse duration, number of pulses, and duration of time between electrical pulses. Voltage is applied across the embryo using tweezer electrodes (BTX Tweezertrodes, Harvard Apparatus, Holliston, MA). For electroporation of both rat and mouse embryos, a 7-mm diameter tweezertrode is sufficient, but for the smaller-sized mouse embryos or to more finely control the placement of the transfection, a smaller-diameter tweezertrode may be used. The voltage generator should be connected to a foot pedal to allow the operator to switch on voltage pulses while keeping hands free to hold the embryo and electrode.

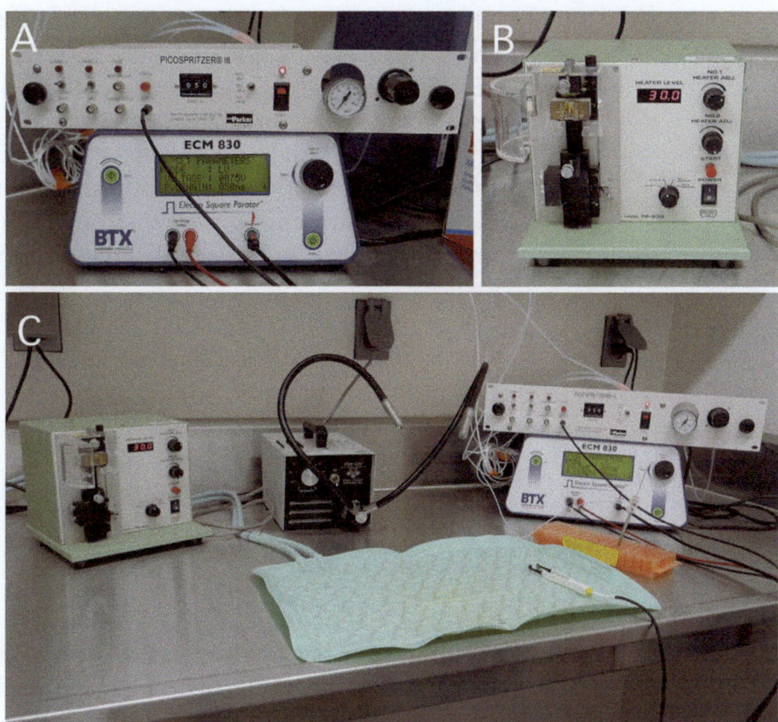

Fig. 1. *IUE equipment.* (**a**) The Picospritzer III allows the experimenter to control the amount DNA that is injected by adjusting the pressure and pulse duration (*top*). The BTX ECM 830 is used to generate and control the current necessary for electroporation. (**b**) Glass injection needles are made using this vertical pipette puller. (**c**) All the equipment needed during surgery should be placed within reach of the sterile field.

We have developed an inexpensive simplified electroporator that can be easily built by most electronic shops (see Appendix). This homemade electroporator is at its core a power capacitor (500 μF, 250 VDC) that is charged to a desired voltage by an external power supply and then discharged by a foot switch. Although this electroporator does not generate the fast square wave typical of commercially available electroporators, from our experience we have found that it electroporates as efficiently as the BTX system with slightly less control of the size of the transfection (4).

2.2. Injection Needle, Pipette, and Puller

The fabrication of the injection needle is critical for successful injection of DNA into the embryo's ventricle. In our experiments, the needles are pulled from borosilicate glass capillary (10-ul microdispenser, Drummond Scientific Company, Broomall, PA). These capillaries have a 0.7-mm inner diameter and 1.19-mm outer diameter.

Injection needles are fabricated with a vertical pipette puller (Fig. 1a, c; PC-10 Puller, Narishige, Tokyo, Japan). The puller uses an electric heating element and gravity to melt and pull glass.

This puller allows for adjustment of the heating element temperature, and weight can be added to increase the gravitational pull on the glass. To make an injection needle, the pipettes are pulled to a point prior to the glass breaking and a tip forming. After the pipette is pulled, we obliquely cut the glass 1 cm from the taper with a small surgical scissors or fine forceps. We find this creates a semirigid needle that is sharp enough to penetrate the uterus. A typical injection needle will have an outside diameter of 25–30 μm at its tip.

2.3. Microinjector

The pressure microinjector system used should allow for repeatable controlled pressure pulses applied to the back of the injection pipette. A Picospritzer III is an ideal instrument for this purpose (Fig. 1a, c; General Valve, Parker Hannifin Corp, Cleveland, OH). In this system, the volume dispensed is approximately linear with respect to both time and pressure. Injection is initiated with a foot pedal, and air pressure is provided by a tank of compressed nitrogen or air. The injection needle is fastened to the pipette holder that has an airtight rubber gasket (Warner MP-S12T, Hamden, CT). The pipette holder is then coupled through another airtight fitting to the picospritzer with tubing. It is important to adjust pressure pulses in the beginning of each experiment because of variability in needle size and viscosity of plasmid DNA mix. For typical injection pipettes and plasmid mixtures, the picospritzer pressure is set to 10–12 psi, and the duration of the pressure pulse can range between 20 and 60 msec. Adjust the pressure pulse by increasing or decreasing the duration of the pressure pulse until a small droplet (approx. 2 mm diameter) of DNA is ejected from the needle tip. Sometimes, it is necessary to place the needle tip in contact with saline while injecting pressure to overcome capillary force inside the tip of the needle. Avoid giving too much pressure, as it is detrimental to the survival of the embryo.

2.4. General Surgical Equipment

The surgery requires creating a sterile surgical field with ample light and nearby access to all the above listed equipment (Fig. 1d). During the procedure, a heat pad that circulates warm water controls the animal's body temperature (Fig. 1c; CSZ Micro Temp LT, Cincinnati, OH). A fiber-optic light is useful to illuminate the surgical area and the embryos inside the uterus. Sterile saline (0.9% NaCl + 1% penicillin/streptomycin) should be prepared and placed in close proximity to the surgical field. All surgical tools must be sterilized. These include a straight scissors (25 mm cutting edge), straight forceps (length 12 cm; teeth 1 × 2), curved hemostat (12 cm), and spatula. The incision is closed with monofilament suture (4/0 monofil).

3. Methods

3.1. Preparation of Plasmid DNA

Plasmid DNA is prepared using Qiagen Plasmid Maxi Kit and must be clean and free of endotoxins to ensure health of embryos and for efficient electroporation. Plasmid DNA mixtures for injection are usually prepared just prior to surgery, but can be made a day in advanced and stored at 4°C. Approximately 1 μL of DNA is injected into each ventricle, and typically a pregnant rodent will have 10–12 embryos (some mouse strains 4–6 embryos). Plasmid DNA is diluted in sterile water and 1 μl of 5% fast green solution to a final concentration of 0.5–2 μg/μl for each plasmid in the mixture. Addition of fast green allows for visualization of plasmid mix as it is injected into the ventricle. Injection needles can be backfilled with mixtures of plasmid DNA. Backfilling glass injection needles is easily done using a Rainin pipetter (p20) with Gell-Well tips (Rainin Instrument, Oakland, CA) inserted into the back of the needle. The filled needle can then be placed into the pipette holder that is attached to the picospritzer.

3.2. Surgery

3.2.1. Preoperative Preparation

Anesthesia is induced with an intraperitoneal (ip) injection of ketamine/xylazine (100/7 mgs/Kg). It is helpful to preanesthetize the animal with 2.5% isoflurane/3% oxygen using an induction chamber and anesthesia vaporizer. Throughout the surgery, the depth of anesthesia should be monitored by occasionally testing for a reflex response to a toe pinch and for steady breathing. If necessary, additional ¼ doses of ketamine/xylazine can be given directly into the intraperitoneal cavity.

Before the surgery begins, the pregnant dam must be completely anesthetized and abdomen prepped for surgery. To protect the dam's eyes during the surgery, Puralube Vet ointment can be applied over each eye with a cotton swab. Hair clippers are used to trim the fur on the abdomen. The site of incision is sterilized with Betadine and cleaned with 70% ethanol. The dam is then placed on top of a sterile drape that covers the heating pad. A small hole, approximately 4 cm in diameter, is cut into another sterile drape and placed over the dam above the abdomen, revealing the incision site and creating a sterile field.

3.2.2. Incision

Correct positioning of the laparotomy is critical to prevent excess bleeding, and care should be taken to avoid cutting mammary tissue. Two incisions are made, first through the skin and then through the underlying abdominal muscle. For rats, the incision is made along the midline and approximately 3–4 cm anterior to the vagina. Using the straight forceps to grasp the overlying skin, a sharp surgical scissors is used to perform a laparotomy. It is helpful

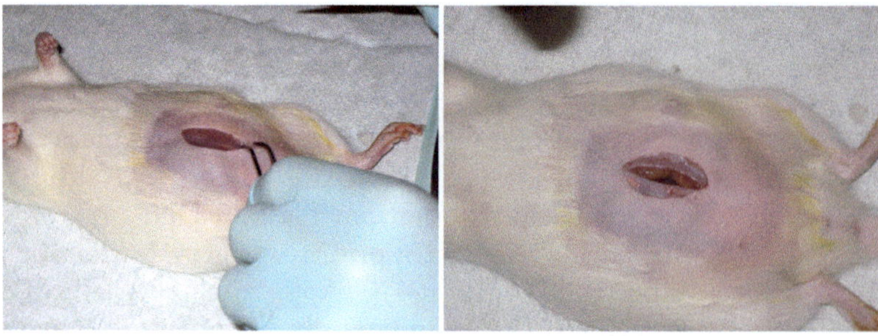

Fig. 2. *Laparotomy on an E15 pregnant rat.* The fur is shaven, and the skin is cleaned with Betadine and ethanol. Two separate incisions are performed, one incision to cut the outer layer of skin (*left*) and another incision to cut the abdominal wall (*right*).

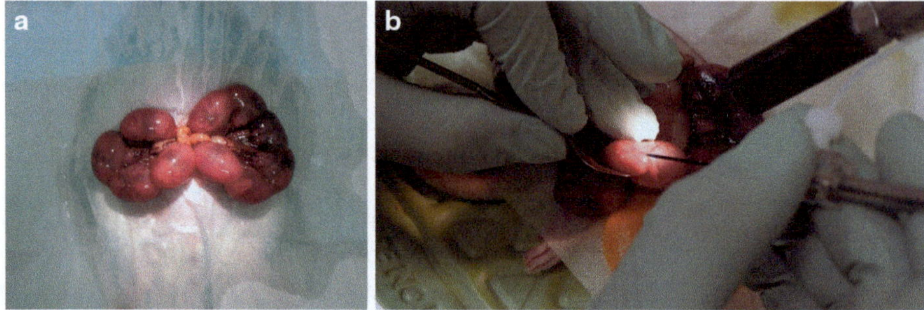

Fig. 3. *Exposure of the uterus and injection of lateral ventricle.* (**a**) Each uterine horn is exposed, and ten embryos can be observed. (**b**) An injection needle penetrates through the uterine wall and into the lateral ventricle. The embryo is stabilized between a spatula and middle finger.

to first use the forceps to pull the skin away from the underlying muscle and then cut approximately a 3–5-cm incision along the midline. This incision will reveal the underlying muscle and mammary tissue. Avoiding the mammary tissue, the muscle layer and fascia are grasped with the forceps, and another incision is made, taking care to avoid cutting internal organs by keeping slight upward pressure on the scissors while cutting (Fig. 2).

3.2.3. Manipulating Uterus and Embryos

If the incision is properly placed, the underlying uterus will be visible. Using a blunt forceps, pull open the incision and insert a spatula under the uterus. Carefully pull each horn of the uterus through the incision with the spatula and place the uterine horns on top of the saline-moistened drape (Fig. 3a). It is important to keep the uterus and internal organs moist by occasionally applying sterile saline (0.9% NaCl + 1% penicillin/streptomycin, 35°C) to the exposed uterus and into the intraperitoneal cavity. Manipulating the embryos carefully will increase survival and improve injections dramatically. Too much pressure on the uterus can burst the amniotic sac surrounding each embryo and is detrimental to survival.

The embryo can be safely rotated inside the uterus by gently sliding the spatula over the surface of the uterus. Proper positioning and handling of the embryo is important for successful injection into the lateral ventricles. To improve visualization of the embryo, it is helpful to bring a fiber-optic light in close proximity to the uterus. For a right-handed surgeon, the embryo's head should point in the direction of the surgeon's right hand. Holding the spatula in the left hand between the thumb and forefinger, the surgeon can grasp the embryo between the spatula and the middle finger (Fig. 3b). Slight pressure is needed to stabilize the embryo and bring the embryo's head close to the inner wall of the uterus. This positioning allows the surgeon to use his or her right hand to approach the embryo with the injection needle.

3.3. Injection and Electroporation

3.3.1. Targeting the Lateral Ventricles for Injection

The lateral ventricles of E13–E16 rats and mice are the easiest to inject into because, at these ages, the ventricular space is large relative to developing cortical and hippocampal layers. The volume of the lateral ventricle decreases with increasing embryonic age, and therefore, surgery on E17–E18 embryos can be more challenging. To locate the lateral ventricle, several landmarks are visible on the surface of the embryo's brain. A good reference point is the intersection of the sagittal and lambdoid sutures. For the injection, the needle should be placed lateral to the sagittal suture and anterior to the lambdoid suture (Fig. 4). For E17–E18 embryos, the needle placement may need to be more anterior from the lambdoid suture. Penetration of the uterine wall can be difficult because of variation in the needle rigidity and sharpness. This can be overcome by slowly increasing the pressure applied to the needle and by applying a slight circular motion to the needle tip. The depth of the injection needle into the brain is also very important but can be difficult to ascertain. Therefore, once the needle is inserted into brain, injection of plasmid mix and observation of dye can help

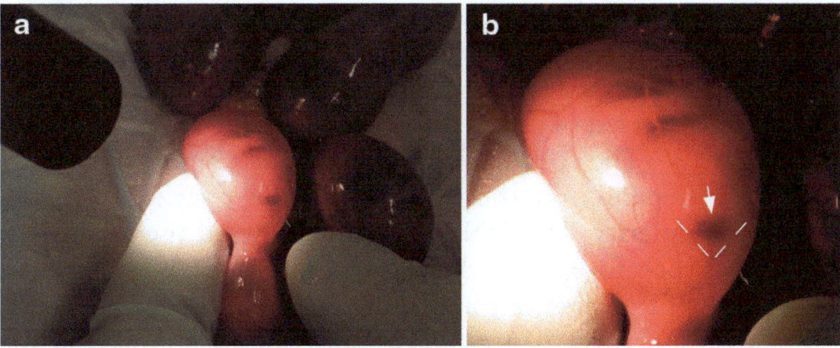

Fig. 4. *Example of a good injection into the lateral ventricle.* (**a**) If the injection is well placed, the ventricle will immediately fill with dye. (**b**) *Magnified view* of the previous image showing the ventricle filled with dye (*arrow*). The *dotted lines* indicate the location of the sagittal and lambdoid sutures.

locate the ventricle. If no dye is observed when the injection is made, then the needle is typically too deep and past the lateral ventricle. In this case, continue injecting more plasmid as the needle is gently pulled back until the ventricle begins to fill with dye. Once the needle is in the ventricle, inject three to five times to completely fill the ventricle space.

3.3.2. Signs of Good and Bad Injection

A good injection of the ventricle completely fills the ventricular space of the lateral ventricles, creating the shape of a right triangle, with the right angle at the intersection of the sagittal and lambdoid sutures (Fig. 4). When the needle tip is inside the ventricle, injection of dye will instantly become visible as it immediately begins to fill and outline the ventricle. On occasion, the needle will not penetrate the brain tissue or pia mater, and plasmid DNA is deposited on the surface of the brain directly underlying the scull. In this case, the dye will appear as a bright dot that will gradually spread over the surface of the brain, outlining the contours of the cortical surface.

3.3.3. Electroporation and Directed Transfection

After injecting DNA, a voltage pulse is given to drive the plasmid into the progenitor cells that line the wall of the ventricle. The intensity, duration, number of pulses, and duration between pulses will determine the extent of transfection, and the orientation of the paddle electrodes and polarity of the pulse will determine largely the location of transfection. The pulse amplitude is dependent on the rodent species and embryonic age, where older embryos require higher voltages and smaller embryos require less voltage. The smaller the animal, the shorter the distance between paddle electrodes, and thus the same voltage will have a higher field density in smaller animals and therefore must be reduced to prevent mortality. In practice, for electroporation in mouse, voltages range between 20 and 60 V, 30–50 ms pulse duration, 1 s pulse intervals, and 3–8 pulses are reported (2, 3, 9). Electroporation of rat embryos generally requires slightly higher voltages with similar pulse durations, intervals, and number of pulses (40–75 V, 30–50 ms pulse duration, 1 s pulse intervals, and 3–5 pulses). For electroporators that deliver a single exponentially decaying pulse, such as that described in Appendix, voltages between 50 and 100 V are used (4).

The specific region of the forebrain that will be transfected depends on the position of the tweezer electrodes and pulse polarity. The DNA will move toward the positive electrode, and therefore, the position of positive pole will determine the area of transfection. Several labs have reported transfection of different cortical areas (10) and hippocampus (9) (see below for details).

3.4. Closing, Recovery, and Post-op Considerations

3.4.1. Closing

After injection and electroporation is complete, the embryos are placed back inside the abdomen. Care in handling is important, and the uterus should be returned in the same way it was taken out. The intraperitoneal cavity is filled with saline prior to closing the incision. Two sets of sutures are used to close both the abdominal wall and then the skin. It is important to cut the tail end of the sutures close to the knot to prevent the dam from chewing open the sutures.

3.4.2. Post-op Recovery and Analgesia

Rodents are not capable of regulating body temperature well while under anesthesia. Therefore, it is important to keep the dam on a heating pad until she has recovered. We typically wait until the dam has rolled over and can maintain an upright posture before returning the animal to the colony. To decrease postoperative pain and inflammation, we provide two subcutaneous doses of Metacam (1.0 mg/Kg), one dose immediately after surgery and another 24 h later. The health of the dam should be monitored daily, and more doses of Metacam may be given without adverse effects to the embryos or dam.

4. Notes

4.1. Vector Systems and Directed Expression

The cell types targeted and the level of expression achieved are dependent on the promoter the plasmid vectors contain, the amount of plasmid injected, the efficiency of the electroporation, the placement of the electroporation paddles, and the developmental time at which the injection and electroporation are made. Many labs have found that the CAG promoter (11), which contains the cytomegalovirus immediate early enhancer with the chicken ß-actin promoter, provides strong, reliable, and stable expression of the majority of transgenes tested. Transgene expression can also be driven using cell-type-specific promoters that take advantage of promoter sequences that are specific to certain types of cells and stages of development (7, 12, 13). By varying the location of transfection, types of promoters in vectors, and the time of transfection, several different cell populations can be targeted for cellular transgenesis.

4.2. Cell-Type-Specific Expression

4.2.1. Time of Electroporation

The temporal order of cortical development allows for transfection of different subsets of progenitors and different layers of cortical neurons depending on the age of the embryo at the time of transfection. Because cells of a particular type and layer are present at the ventricular surface at defined times in development, and only these surface cells are electroporated, it is possible to create transgenesis in subsets of cells. By comparing the population of cells labeled by IUE versus BrdU at different times during corticogenesis from E12.5 to E15, Langevin et al. (12) showed specific labeling of neurons occupying different cortical lamina depending on the time of transfection, where E12.5 transfections primarily labeled layer VI and E15 transfections labeled layer II/III (14). Taking advantage of this temporally determined specificity of cell-type targeting, Hatanaka et al. (2004) determined that Cdk5 is important for migration of upper layer neurons but not for the earlier born deeper layer neurons. If IUE is performed on E18 or later embryos, we observe expression of transgenes in both upper layer 2 neurons and astrocytes (13).

4.2.2. Spatially Directed Transfection

Using IUE, it is possible to specifically transfect different brain regions by directing the flow of current into specific progenitor populations. Different regions of cortex can be transfected by placing the positive pole of the tweezer electrode over the cortical region of interest. For example, to transfect visual cortex, the current flow should be directed toward the developing visual cortex, such that the positive pole is placed over the posterior cortex. Niwa et al. (2010) showed bilateral transfection of the medial prefrontal cortex was accomplished by orienting the positive electrodes at an approximately 20° outward angle from the midline and an approximately 30° angle downward from an imaginary line from the olfactory bulbs to the caudal side of the cortical hemisphere.

Transfection of hippocampus is achieved by directing the transfection into the medial surface of the lateral ventricle. Transfection of all hippocampal regions (dentate gyrus to CA1) is accomplished by sweeping the tweezer electrode from 0 (interaural plane) to 45° during five consecutive voltage pulses (9).

In order to transfect cortical interneurons, transfection of the progenitors located in the ganglionic eminence is required. The positive pole of the tweezer electrode should be placed approximately 30° below the brain's horizontal plane. This type of transfection will not only transfect cortical interneurons but also hippocampal interneurons and striatal neurons (15).

4.2.3. Cell-Type-Specific Promoters

To label different populations of neocortical progenitor cells that may be generated or present at the same time on the ventricular surface, several studies have taken advantage of cell-type-specific promoters to selectively drive transgene expression in subsets of progenitors. Using IUE, Wang et al. (7) identified and characterized the promoter region of DCX and showed IUE of DCX-dsRED faithfully recapitulated endogenous doublecortin expression. Similarly, neuronal subpopulations that express the transcription factors ER81 and NGN2 are labeled by controlling the expression of fluorescent proteins with the specific promoters for ER81 and NGN2 (12). To selectively label neuronal progenitors and radial glia simultaneously, Gal et al. (2006) used the neuronal progenitor promoter Talpha1 and the radial glia promoters

BLBP and GLAST to drive GFP or DsRED2, respectively. By coelectroporating these cell-specific constructs, the authors showed a lack of overlap between cells transfected with the different promoter constructs and therefore were able to genetically separate these two cell types.

4.3. Cre-Mediated Expression Systems

Cre-mediated conditional expression systems are now developed for use in IUE experiments and allow for cell- and time-specific expression of shRNAs or transgenes (6). This two-plasmid approach uses cre recombination to gate the expression of a target plasmid that contains a neo-stop-polyA cassette that is flanked by loxP sites. The cre recombination is required to remove and recombine the target plasmid before the transgene is expressed. If cre recombinase is under the control of a cell-type-specific promoter, then expression of the transgene will only occur in cells that recognize the cell-type-specific promoter. This type of approach was used to demonstrate that subsets of progenitor cells in the ventricle zone generate phenotypically different progeny (16).

Conditional cre recombinase gated by two estrogen receptors (ER-cre-ER) has also been shown to be effective when delivered by IUE. Activation of the estrogen receptors by intraperitoneal injection of tamoxifen (2 mg/100 g body weight) results in nuclear localization, cre recombination, and expression of the transgene (6, 8). This inducible expression system allows experimental changes in expression during selected times. We currently use this expression system to induce expression of a mutant version of DISC1 (DISC1ΔC-GFP) in postnatal rats, in order to circumvent migration defects that occur when this protein is expressed during embryogenesis (Fig. 5). This system was also used to rescue a migration phenotype that is the result of embryonic knockdown of DCX by conditionally reexpressing the target gene DCX (8).

4.4. RNA Interference (RNAi)

IUE is also suitable for loss-of-function experiments using RNAi knockdown of gene expression. Constitutive knockdown is effectively achieved using the mU6 pro expression system (17). This vector system uses the mouse U6 promoter and is a polymerase III system. The use of RNAi to knockdown specific genes during corticogenesis has shed light on the molecular mechanisms important for the regulation of neuronal migration, a process responsible for several neurological disorders including lissencephaly, double cortex, and periventricular heterotopia (18, 19). For instance, our lab showed embryonic knockdown of DCX resulted in a double cortex phenotype that is reminiscent of the cortical malformations observed in humans that carry a mutation in this gene (4). RNAi knockdown of LIS1, a gene associated with lissencephaly, inhibited centrosomal movement and migration along radial glial fibers (20). In addition, this technology has resulted in the identification of several other molecular pathways important

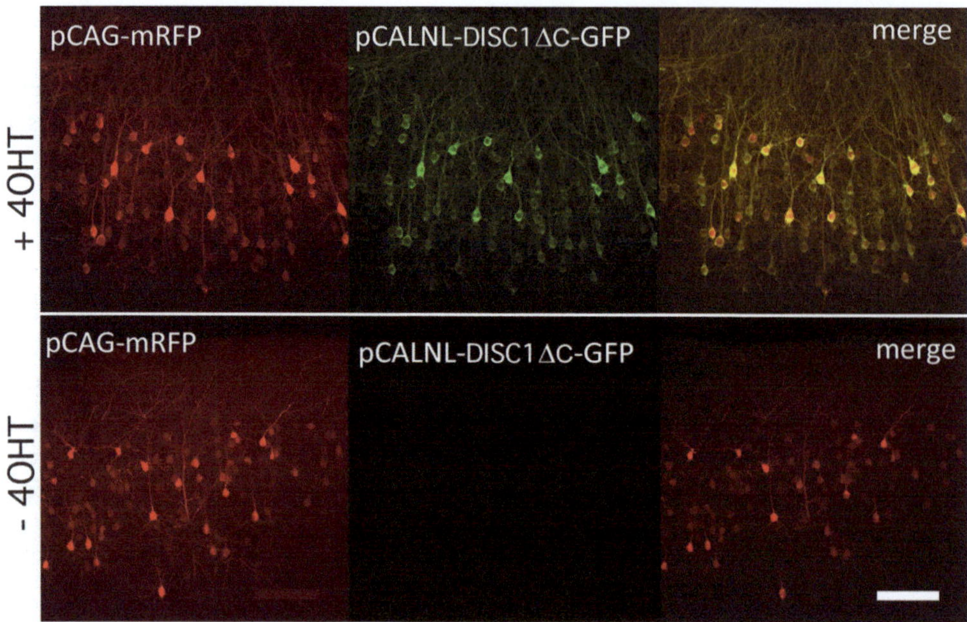

Fig. 5. *Inducible expression of a transgene using IUE.* Expression of DISC1ΔC is induced by postnatal administration of 4OHT (P5–7) as seen by expression of GFP fused to DISC1ΔC in this P28 brain slice. No GFP expression is observed in vehicle treated animals (−4OHT). Scale bar equals 100 μm.

for neuronal migration, including the small GTP-binding protein Rnd2 (21) and two gap junction proteins connexin 26 and connexin 43 that are required for proper cell adhesion between radial glial fibers and migrating neurons (22).

4.5. IUE in Combination with Transgenic Mice

IUE can also be used in combination with transgenic mouse lines that contain floxed alleles to create mosaic knockouts. IUE of cre recombinase will knock out the floxed gene only in the subset of cells that are transfected. This approach is valuable to determine the cell-autonomous functions of a particular gene and to validate results obtained from knocking down gene expression with RNAi. For instance, Chen et al. (2008) showed RNAi knockdown of NP1 impaired radial migration of neocortical neurons. This phenotype was recapitulated by IUE of cre recombinase into the ventricular zone of NP1 conditional knockout mice, in which loxP sites flanked the second exon of the *Nrp1* gene (floxed NP1). A similar migration defect was observed when cre recombinase was transfected into ventricular progenitor cells of the alpha5 integrin conditional knockout mouse (23). As more transgenic mouse containing floxed alleles are created, this technology will become a common strategy to assay the function of genes involved in neurodevelopment.

Appendix: Parts List

1. Aluminum Enclosure, 5.25″ D x 3″ W x 2.13″ H (BUD Part No. CU-3006-A) $5.00
2. Footswitch, (Linemaster Treadlite II, Model T-91-SC36) $25.00
3. Capacitor, 500 μF at 250 VDC,(Vishay Sprague Powerlytic (R) Part No. 36DX501F250AB2A) $7.17
4. Banana Jack, Black, package of 10 (Pomona Part No. 1581-0) $11.00
5. Banana Jack, Red, package of 10 (Pomona Part No. 1581-2) $11.00
6. Power Resistor 0,000 at 12 W (Ohmite 210 series DIVIDOHM Part No. D12K3K0) $10.00
7. Thru-Bolt, #8 size, 2.5″ length, for mounting 210 series(Ohmite Part No. 7PA10) $5.00
8. Switch, 2PDT, On-none-On (Eaton Cutler-Hammer 7312K38) $17.49
9. Wire, zip cord, 2 conductor, 18 AWG, PVC Jacket, 250 foot roll (Belden Part No. 19122-250-10) $57.27
10. Wire, hookup, black, 22AWG, solid (Belden Part No. 8530) $14.33
11. Wire, hookup, red, 22AWG, solid (Belden Part No. 8530) $14.33
12. Wireties, package of 100, (Panduit PAN-TY Part No. PLC2S-S10-C) $30.00
13. Screws, machine, 6-32, stainless steel, phillips head
14. Nuts, 6-32, stainless steel
15. Strain Relief (Voltrex Part No. 2463) $5.42
16. Strain Relief (Voltrex Part No. 2467) $9.77
17. Kelly Hemostatic forceps, 14 cm, straight, serrated, (Fine Science Tools Part No. 13018-14) $33.50
18. Heat Shrink Tubing
19. Gold plated copper pins Contact us

Schematic Diagram

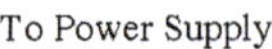

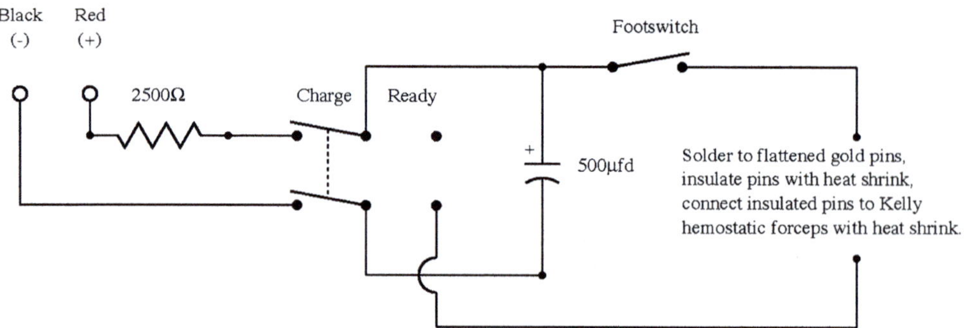

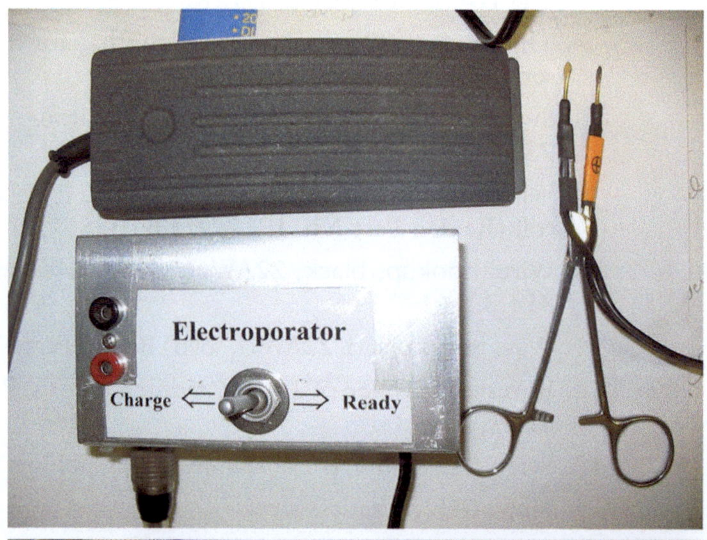

References

1. Fukuchi-Shimogori, T., and Grove, E. A. (2001) Neocortex patterning by the secreted signaling molecule FGF8, Science 294, 1071–1074.
2. Saito, T., and Nakatsuji, N. (2001) Efficient gene transfer into the embryonic mouse brain using in vivo electroporation, Dev Biol 240, 237–246.
3. Tabata, H., and Nakajima, K. (2001) Efficient in utero gene transfer system to the developing mouse brain using electroporation: visualization of neuronal migration in the developing cortex, Neuroscience 103, 865–872.
4. Bai, J., Ramos, R. L., Ackman, J. B., Thomas, A. M., Lee, R. V., and LoTurco, J. J. (2003) RNAi reveals doublecortin is required for radial migration in rat neocortex, Nat Neurosci 6, 1277–1283.
5. Kawauchi, T., Chihama, K., Nabeshima, Y., and Hoshino, M. (2003) The in vivo roles of STEF/Tiam1, Rac1 and JNK in cortical neuronal migration, EMBO J 22, 4190–4201.
6. Matsuda, T., and Cepko, C. L. (2007) Controlled expression of transgenes introduced by in vivo electroporation, Proc Natl Acad Sci USA 104, 1027–1032.
7. Wang, X., Qiu, R., Tsark, W., and Lu, Q. (2007) Rapid promoter analysis in developing mouse brain and genetic labeling of young neurons by doublecortin-DsRed-express, J Neurosci Res 85, 3567–3573.
8. Manent, J. B., Wang, Y., Chang, Y., Paramasivam, M., and LoTurco, J. J. (2009) Dcx reexpression reduces subcortical band heterotopia and seizure threshold in an animal model of neuronal migration disorder, Nat Med 15, 84–90.
9. Navarro-Quiroga, I., Chittajallu, R., Gallo, V., and Haydar, T. F. (2007) Long-term, selective gene expression in developing and adult hippocampal pyramidal neurons using focal in utero electroporation, J Neurosci 27, 5007–5011.
10. Niwa, M., Kamiya, A., Murai, R., Kubo, K., Gruber, A. J., Tomita, K., Lu, L., Tomisato, S., Jaaro-Peled, H., Seshadri, S., Hiyama, H., Huang, B., Kohda, K., Noda, Y., O'Donnell, P., Nakajima, K., Sawa, A., and Nabeshima, T. Knockdown of DISC1 by in utero gene transfer disturbs postnatal dopaminergic maturation in the frontal cortex and leads to adult behavioral deficits, Neuron 65, 480–489.
11. Niwa, H., Yamamura, K., and Miyazaki, J. (1991) Efficient selection for high-expression transfectants with a novel eukaryotic vector, Gene 108, 193–199.
12. Langevin, L. M., Mattar, P., Scardigli, R., Roussigne, M., Logan, C., Blader, P., and Schuurmans, C. (2007) Validating in utero electroporation for the rapid analysis of gene regulatory elements in the murine telencephalon, Dev Dyn 236, 1273–1286.
13. LoTurco, J., Manent, J. B., and Sidiqi, F. (2009) New and improved tools for in utero electroporation studies of developing cerebral cortex, Cereb Cortex 19 Suppl 1, i120–125.
14. LoTurco, J. J., and Bai, J. (2006) The multipolar stage and disruptions in neuronal migration, Trends Neurosci 29, 407–413.
15. Borrell, V., Yoshimura, Y., and Callaway, E. M. (2005) Targeted gene delivery to telencephalic inhibitory neurons by directional in utero electroporation, J Neurosci Methods 143, 151–158.
16. Stancik, E. K., Navarro-Quiroga, I., Sellke, R., and Haydar, T. F. Heterogeneity in ventricular zone neural precursors contributes to neuronal fate diversity in the postnatal neocortex, J Neurosci 30, 7028–7036.
17. Yu, J. Y., DeRuiter, S. L., and Turner, D. L. (2002) RNA interference by expression of short-interfering RNAs and hairpin RNAs in mammalian cells, Proc Natl Acad Sci USA 99, 6047–6052.
18. Feng, Y., and Walsh, C. A. (2001) Protein-protein interactions, cytoskeletal regulation and neuronal migration, Nat Rev Neurosci 2, 408–416.
19. Kato, M., and Dobyns, W. B. (2003) Lissencephaly and the molecular basis of neuronal migration, Hum Mol Genet 12 Spec No 1, R89–96.
20. Tsai, J. W., Bremner, K. H., and Vallee, R. B. (2007) Dual subcellular roles for LIS1 and dynein in radial neuronal migration in live brain tissue, Nat Neurosci 10, 970–979.
21. Heng, J. I., Nguyen, L., Castro, D. S., Zimmer, C., Wildner, H., Armant, O., Skowronska-Krawczyk, D., Bedogni, F., Matter, J. M., Hevner, R., and Guillemot, F. (2008) Neurogenin 2 controls cortical neuron migration through regulation of Rnd2, Nature 455, 114–118.
22. Elias, L. A., Wang, D. D., and Kriegstein, A. R. (2007) Gap junction adhesion is necessary for radial migration in the neocortex, Nature 448, 901–907.
23. Marchetti, G., Escuin, S., van der Flier, A., De Arcangelis, A., Hynes, R. O., and Georges-Labouesse, E. Integrin alpha5beta1 is necessary for regulation of radial migration of cortical neurons during mouse brain development. Eur J Neurosci 31, 399–409.

24. Hatanaka, Y., Hisanaga, S., Heizman, C.W., Murakami, F. (2004) Distinct migratory behavior of early- and late-born neurons derived from the cortical ventricular zone, J Comp Neurol 479, 1–14.

25. Gal, J.S., Morozov, Y.M., Ayoub, A.E., Chatterjee, M., Rakic, P., Haydar, T.F. (2006) Molecular and Morphological Heterogeneity of Neural Precursors in the Mouse Neocortical Proliferative Zones, J Neurosci 26, 1045–1056.

26. Chen, G., Sima, J., Jin, M., Wang, K., Xue, X., Zheng, W., Ding, Y., Yuan, X., Semaphorin-3A guides radial migration of cortical neurons during development, Nat Neurosci 11, 36–44

Chapter 7

Single-Cell Electroporation of siRNA in Primary Neuronal Cultures

Masahiko Tanaka

Abstract

Single-cell electroporation is a recently developed method to introduce polar and charged molecules such as dyes, drugs, peptides, proteins, and nucleic acids into single and identified cells. This feature is advantageous, especially in investigations of the nervous system, because the nervous system is composed of various types of cells. Using this method, we transferred siRNA into cerebellar neurons in primary cultures. The gene-silencing effects of siRNA introduced by this method were evaluated by real-time monitoring using GFP imaging over several days after electroporation. The experimental results indicate that this technique could be a simple but effective tool for silencing gene expression in primary neuronal cultures.

Key words: Single-cell electroporation, Small interfering RNA (siRNA), Primary neuronal culture, Cerebellum, Purkinje cell, Golgi cell, Green fluorescent protein (GFP).

1. Introduction

Electroporation is a widely used method to introduce polar and charged molecules such as dyes, drugs, peptides, proteins, and nucleic acids into cells, using an electric field (1–3). When short-duration electric pulses are applied, cell membranes are transiently permeabilized by the formation of minute pores, and molecules are electrophoretically transferred from extracellular solution into cells through these pores. After pulse application, pores close and the cell membrane continuity recovers. However, it is impossible to introduce molecules of interest into selected or single cells by such traditional bulk electroporation.

Single-cell electroporation is a recently developed method to introduce polar and charged molecules into single and identified cells (4). In contrast to traditional bulk electroporation using large plane electrodes, single-cell electroporation uses carbon fiber

Alexei Morozov (ed.), *Controlled Genetic Manipulations*, Neuromethods, vol. 65,
DOI 10.1007/978-1-61779-533-6_7, © Springer Science+Business Media, LLC 2012

microelectrodes (5), electrolyte-filled capillaries (6), micropipettes (patch-clamp electrodes) (7–10), or chip structures (11–13) to transfer the molecules of interest into single cells.

RNA interference (RNAi) is a powerful means to elucidate gene function in molecular and cellular biology. Among several methods for RNAi, small interfering RNA (siRNA)-based RNAi has the advantage of simplicity. Using single-cell electroporation of siRNA, it is possible to transfer siRNA into microscopically identified single cells (14, 15). This enables us to inhibit expression of target genes in specific cells within a heterogeneous cell population. This feature is advantageous, especially in investigations of the nervous system, because the nervous system is composed of various types of cells.

In this chapter, I describe our method of single-cell electroporation of siRNA in cerebellar neurons in primary cultures (15). The gene-silencing effects of siRNA introduced by this method were evaluated by real-time monitoring using green fluorescent protein (GFP) imaging over several days after electroporation. The experimental results indicate that this technique could be a simple but effective tool for silencing gene expression in primary neuronal cultures.

2. Primary Neuronal Culture

2.1. Mice

As primary neuronal cultures, we use cerebellar dissociated cell cultures derived from neonatal (postnatal day 0 or 1) C57BL/6 mice. Usually, 7 to 9 pups are used to prepare 9 to 11 dishes of dissociated cell cultures for one experiment. For real-time monitoring of the gene-silencing effects of electroporated siRNA using GFP imaging (15), we used GAD67-GFP knock-in mice, in which GFP is selectively expressed in Purkinje cells and inhibitory interneurons such as stellate, basket, and Golgi cells (16). In this case, we mated a mouse heterozygous for the GAD67-GFP allele ($GAD67^{GFP/+}$) with a wild-type mouse and used both $GAD67^{GFP/+}$ and $GAD67^{+/+}$ pups obtained from the breeding pair because seven or more pups per experiment are needed to maintain the viability of Purkinje cells in dissociated cell cultures.

2.2. Dissociated Cell Culture

Cerebellar tissues are dissociated into cells by the proteolytic reaction with papain (Worthington, Lakewood, NJ, USA) and trituration with a Gilson-compatible pipette tip (blue type; Watson/Nippon Genetics, Tokyo, Japan) (15, 17). Seventy-five microliters of cell suspension at a density of $4–5 \times 10^6$ cells per ml is plated on a plastic bottom dish (μ-Dish, low with Grid-500, ibiTreat; plating area, 21 mm in diameter; ibidi/Nippon Genetics, Tokyo, Japan), the bottom of which is coated with poly-L-lysine. This results in a cell density of $3–4 \times 10^5$ cells per dish ($1.5–2.0 \times 10^5$ cells per cm^2).

Such a high cell density is needed to maintain the viability of Purkinje cells. After 1.5 h incubation for cell attachment, 0.5 ml of the Sumitomo nerve-cell culture medium (Sumitomo Bakelite, Tokyo, Japan) is added per dish.

The μ-Dish with grid has advantages such as good optical conditions due to a thin (175 μm) plastic bottom and easy identification of electroporated cells using the grid on the bottom. In addition, after culture and fixation, cells can be used for immunocytochemistry by cutting out the thin plastic bottom with fine scissors. The Sumitomo nerve-cell culture medium is good for the maintenance and differentiation of Purkinje cells.

3. Single-Cell Electroporation of siRNA

3.1. Equipment, Electrodes, Internal Solution

Equipment for single-cell electroporation is shown in Fig. 1a, b. Among several types of devices for single-cell electroporation, we use micropipettes because they are easy to make and most effective to restrict the electric fields to single cells (high spatial resolution). During electroporation, the tip of the micropipette never penetrates but touches the cell membrane as in patch clamp (Fig. 1c), which makes this technique less invasive to cells than microinjection.

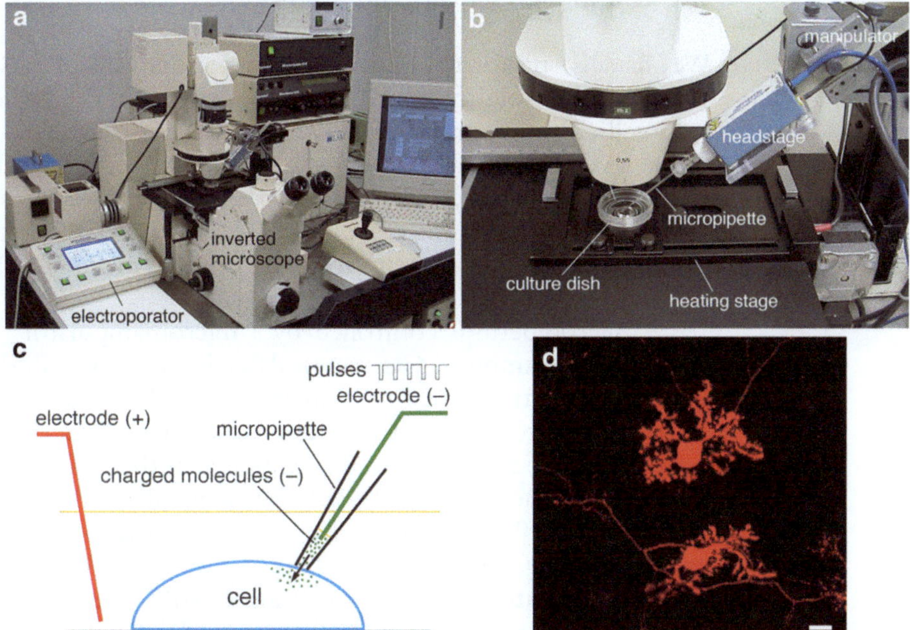

Fig. 1. (**a**, **b**) Equipment for single-cell electroporation. (**c**) A schema of single-cell electroporation. (**d**) AF594 fluorescence of the electroporated Purkinje cells at 17 DIV. The fluorescent image was acquired 30 min after electroporation. Note that the cells with stronger fluorescence than our usual levels, like in Fig. 2A2, D2, are selected for this figure to show the cell morphology clearly. Scale bar, 20 μm. (**a**, **b** were reproduced from Tanaka et al. (15). With permission from Elsevier).

Micropipettes are pulled on a micropipette puller (PB-7; Narishige, Tokyo, Japan) using 1.2 mm O.D. glass capillaries with filament (GD-1.2; Narishige). The internal glass filament facilitates tip filling. Micropipette tips are roughly 0.5 μm in diameter. Although the glass capillaries we use are standard-wall type, others favor thin-wall type, which allows a lower resistance without producing larger tips (8).

Micropipette tips are filled with 2–3 μl (or less) of a solution containing 0.1–3 μM siRNA, 0.25 mM Alexa Fluor 594 hydrazide (AF594; Molecular Probes/Invitrogen, Eugene, OR, USA), 150 mM potassium methanesulfonate, and 5 mM Hepes (pH 7.2), using the Microloader pipette tips (Eppendorf, Hamburg, Germany). The internal solution does not contain sodium, calcium, and chloride ions because the entry of these ions into the cells should be minimized. AF594 is added to the solution to monitor electroporation and visualize electroporated cells. In our experience, high concentrations of siRNA (above 4 μM) and AF594 (above 0.5 mM) should be avoided for the viability of electroporated cells. If air bubbles form in the solution in the micropipette, they are removed by tapping the micropipette. A silver wire electrode (cathode) is inserted in the micropipette. The resistance of the micropipettes is 30–40 MΩ. The micropipette is connected to the headstage of the electroporator (Axoporator 800A; Molecular Devices, Sunnyvale, CA, USA), which has a simple voltage clamp circuit with high stability and bandwidth.

3.2. Single-Cell Electroporation

We usually perform single-cell electroporation of siRNA in Purkinje and Golgi cells at 9–11 days in vitro (DIV). The dissociated cell culture is transferred to the stage of an inverted microscope (Axiovert; Zeiss, Jena, Germany) equipped with a heating stage and epifluorescence. A ground electrode (anode) is placed in the culture medium (red line in Fig. 1c). Purkinje and Golgi cells are identified by their large cell bodies (maximal diameter, ≥18 μm).

A micropipette is controlled by a micromanipulator (#5170; Eppendorf, Hamburg, Germany). After identifying a Purkinje or Golgi cell, the tip of the micropipette is placed in contact with the center of the cell body of the target cell. Then, it is driven very gently to push against the cell membrane so the resistance of the micropipette increases by 20–30%. Subsequently, after switching a microscope to fluorescence mode, square electric pulses (standard protocols: –4 V, 1 ms width, 100 Hz, 2-s duration) generated with the electroporator are applied to transfer siRNA and AF594 into the cell. In our experience, high voltage (above 8 V) and long duration (above 4 s) of electric pulses may induce cell damage, fusion of the micropipette tip to the cell, or plugging of the micropipette. During pulse application, AF594 is observed to enter the cell, which diffuses rapidly in the cell body, dendrites, and axon (Figs. 1d, 2A2, D2).

Electroporation should be performed as quickly as possible to maintain the viability of the cells. We usually electroporate about 10–20 cells per dish within 20–30 min.

4. Real-Time Monitoring of the Gene-Silencing Effects of Electroporated siRNA Using GFP Imaging

4.1. Procedures

To monitor the gene-silencing effects of electroporated siRNA in real time, siRNA against GFP was introduced in GFP-expressing Purkinje and Golgi cells in cerebellar dissociated cell cultures derived from GAD67-GFP knock-in mice, and the temporal changes in the intensity of GFP fluorescence were quantified in the same cells (15). Single-cell electroporation was usually performed at 9–11 DIV because GFP fluorescence of Purkinje and Golgi cells was so weak that correct identification of these cells was difficult before 9 DIV. GFP and negative control siRNAs were used at various concentrations (0.01–10 μM) in micropipettes. Fluorescent images of the electroporated cells were acquired 0, 1, and 4 days after electroporation using a confocal laser scanning microscope (CFLSM) (LSM510 META; Zeiss) equipped with a heating stage. The electroporated cells were easily identified by AF594 fluorescence 1 h after electroporation (Fig. 2A2, D2). However, AF594 fluorescence gradually decreased 1 and 4 days after electroporation (Fig. 2B2, C2, E2, F2). Therefore, the electroporated cells were identified not only by AF594 fluorescence but also by the grid on the bottom of μ-Dish at these stages. For quantification of GFP fluorescence, the intensity of GFP fluorescence in each cell in CFLSM images was measured using ImageJ 1.40 (NIH, Bethesda, MA, USA). The fluorescent intensity was normalized by that 1 h (0 day) after electroporation.

4.2. Results

Electroporation with negative control siRNA at 0.01–3 μM in Purkinje cells (Fig. 2A1–C1; white circles in Fig. 3a) and at 0.01–10 μM in Golgi cells (white circles in Fig. 3b) did not influence GFP fluorescence (cf. nonelectroporated cells shown by gray diamonds in Fig. 3). However, electroporation with 10 μM negative control siRNA in Purkinje cells had inhibitory effects on GFP fluorescence 4 days after electroporation (white circles in Fig. 3a, 10 μM), showing that high concentrations of siRNA exert off-target effects.

In contrast, electroporation with GFP siRNA at 0.01–10 μM reduced GFP fluorescence almost dose dependently in both Purkinje (Fig. 2D1–F1; black circles in Fig. 3a) and Golgi (black

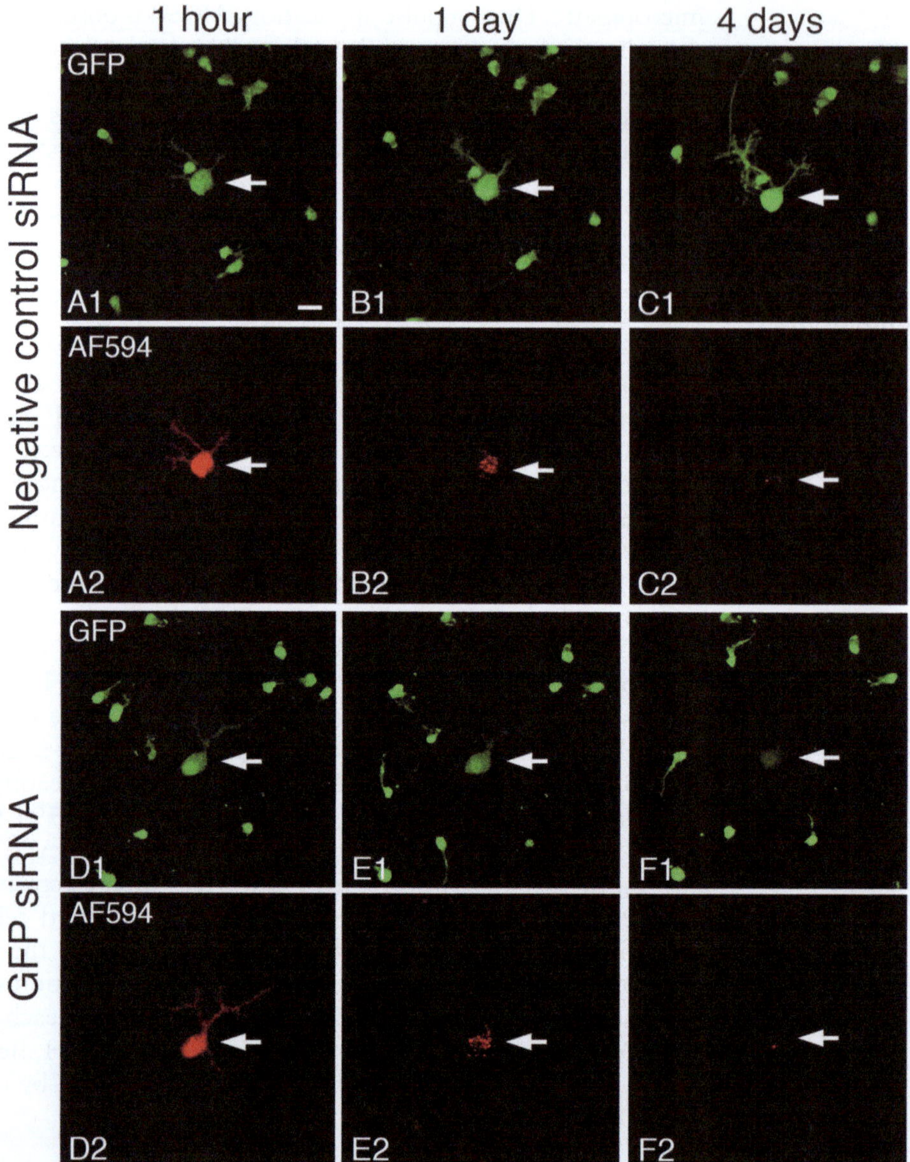

Fig. 2. GFP imaging after single-cell electroporation of negative control or GFP siRNA in Purkinje cells. Single-cell electroporation was performed in GFP-expressing Purkinje cells at 10 DIV using micropipettes loaded with negative control (**A–C**) or GFP (**D–F**) siRNA (0.1 μM) in addition to AF594. Fluorescent images of the electroporated cells (*arrows*) were acquired 1 h, 1 day, and 4 days after electroporation using a CFLSM. Scale bar, 20 μm (Reproduced from Tanaka et al. (15). With permission from Elsevier).

circles in Fig. 3b) cells, showing that RNAi was achieved effectively by this method. Our recent study has shown that sufficient reduction in GFP fluorescence continued for at least 14 days after electroporation (18).

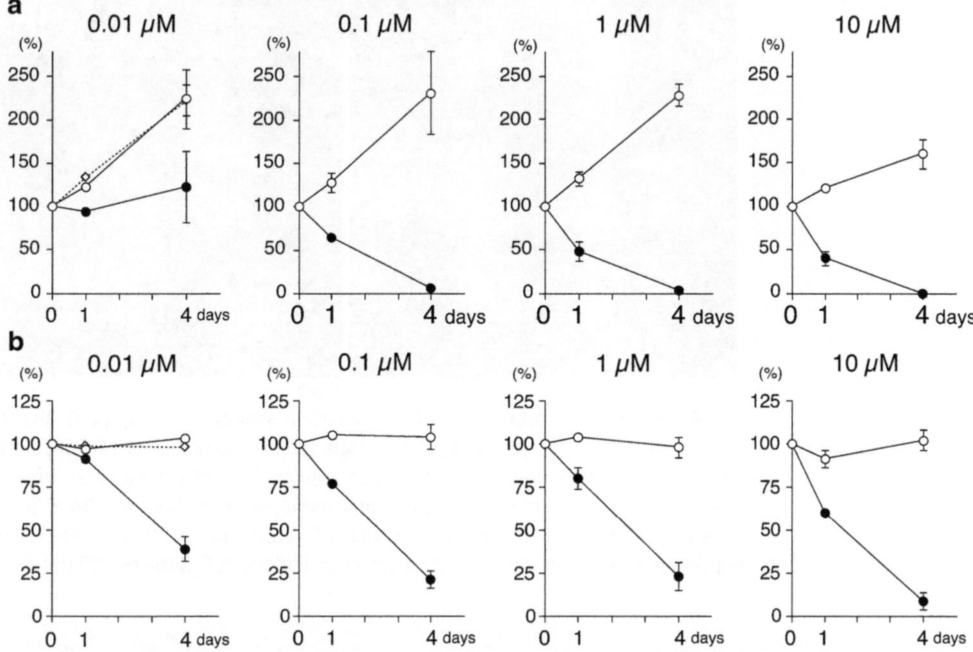

Fig. 3. Quantification of temporal changes in the intensity of GFP fluorescence after single-cell electroporation of siRNA in Purkinje (**a**) and Golgi (**b**) cells. Single-cell electroporation was performed in GFP-expressing Purkinje (n = 8–12) and Golgi (n = 10–17) cells at 9–11 DIV, using micropipettes loaded with various concentrations (0.01–10 μM) of negative control (*white circles*) or GFP (*black circles*) siRNA. Fluorescent images of the electroporated cells were acquired 1 h (0 day), 1 day, and 4 days after electroporation using a CFLSM. The intensity of GFP fluorescence in each cell was normalized by that 0 day after electroporation. The data of randomly selected Purkinje (n = 20) and Golgi (n = 44) cells that were not electroporated are also represented (*gray diamonds*). Note that GFP fluorescence in Purkinje cells increases owing to neuronal maturation (*gray diamonds* and *white circles* in (**a**)). Error bars represent S.E.M (Modified from Tanaka et al. (15). With permission from Elsevier).

5. Immunocytochemistry of Electroporated Cells

5.1. Procedures

To examine the gene-silencing effects of siRNAs against endogenous genes and the morphology of electroporated cells, immunocytochemistry can be performed several days after electroporation (15). After culture for a necessary number of days after electroporation, dissociated cell cultures are fixed for an appropriate time (e.g., 10–40 min) and in an appropriate fixative (e.g., 4% paraformaldehyde in PBS) for target proteins. The plastic bottom of μ-Dish is cut out and used as a specimen. Specimens are preincubated in a blocking solution (e.g., 10% normal serum and 0.3% Triton X-100 in PBS), then incubated in a solution containing a primary antibody. Immunoreactivity is visualized by a fluorescent dye-conjugated secondary antibody or by a horseradish peroxidase (HRP)-conjugated secondary antibody and subsequent development with a substrate for HRP (e.g., 3,3′-diaminobenzidine, 3-amino-9-ethylcarbazole).

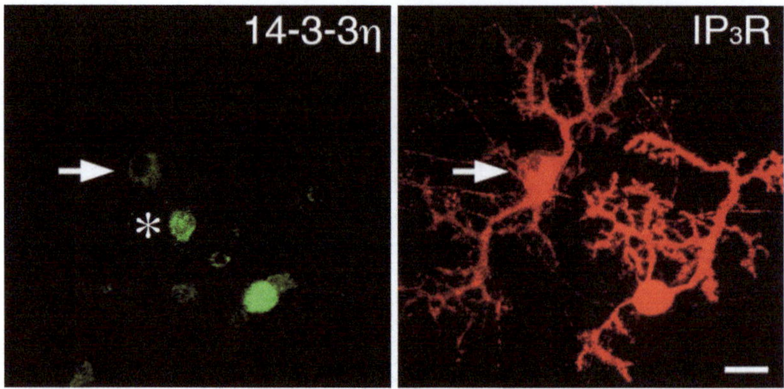

Fig. 4. Double fluorescent immunocytochemistry against 14-3-3η (*left*) and inositol 1,4,5-trisphosphate receptor (IP_3R; *right*). Single-cell electroporation of 14-3-3η siRNA was performed in Purkinje cells in cerebellar cell cultures. The cell cultures were fixed 4 days after electroporation and processed for immunocytochemistry. Note that 14-3-3η signals were reduced in the electroporated Purkinje cell (*arrow*), but not in an adjacent nonelectroporated Purkinje cell. An *asterisk* shows a non-Purkinje cell. Scale bar, 20 μm.

5.2. Applications

In many cases, we perform double fluorescent immunocytochemistry using antibodies against the target protein of siRNA and cell markers of Purkinje (e.g., inositol 1,4,5-trisphosphate receptor, calbindin-D-28K) or Golgi (e.g., neurogranin) cells to confirm that siRNA has been introduced into the target cells or to examine the correlation between the gene-silencing effects of siRNA and cell morphology (Fig. 4). By such immunocytochemical study, it was shown that Purkinje cells electroporated with negative control or GFP siRNA (≤1 μM in micropipettes) had normal morphology, including their elaborated dendrites (15). However, inhibitory effects on dendritic extension were observed by electroporation of 10 μM siRNA. Thus, single-cell electroporation appeared not to exert any negative effects on the development of Purkinje cell dendrites if the concentration of siRNA is properly controlled. Recent study has shown that sufficient reduction in immunocytochemical signals for endogenous genes continued for at least 14 days after electroporation (18).

6. Single-Cell Electroporation of Other Molecules

Other than siRNA, we have tried to transfer several types of molecules into Purkinje cells in cerebellar cell cultures.

6.1. Fluorescent Dyes

Fluorescent dyes such as Lucifer yellow and calcium ion indicators such as Oregon green 488 BAPTA-1 were easily transferred, although the fluorescence of these dyes became very weak several hours to 1 day after electroporation. In contrast, dextran conjugates of Alexa Fluor dyes were difficult to transfer in comparison to nonconjugated dyes.

6.2. Expression Plasmids

So far, we have not succeeded in the expression of GFP by the electroporation of GFP expression plasmids in Purkinje cells even when the voltage of electric pulses was increased up to –10 V. The large size of expression plasmids may prevent our success in this experiment. Another problem is that the high voltage around –10 V often induced cell damage, fusion of the micropipette tip to the cell, or plugging of the micropipette, as mentioned above. Because Purkinje cells are susceptible neurons, it may be difficult to apply enough power of electric pulses for expression plasmids without inducing cell damage. Others succeeded in GFP or DsRed expression by introducing expression plasmids into hippocampal and cortical pyramidal neurons in slice cultures (7, 9) and into cortical interneurons in vivo (19) by applying electric pulses of –10 V or higher.

7. Concluding Remarks

As shown by the real-time monitoring using GFP imaging, RNAi was achieved effectively by single-cell electroporation of siRNA. Recent study has shown that sufficient gene-silencing effects continued for a long period. Thus, this method is useful for examining roles of genes involved in neuronal differentiation, function, and degeneration in primary cultures.

This method allows the introduction of siRNA into specific cells, which is an important advantage over the other methods for introducing siRNA, especially in primary neuronal cultures. In general, primary neuronal cultures are composed of different types of cells (e.g., neurons and glia, projection neurons and local interneurons, excitatory and inhibitory neurons). By widely used transfection methods such as lipofection and bulk electroporation, siRNA is introduced nonspecifically into several types of cells. In contrast, the single-cell electroporation realizes introduction of siRNA into specific cells, enabling us to determine functions of target genes clearly, without using cell-type-specific promoters.

In addition, this method can easily realize the introduction of multiple siRNAs into a cell simply by loading them together in one micropipette. Furthermore, this method does not require cell harvesting and resuspension, which reduce the cell viability of neuronal primary cultures. Cell damage by this method is less than that by microinjection because the tip of the micropipette does not penetrate the cell membrane.

Unfortunately, in our experience, it appears to be impossible at present to express foreign genes in Purkinje cells by single-cell electroporation of expression plasmids probably because Purkinje cells are susceptible neurons. In general, the transfection efficiency by widely used transfection methods such as lipofection, bulk electroporation, and gene gun biolistics is extremely low in Purkinje cells except when using viral vectors. It is desirable to overcome the susceptible nature of Purkinje cells and realize expression of foreign genes introduced by single-cell electroporation in Purkinje cells in the future.

Because single-cell electroporation must be performed as quickly as possible to maintain the viability of susceptible neurons such as Purkinje cells, the number of cells electroporated by this method is limited (about 10–20 cells per dish in our usual experiments), which prevents application of this method to large-scale analysis. To overcome this limitation, chip-type single-cell electroporation has been developed recently (11–13).

Acknowledgments

I thank Dr. Naohide Hirashima (Nagoya City University) for his generous support of this study and Dr. Yuchio Yanagawa (Gunma University) for his kind gift of GAD67-GFP knock-in mice.

References

1. Weaver, J. C. (1993) Electroporation: a general phenomenon for manipulating cells and tissues. *J. Cell. Biochem.* 51, 426–435.
2. Ho, S. Y., and Mittal, G. S. (1996) Electroporation of cell membranes: a review. *Crit. Rev. Biotechnol.* 16, 349–362.
3. Neumann, E., Kakorin, S., and Toensing, K. (1999) Fundamentals of electroporative delivery of drugs and genes. *Bioelectrochem. Bioenerg.* 48, 3–16.
4. Olofsson, J., Nolkrantz, K., Ryttsén, F., Lambie, B. A., Weber, S. G., and Orwar, O. (2003) Single-cell electroporation. *Curr. Opin. Biotech.* 14, 29–34.
5. Lundqvist, J. A., Sahlin, F., Aberg, M. A. I., Strömberg, A., Eriksson, P. S., and Orwar, O. (1998) Altering the biochemical state of individual cultured cells and organelles with ultramicroelectrodes. *Proc. Natl. Acad. Sci.* 95, 10356–10360.
6. Nolkrantz, K., Farre, C., Brederlau, A., Karlsson, R. I. D., Brennan, C., Eriksson, P. S., Weber, S. G., Sandberg, M., and Orwar, O. (2001) Electroporation of single cells and tissues with an electrolyte-filled capillary. *Anal. Chem.* 73, 4469–4477.
7. Haas, K., Sin, W.-C., Javaherian, A., Li, Z., and Cline, H. T. (2001) Single-cell electroporation for gene transfer in vivo. *Neuron* 29, 583–591.
8. Rae, J. L., and Levis, R. A. (2002) Single-cell electroporation. *Pflügers Arch* 443, 664–670.
9. Rathenberg, J., Nevian, T., and Witzemann, V. (2003) High-efficiency transfection of individual neurons using modified electrophysiology techniques. *J. Neurosci. Meth.* 126, 91–98.
10. Lovell, P., Jezzini, S. H., and Moroz, L. L. (2006) Electroporation of neurons and growth cones in *Aplysia californica. J. Neurosci. Meth.* 151, 114–120.
11. Huang, Y., and Rubinsky, B. (1999) Microelectroporation: improving the efficiency and understanding of electrical permeabilization of cells. *Biomed. Microdev.* 2, 145–150.
12. Khine, M., Lau, A., Ionescu-Zanetti, C., Seo, J., and Lee, L. P. (2005) A single cell electroporation chip. *Lab Chip* 5, 38–43.
13. Vassanelli, S., Bandiera, L., Borgo, M., Cellere, G., Santoni, L., Bersani, C., Salamon, M.,

Zaccolo, M., Lorenzelli, L., Girardi, S., Maschietto, M., Dal Maschio, M., and Paccagnella, A. (2008) Space and time-resolved gene expression experiments on cultured mammalian cells by a single-cell electroporation microarray. *N. Biotechnol.* 25, 55–67.

14. Boudes, M., Pieraut, S., Valmier, J., Carroll, P., and Scamps, F. (2008) Single-cell electroporation of adult sensory neurons for gene screening with RNA interference mechanism. *J. Neurosci. Meth.* 170, 204–211.

15. Tanaka, M., Yanagawa, Y., and Hirashima, N. (2009) Transfer of small interfering RNA by single-cell electroporation in cerebellar cell cultures. *J. Neurosci. Meth.* 178, 80–86.

16. Tamamaki, N., Yanagawa, Y., Tomioka, R., Miyazaki, J., Obata, K., and Kaneko, T. (2003) Green fl uorescent protein expression and colocalization with calretinin, parvalbumin, and somatostatin in the GAD67-GFP knock-in mouse. *J. Comp. Neurol.* 467, 60–79.

17. Tanaka, M., Yanagawa, Y., Obata, K., and Marunouchi, T. (2006) Dendritic morphogenesis of cerebellar Purkinje cells through extension and retraction revealed by long-term tracking of living cells *in vitro*. *Neuroscience* 141, 663–674.

18. Tanaka, M., Asaoka, M., Yanagawa, Y., and Hirashima, N. (2011) Long-term gene-silencing effects of siRNA introduced by single-cell electroporation into postmitotic CNS neurons. *Neurochem. Res.* 36, 1482–1489.

19. Kitamura, K., Judkewitz, B., Kano, M., Denk, W., and HÐsser, M. (2008) Targeted patchclamp recordings and single-cell electroporation of unlabeled neurons *in vivo*. *Nat. Methods* 5, 61–67.

Chapter 8

Development and Application of Membrane-Tethered Toxins for Genetic Analyses of Neuronal Circuits

Sebastian Auer and Inés Ibañez-Tallon

Abstract

Peptide toxins derived from venomous animals are widely employed in neuroscience research because of their ability to manipulate specific ionic currents. However, the detailed characterization of particular classes of ion channels and their contributions to given neuronal circuit is often hampered by the fact that the action of soluble neurotoxins cannot be restricted to single cell populations. Here, we review a recombinant strategy based on genetically encoded membrane-tethered toxins (t-toxins) for long-term inhibition of ion channels and receptors. Due to their membrane attachment, the activity of fused toxins is restricted to the genetically targeted cells, without affecting identical channels on neighboring cells that do not express the t-toxin. This review describes the development and modular design of t-toxins and their application to studies of neurocircuitry using transgenic and viral constructs to target expression to specific CNS cell types in vitro and in vivo.

Key words: Tethered toxin, Ion channels, Lentivirus, Neurotransmission, Recombinant antagonists

1. Introduction

The complexity of higher brain functions greatly depends on the enormous diversity of ion channels that control ion transport and cell communication between distinct cell populations. Hence, the dissection of neuronal circuits underlying a specific function requires both the identification of the cell populations and the profiling and characterization of the ion channels that control the excitability of those cell types. Because of their unique and essential roles in cell signaling, ion channels and receptors belong to the most promising molecular targets for therapeutic interventions in many diseases. Until recently, traditional methods to decode neuronal circuits were based on electrolytic lesions and electrical stimulation of brain regions, or used local administration of neurotoxic and pharmacologic agents. Besides being highly

Alexei Morozov (ed.), *Controlled Genetic Manipulations*, Neuromethods, vol. 65,
DOI 10.1007/978-1-61779-533-6_8, © Springer Science+Business Media, LLC 2012

invasive, these manipulations are usually not cell-type or even ion channel specific and often disrupt more than one neuronal circuit in an irreversible manner. Recently, new genetic approaches have provided powerful opportunities for more precise manipulations of ion channels in vivo. For example, targeted gene deletion of ion channel subunits, for example, in knockout mice, or downregulation by siRNAs can be used to interfere with ion channel function. Yet, since the majority of receptors and ion channels are multimeric, gene deletion of one subunit might be accompanied by compensatory upregulations of closely related receptors that share common subunits or compete for the same binding partners (1–3). Another important development in the study of ion channels has been the identification of gene promoters that allow cell-type-specific manipulations of functionally related neuronal populations, for example, in BAC (bacterial artificial chromosome) transgenic mice (4). These mouse models can be used to trace neuronal connections, achieve cell-specific conditional mutagenesis, or drive functional changes within circuits in vivo by injections with viral vectors (5–7).

In this chapter, we describe a genetic approach termed "tethered toxins" that extends our ability to modify singular ionic currents in neurons in vivo. This method makes use of natural peptide toxins which have originally been isolated from venoms of snakes, scorpions, spiders, sea anemones, and cone snails. Being complex mixtures of peptide toxins, these venoms allow poisonous animals to capture prey or to protect themselves against predators by acting on different ion channels and receptors for most efficient paralysis or deterrence.

Adaptive evolution, driven by natural selection, has led to strong variations in venom composition, providing a vast array of naturally occurring peptide toxins with different affinities for specific ion channel subclasses. During the last decade of peptide toxin research, an important number of specific inhibitors and modulators of ion channels and receptors have been identified. Thus, unique peptide toxins with characteristic cysteine backbones and selective affinities for voltage-gated sodium (Na_v), calcium (Ca_v), and potassium (K_v) ion channels, and ligand-gated receptors, including nicotinic acetylcholine receptors (nAChRs), N-methyl-D-aspartate (NMDA), and G-protein-coupled receptors (GPCRs), have been isolated and characterized (for reviews see: (8–11)). Their high specificity makes them ideal tools for deciphering the contribution of ionic currents in neuronal networks. However, their activity cannot be restricted to a single cell population in brain slices or in a living organism, and their application usually requires constant administration to compensate for degradation and diffusion effects (Fig. 1a). To bypass these

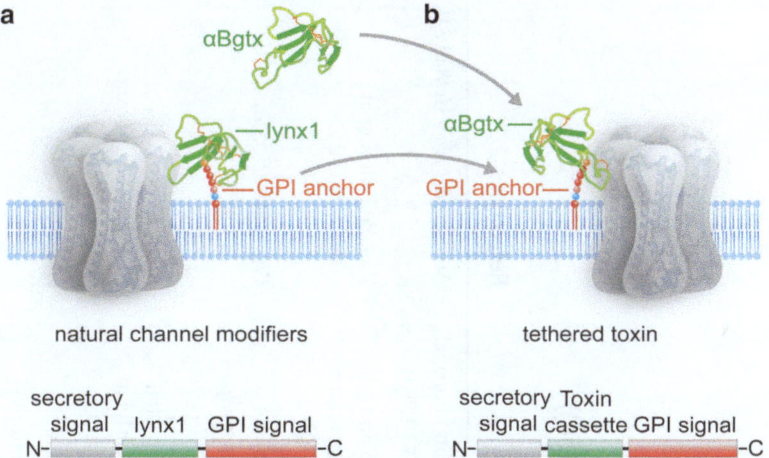

Fig. 1. Schematic representation of prototoxin channel modifiers and engineered tethered toxins (t-toxins). (**a**) Examples of the Ly-6/uPAR superfamily: snake α-bungarotoxin (αBgtx) and the GPI-anchored, cell-membrane bound lynx1. The schematic shows the coding sequence of lynx1 (N-terminal secretory signal, lynx1 peptide, C-terminal GPI anchor). (**b**) The structural homology of lynx1 with αBgtx gave rise to the t-toxin strategy of using the biological scaffold of lynx1 (secretory signal and GPI signal) to generate recombinant membrane-bound toxins, such as the illustrated t-aBgtx, by substitution of the lynx 1 peptide with the toxin sequence, Flag-tag, and linker regions (Adapted from Holford et al. (16)).

limitations, we developed genetically encoded tethered toxins (t-toxins) that are bound to the cell surface by membrane tethers and act only on ion channels and receptors of the cell population that expresses the t-toxin, and not on identical receptors present in neighboring cells that do not express the t-toxin (12–14) (Fig. 1b). In this chapter, we discuss the development of modular t-toxins with preserved activity and specificity and their application to studies of neurocircuitry. Examples include expression of various t-toxins for manipulation of nAChR functions, as well as for the inhibition of voltage-gated Na^+ and Ca^{2+} channels in order to suppress neuronal excitability by blockade of action potentials or presynaptic vesicle fusion (Table 1). Combining the extremely high affinity of peptide toxins with the ability to restrict their action by membrane tethering, while allowing complementation with cell-specific and inducible genetic methodologies, makes the t-toxin approach a powerful tool for investigating ion channel function in vitro and in vivo. We believe that t-toxins can be broadly applied to the genetic dissection of specific channels and pathways that control the development and function of the CNS across various species.

Table 1
Overview of peptide toxins used in tethered toxin studies (From Auer and Ibañez-Tallon (54))

Toxin name	Species	Length (aa)	Tethered toxin construct	Activity	References
Agatoxin IVA	*Agenelopsis aperta*	48	Venus-flag-linker-AgaIVA-GPI anchor	Inhibition of $Ca_v2.1$ in neuronal culture and in mice in vivo	Auer et al., *Nat Meth* (13)
Bungarotoxin	*Bungarus multicinctus*	74	αBgtx-linker-GPI anchor	Inhibition of α1β1γδ and α7nAChR in *Xenopus laevis* oocytes and zebra fish	Ibañez-Tallon et al., *Neuron* (12)
δ-ACTX-Hv1a	*Hadronyche versuta*	42	δ-ACTX-Hv1a-GPI anchor	Inhibition of Na_v channel inactivation in *Drosophila melanogaster*	Wu et al., *PLoS Biol* (20)
GID	*Conus geographus*	19	GID-flag-linker-TM domain-EGFP	Inhibition of α7 and partial block of α3β4 nAChRs in *Xenopus laevis* oocytes	Holford et al., *Front Mol Neurosci* (16); Auer (26)
MVIIA	*Conus magus*	25	MVIIA-flag-linker-TM domain-EGFP	Inhibition of $Ca_v2.2$ in neuronal culture and in mice in vivo	Auer et al., *Nat Meth* (13)
MrVIA	*Conus marmoreus*	31	MrVIA-flag-linker-GPI anchor	Partial inhibition of Na_v 1.8 in DRG culture and in mice in vivo	Stürzebecher et al., *J Physiol* (14)

2. Origin and Development of Tethered Toxins

The engineering of tethered toxins derived from the discovery of endogenous, cell-membrane bound prototoxins of the lynx1 family (15). *Lynx1* belongs to the Ly-6/neurotoxin gene family and is an evolutionary precursor to snake venom toxins with structural homologies to the nicotinic acetylcholine receptor (nAChR) antagonists α- and κ-bungarotoxin. Lynx1 is attached to the cell surface by a glycosylphosphatidylinositol (GPI) anchor and shows a complex disulfide fold (cysteine knot), a characteristic feature of most peptide toxins (Fig. 1a) (15, 16). Functional analyses indicated that lynx1 and the closely related molecule lynx2 are not ligands or neurotransmitters but directly assemble with nAChRs at the cell membrane and modulate their functions in the presence of acetylcholine or nicotine (15, 17–19). The first recombinant membrane-bound toxins were designed by replacing lynx1 with the sequences encoding for bungarotoxins and α-conotoxins downstream of the secretory signal sequence, followed by a short linker and the GPI anchor signal sequence (12). This design directs the toxin peptide to the secretory pathway, where the signal sequence is cleaved and the GPI targeting sequence is substituted by a covalent bond to GPI, thereby anchoring the peptide to the extracellular side of the plasma membrane of the cell in which it is expressed (Fig. 1b).

Since then, the t-toxin design has been further optimized by introducing other membrane tethers, that is, the transmembrane domain of the PDGF receptor (13), as well as fluorescent markers (EGFP, mCherry, and EBFP2) and immunotags (i.e., Flag-tag) (Fig. 2b). These modifications have greatly increased the ability to monitor the expression levels and subcellular localization of the recombinant molecules, which are important prerequisites for their use in neurocircuitry. So far, approximately 40 different t-toxin constructs have been cloned in our group, and their activity has been characterized on voltage- and ligand-gated ion channels (12–14, 16). Studies in vivo have been possible using different genetic approaches to drive their cell-autonomous action. These include transgenesis in zebra fish (12), Drosophila (20), and mouse (14), as well as recombinant viral systems (13, 21) (Table 1). In particular, the possibility to encode t-toxins in viral vectors has further allowed the stable genetic delivery of t-toxins to a variety of mammalian cells, including neurons in vitro and in vivo. Furthermore, the use of viral vectors has provided the possibility to implement inducible and Cre recombinase-dependent approaches for regulated and cell-specific expression of t-toxins in mice that we will discuss later (13).

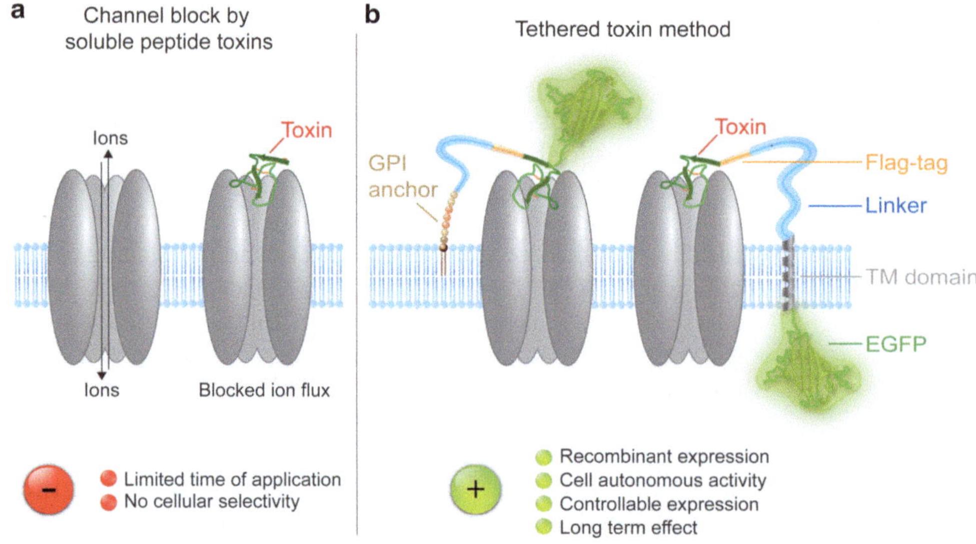

Fig. 2. Illustration of ion channel inhibition by soluble peptide toxins or genetic expression of recombinant tethered toxins. (**a**) Soluble peptide toxins cause instantaneous block of target ion channels, but their use in vivo is limited by the accessibility of the brain structure to be targeted, the necessity of constant readministration, and the lack of cellular selectivity within a circuit or cell network. (**b**) Genetically encoded toxins tethered to the membrane via a glycosylphosphatidylinositol (*GPI*) anchor or a transmembrane (*TM*) domain allow cell-autonomous block of ion channels and the integration of domains (i.e., Flag or other epitopes, EGFP or other markers) to monitor their long-term expression in targeted cells (From Auer and Ibañez-Tallon (54)).

3. Modular Components to Engineer t-Toxins

The biological activity of each tethered construct has to be individually evaluated and optimized for every toxin/channel combination. This is due to several factors, including the variability in the nature of the bioactive peptide and ion channel or receptor being targeted, and in the constraints imposed by the modular architecture of the tethers, linkers, and epitope tags. The vast amount of available modular choices for constructing t-toxins indicate the feasibility of tailor-made tethered manipulators for a wide range of different receptors/channels, and point to their broad applicability in various fields of ion channel and neurocircuitry studies.

3.1. Characteristics of the Bioactive Peptide

Expression and functional assays have revealed that several elements are critical to achieve robust expression on the cell surface and rotational flexibility of the peptide toxin in order to be able to bind to its target region. The affinity of the bioactive peptide for its cognate ion channel or receptor has to be taken into account, and toxins with a strong affinity per se are potentially more effective when applied as t-toxins compared to less potent ones. Furthermore, it is also important to consider the composition and length of the peptides to be tethered, that is, charges and hydrophobicity of the amino acid residues, number of cysteine bonds in the case of toxins, and existence of noncanonical residues, or terminal amidations. For example, substitution of hydroxylated or carboxylated amino acids with nonmodified residues in highly posttranslational modified conotoxins, such as GVIA (22), RIIIK (23), and PIIIA (24), failed to yield satisfactory activity in their tethered constructs (Ibañez-Tallon, unpublished), except for the case of GID (16, 25). However, when tested as tethered toxin, t-GID showed block of α7 and α3β4 nAChRs, while, contrary to findings with the soluble GID (25), α4β2 nicotine–induced currents were not altered, irrespective of the linker length (26). Most probably, the substitution of the nonproteinogenic γ-carboxyglutamic acid with glutamic acid (γ4E) in t-GID could account for the selective loss of activity on α4β2 nAChRs. This hypothesis is supported by the report of a lack of tolerance to sequence variation in the N-terminal region and the finding that deletion of the N-terminal four residues of GID (GIDΔ1–4) leads to ~4-fold decrease in activity at the α4β2 receptor, without affecting the functionality at α3β2 and α7 nAChRs (25, 27). In general, possible variations in the efficacy of t-toxins with modified amino acids are likely to be dependent on the location of the posttranslational changes in the toxin sequence and have to be individually tested. Furthermore, the makeup of the peptide or toxin to be tethered has to be considered, but it is not predictive of the expression and activity of its tethered form. For instance, the conotoxins MrVIA and MrVIB are well expressed at the cell membrane and functionally active when tethered (12, 14, 20) despite their high number of hydrophobic amino acids and difficulties to be chemically synthesized (28, 29). Other examples are the tethered forms of α- and κ-bungarotoxins, which were well expressed and functional (12, 30), although being relatively long peptides (68–82aa), while shorter conotoxins such as SmIIIA (30aa) or MVIIC (26aa) did not lead to active t-toxins, possibly due to folding disturbances or high proportion of charged amino acids (Ibañez-Tallon, unpublished).

3.2. Linker Length and Composition

Another relevant feature when designing t-peptide constructs is the linker sequence bridging the toxin peptide to the GPI anchor or TM domain (Figs. 1b, 2b). Most of the t-toxins generated so far contain a stretch of 20 alternating glycine and asparagine residues. This $(\text{Gly-Asn})_{20}$ linker preceding the membrane tether provides the rotational flexibility and distance necessary for the peptide toxin to bind to its cognate ion channel (12). Similarly, tethered peptide ligands of G-protein-coupled receptors (GPCRs), which have been engineered using the same scaffold as the t-toxins, also required this $(\text{Gly-Asn})_{20}$ flexible linker for effective binding to its receptor (31, 32). A number of tethered constructs including α- and κ-bungarotoxins have been cloned using a very short $(\text{Ala-Gly})_6$

linker which was sufficient for efficient binding and activity, most likely because of the large size of these toxins (74/66 aa for α- and κ-bungarotoxin in comparison to 15–30 aa for most conotoxins). In some other cases, we have used Gly-Asn repeats with lengths varying from 40 aa (2× long) to 60 aa (3× long) (12, 16). Theoretically, longer flexible linkers may provide further freedom for the t-toxin to bind within the vestibule of voltage-gated channels or allosteric binding sites on other ion channels and receptors. However, we found that t-GID efficiently blocked α7 nAChRs in *Xenopus* oocytes when fused to the $(\text{Ala-Gly})_6$ and $(\text{Gly-Asn})_{20}$ linkers, but its activity was progressively reduced when longer linkers were tested (16, 26). Taken together, a distance of the peptide toxin from the membrane tether of 6–20 aa resulted in the highest activities of the so far tested t-toxins; however other t-toxin/ion channel combinations might require empirical adjustment of the linker length.

3.3. Membrane Tethers for Cell-Surface Expression

Membrane tethers allow the restriction of peptide toxin activity to the expressing cell and also enhance the working concentration of the toxin in its target region, the plasma membrane. The choice of membrane tether depends on the characteristics of the peptide, as well as on the epitope tags and markers to be used in combination. GPI anchors, which are less bulky than transmembrane (TM) domains, may facilitate the mobility of the t-toxin in close proximity to the receptor or ion channel to be targeted within the plasma membrane. If the toxin or peptide does not require a free N-terminus for interacting with its cognate receptor, GPI-anchored versions of EGFP and peptide toxin fusions may be used. However, when using GPI anchors, it has to be taken into account that they are susceptible to cleavage by endogenous phospholipases, such as PI-PLC and phospholipase D (33). This has been suggested as a mechanism used by cells for selective and rapid release of certain GPI-anchored proteins at specific times. To avoid this potential problem, the GPI anchor of t-toxins can be replaced by a transmembrane domain, for example, the PDGF receptor TM domain (13). TM domains can be used to retain t-toxins at the cell surface, while allowing the fusion of fluorescent markers on the cytoplasmatic side of the plasma membrane, which helps to avoid possible hindrances with the peptide toxin. Furthermore, the fluorescence intensity was found to be significantly higher for intracellular EGFP fusions compared to extracellular localization of the fluorescent marker (Auer et al., unpublished). On the contrary, GPI-anchored t-toxins showed improved antibody detection of fused extracellular immunotags, compared to TM containing t-toxins (Auer et al., unpublished). Thus, the choice of the membrane tether depends on the respective t-toxin/ion channel combination and the t-toxin application requirements.

4. Genetic Approaches for t-Toxin Delivery to Neuronal Circuits

The fact that peptide toxins maintain their functionality when expressed exogenously in mammalian neurons opens up the possibility to implement an important number of genetic strategies for studies of neurocircuitry in vivo by using genetically encoded recombinant t-toxins. These include BAC transgenesis, Cre-dependent knock-in constructs, and viral vectors. Here, we discuss some examples of the use of t-toxins with these genetic approaches to target specific neuronal populations in the mouse nervous system.

4.1. BAC Transgenesis

Mouse transgenesis using bacterial artificial chromosomes (BAC) provides a powerful tool to study neuronal circuitry in vivo. The large genomic inserts of the BAC carry a whole transcription unit with all the regulatory elements and therefore enable precise targeting of cell types and tissue-specific expression of the transgene in certain neuronal circuits. Several studies have shown that BAC transgenesis presents a highly reliable way to obtain accurate transgene expression in rodents (34, 35). We have employed this approach to express two classes of t-toxins specifically in primary afferent nociceptors in the pain pathway. BAC transgenic mice were generated using the two-step modification system described by Gong (4). This approach takes advantage of the precise targeting of the transgene into the genomic locus of the BAC via homologous recombination. To specifically target small and medium nociceptive fibers, the expression cassette containing the t-toxin was cloned into the Scn10a regulatory sequences (Fig. 3), which drive expression of the $Na_v1.8$ voltage-gated sodium channel (VGSC) α-subunit only in a subset of sensory neurons, of which more than 85% are nociceptors (36, 37). The first study using this approach targeted the tethered ω-conotoxin MrVIA, which preferentially inactivates TTX-R VGSC (14), while the second study used the ω-conotoxin MVIIA that selectively inactivates $Ca_v2.2$ voltage-gated calcium channels (VGCC) (13). The functional analyses of these two mouse models are described in further detail in Sect. 5.3.

4.2. Lentiviral Vectors

HIV-1 lentiviral vectors (LV) have been extensively used as gene delivery tools because of their ability to infect dividing and nondividing cells (e.g., neurons) with stable integration of the gene of interest into the host genome, but without inducing immunogenic responses, unlike adenoviral vectors (38, 39). Furthermore, generation and purification of pseudolentiviral particles is fast and relatively easy, with a typical batch production time of 1 week. Pseudolentiviral particles are also able to incorporate large inserts (~8 kb), whereas the size of transgenes can be a limiting factor, permitting the use of adeno-associated virus (AAV, ~3–4 kb). For

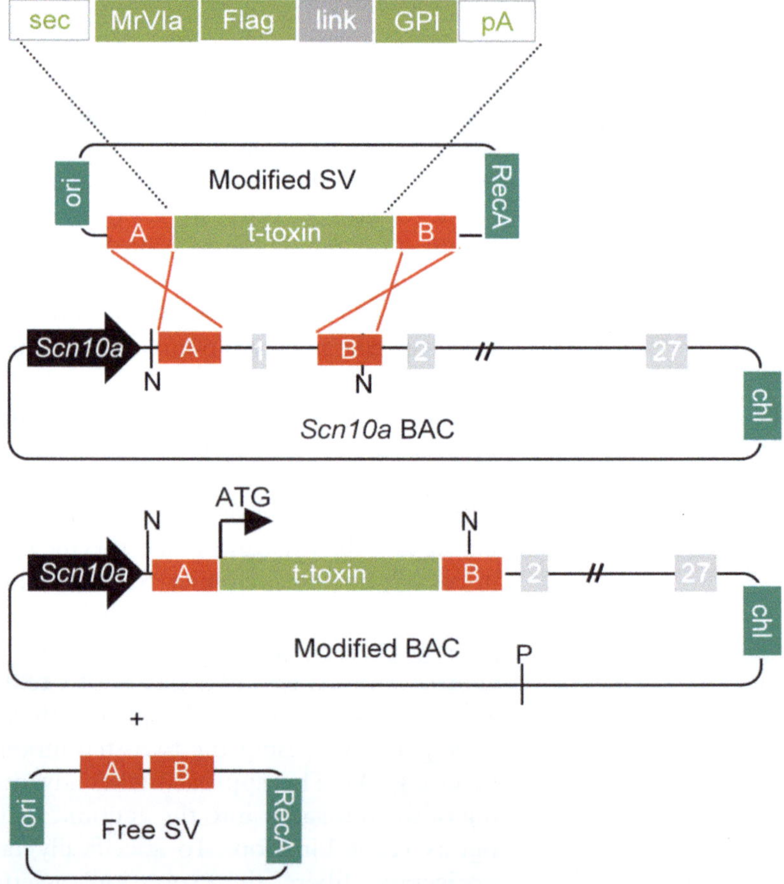

Fig. 3. Generation of BAC transgenic mice encoding membrane-tethered MrVIa toxin. Schematic overview of the BAC modification by two-step homologous recombination. The modified shuttle vector (*SV*) contains the t-toxin cassette (*green*), flanked by a secretion signal (*sec*) and a polyA (*pA*), inserted between two recombination boxes (A and B in *red*) corresponding to the sequences flanking the initiator methionine of the *Scn10a* gene encoding the Nav1.8 VGSC subunit. Recombination (*red lines*) with the *Scn10a* BAC results in the modified BAC and the free shuttle vector. *Scn10a* gene promoter (*black arrow*), *Scn10a* exons (*grey boxes*), selection markers (*dark green boxes*), *NcoI* restriction sites for Southern blot (*N*), PI-SceI restriction site for BAC linearization (*P*) (From Stürzebecher et al. (14)).

the generation of pseudolentiviral particles as vectors, the accessory genes of native HIV-1, which encode crucial virulence factors and are essential for HIV pathogenicity, have been eliminated (40). Furthermore, cis-acting elements in the native HIV-1 genome (noncoding sequence elements required for vector RNA synthesis, packaging, reverse transcription, and integration) have been separated from the trans elements (encoding structural, enzymatic, and accessory proteins) by subcloning to different plasmids (40, 41). Division of these components to four different plasmids in the third generation of LV vectors minimizes the likelihood of generating replication competent lentivirus (RCL), as this would require

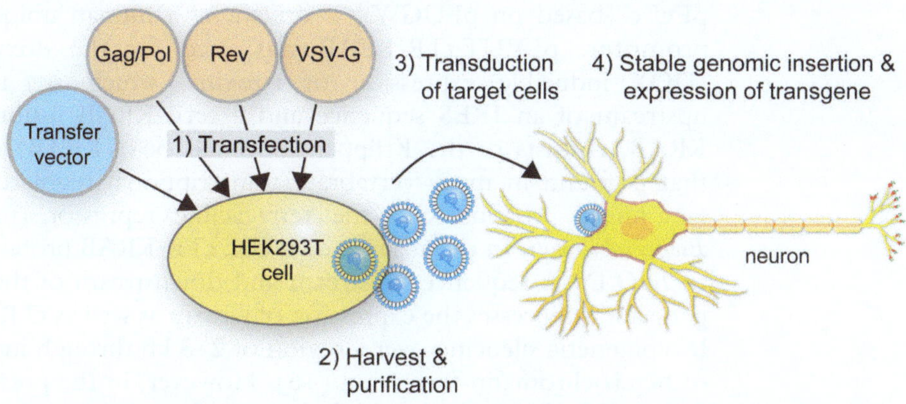

Fig. 4. Schematic representation of lentivirus production. Pseudolentiviral particles are generated by cotransfection of transfer vector (contains gene of interest) and the packaging plasmids into HEK293T cells. *Gag* and *Pol* encode for the structural and enzymatic proteins required to form functional vector particles, *Rev* encodes for the Rev protein, which is necessary for nucleocytoplasmic transport of full-length vector RNA, and *VSV-G* encodes for vesicular stomatitis virus G-protein envelope. Lentiviral particles are harvested ~48 h after transfection, purified and concentrated by ultracentrifugation, and stored at −80°C. Target cells can be transduced by direct addition of LV suspension to the culture medium (in vitro), or by injection into the desired tissue area (in vivo). Upon transduction, the gene of interest is stably integrated into the host genome and subsequently expressed, depending on the chosen promoter and vector design (Adapted from Auer (26)).

at least three recombination events (Fig. 4). As an additional measure of safety, replication defective, self-inactivating (SIN) vectors have been generated, which use a CMV promoter with a deletion in the promoter/enhancer region in the 3′ LTR, that is transposed onto the 5′ LTR following the transduction of a target cell and thus prevents expression of full-length vector RNA in infected cells (38, 40). Additionally to these improvements in vector safety, the efficiency of LV production and expression has been enhanced by insertion of WPRE (woodchuck hepatitis virus posttranscriptional regulatory element) at the 3′end of the insert, which increases the transgene expression more than 5-fold by stabilizing the mRNA (42). In contrast to wild-type HIV-1, most pseudolentiviral particles today use the vesicular stomatitis virus G protein (VSV-G) as envelope instead of the parental HIV-1 envelope (41, 43). The VSV-G envelope introduces three novel features to lentivirus vector particles: (1) it stabilizes the vector particles, thereby allowing vector concentration by ultracentrifugation, (2) it broadens the vector tropism to diverse tissue and cell types of various species, and (3) it changes the vector entry to an endocytic pathway, which reduces the requirements for viral accessory proteins (38, 44).

Our group has employed three different lentiviral vectors allowing constitutive, inducible, or Cre recombinase-dependent expression of t-toxins. The expression in all three vectors, namely the constitutive lentiviral vector pFUGW (45), inducible pLVUT-tTR-KRAB (46), and the newly generated Cre-dependent vector pFU-c (based on pFUGW), is driven by a human ubiquitin C promoter. pLVUT-tTR-KRAB was used for the doxycycline (DOX)-inducible expression of t-toxins, which are inserted upstream of an IRES sequence and a tetR-KRAB fusion. tTR-KRAB, consists of the Krüppel-associated box (KRAB) domain that is found in many vertebrate transcriptional regulators, like zinc-finger proteins, and the tetracycline repressor (tetR) of *Escherichia coli*. In the absence of DOX, tTR-KRAB protein binds to *tetO* DNA sequences upstream and downstream of the transgene and suppresses the expression of t-toxin as well as tTR-KRAB by epigenetic silencing over a region of 2–3 kb through induction of heterochromatin formation (46). However, in the presence of DOX, tTR-KRAB is released from *tetO*, thus allowing temporally controlled expression of the t-toxin. Our group was able to demonstrate the effectiveness of inducible t-toxins in vitro in several cell culture systems, as well as in vivo by stereotaxic delivery of lentiviral vectors to distinct brain areas, followed by administration of DOX in the drinking water of the mice for 1 week (13).

In general, our experiments showed that a lentiviral titer above 6.0×10^{8} TU/ml was important for an efficient delivery of t-toxins to a high number of cells after stereotaxic injection. Comparable titers were routineously reached by resuspending the viral pellet after ultracentrifugation in 1/600 of the original volume (e.g., 50 μl for 30 ml supernatant). Alternatively, a second round of ultracentrifugation can be applied to pool several resuspensions and thereby increase the titer. In order to prevent potential immunogenic reactions after stereotaxic delivery, we used sterile filtered 1×PBS for virus resuspension instead of protein-containing media and buffers, which are believed to increase viral solubility. The detailed protocol for lentivirus production can be downloaded from http://www.mdc-berlin.de/en/research/research_teams/molecular_neurobiology/Research/index.html.

To selectively express t-toxins only in a specific cell population within a given brain structure, we have applied the Cre/loxP recombination system. The Cre recombinase enzyme was originally isolated from the P1 bacteriophage and belongs to the integrase family of site-specific recombinases. It catalyzes the recombination between two of its DNA recognition sites, called loxP sites (47, 48). These loxP sequences consist of a 34-bp consensus sequence, with a core spacer sequence of 8 bp and two 13-bp palindromic flanking sequences. The Cre recombinase binds to the loxP sites, then a tetramer is formed, thus bringing two loxP sites together, and subsequently, the recombination occurs within the spacer area of the loxP sites (47, 48). As a result, the sequence in between gets excised or inverted, depending on the direction of the two loxP sites. Since the first isolation and characterization of the Cre/loxP recombination system, this approach has been used to generate hundreds of different transgenic mouse lines, each one expressing Cre recombinase in a different subset of neurons (49). This renders Cre-selective expression or conditional knockout, a powerful tool to study cell functions in defined cell types. To apply the Cre recombinase approach to t-toxins, we have generated Cre-dependent lentiviral vectors (pFU-c) in which expression of the t-toxin is regulated by a loxP cassette flanking DsRed2 (13). In conditions without Cre recombinase, only DsRed2 is expressed, and generation of functional t-toxin is inhibited by an introduced frameshift in the DNA reading frame after DsRed2, which generates several translational stop codons throughout the t-toxin sequence. Thus, the expression of functional t-toxin is possible only after successful excision of the loxP cassette by Cre recombinase. The proof-of-function of Cre-dependent t-toxins was shown in vitro in rat hippocampus culture (13), and our group is currently expanding their usage to several Cre-transgenic mouse lines for the dissection of neuronal pathways.

5. Analyses of Neuronal Circuits with t-Toxins

5.1. Cell-Autonomous Inhibition of Nicotinic Acetylcholine Receptors in Zebra Fish and Chick

Early studies in the 1970s took advantage of the irreversible binding of α-bungarotoxin to the nicotinic acetylcholine receptor at the motor end plate to characterize and purify it (50). Since then, a great number of other snake venom toxins and α-conotoxins with specific affinities for different nAChR combinations have been identified (8, 51, 52). Given the diversity of pentameric nAChRs composed of different subunits (α1–7, α9–10, β1–4, γ, δ, ε) and the fact that most peptide toxins bind to nAChR combinations and not to single subunits, these have been broadly used to localize and functionally identify the composition of the nAChR pentamers present in different neuronal and nonneuronal populations. For example, α-bungarotoxin binds to muscle α1β1γδ and α7 nAChRs, and not to other neuronal nAChR, while the closely related peptide, κ-bungarotoxin, that is coexisting in the venom of *Bungarus multicinctus*, does not affect those nAChR combinations. Other examples of subtype specificity within nAChR combinations are found in the large family of α-conotoxins (53). The t-toxin strategy has been applied to several bungarotoxins and conotoxins, which retained their selective antagonistic properties against specific nAChR combinations in experiments using *Xenopus* oocytes (12). Furthermore, transgenic zebra fish in which expression of either tethered alpha (t-α-Btx) or kappa bungarotoxin (t-κ-Btx) was driven by a cell-type-specific promoter to muscle fibers demonstrated targeted cell-autonomous elimination of nAChR function in muscle fibers expressing t-α-Btx, but not t-κ-Btx, consistent with the specific activity of the two toxins (Table 1) (54). These in vivo experiments indicated that t-α-Btx-mediated inhibition of nAChR currents results in silencing of neuromuscular neurotransmission without perturbing synapse formation (12). Studies to inhibit α7 receptors in vivo have also been carried out by using retroviral vectors to express t-α-Btx in chick ciliary ganglia (Table 1). By blocking the intracellular calcium influx with t-α-Btx, these studies demonstrated the critical role of α7 nAChRs in the programmed cell death that occurs during early development of the autonomic nervous system (30). These findings illustrate some of the possible uses of t-toxins to dissect the contribution of nAChR combinations to cell network functions in different living organisms.

5.2. Silencing Neurotransmission with Calcium Channel Specific t-Toxins in the Mouse Brain

At presynaptic nerve terminals, the two voltage-gated calcium channels $Ca_v2.1$ and $Ca_v2.2$ play an essential and joint role in the electrochemical signal conversion by coupling the arriving presynaptic action potential to neurotransmitter release. The first step in this complex process is the opening of the voltage-gated $Ca_v2.1$ and $Ca_v2.2$ channels due to a rapid membrane depolarization that is caused by an arriving action potential. The resulting Ca^{2+} influx into the presynapse then enables the multimeric vesicle fusion machinery to fuse neurotransmitter (NT)-filled vesicles with the synaptic membrane, thereby releasing NT to the synaptic cleft (55, 56) (Fig. 5). As the NT release is proportional to the third or

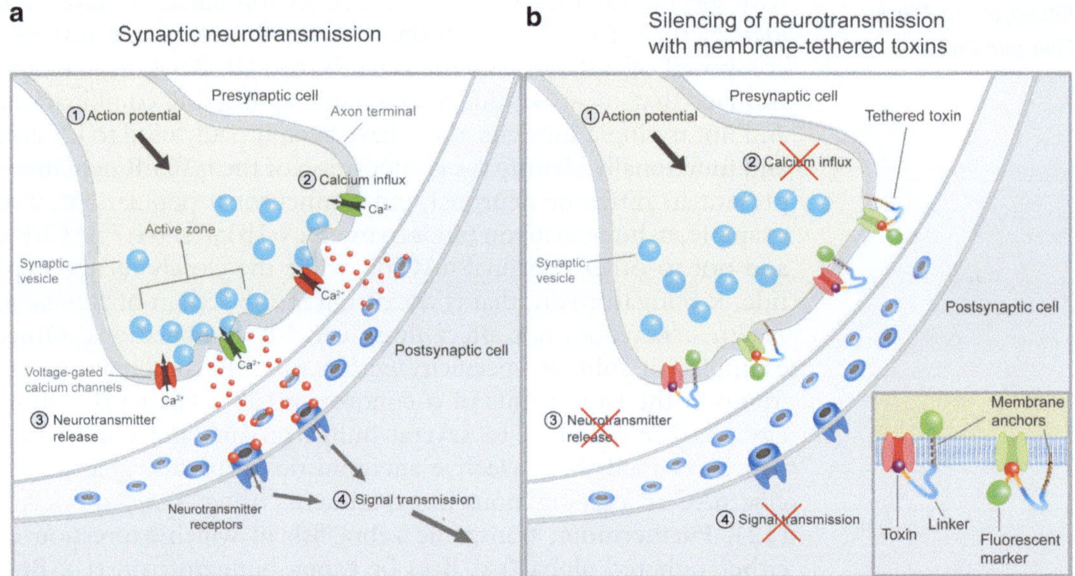

Fig. 5. Schematic of tethered-toxin mediated silencing of neurotransmission by inhibition of voltage-gated calcium channels $Ca_v2.1$ and $Ca_v2.2$ at the presynapse. The block of calcium influx upon binding and inactivation of the calcium channels $Ca_v2.1$ and $Ca_v2.2$ by the t-toxins prevents the vesicle fusion machinery from releasing neurotransmitter to the synaptic cleft, thereby interrupting neurotransmission. (**a**) Synaptic neurotransmission. (**b**) Silencing of neurotransmission with membrane-tethered toxins (Adapted from Auer and Ibañez-Tallon (54)).

fourth power of Ca^{2+} influx, a 2-fold change in presynaptic Ca^{2+} influx results in an 8- to 16-fold change in NT exocytosis (57). Thus, regulation of presynaptic calcium channels is an efficient way to control synaptic transmission. Besides deciphering distinct neuronal connections and their physiological functions, controlling the activity of these channels enables detailed studies of contributions of individual channels to neuronal circuits. With these aims in view, our group generated recombinant t-toxins that are able to block $Ca_v2.1$ and $Ca_v2.2$ channels by integration of the ω-agatoxins AgaIIIA and AgaIVA, as well as ω-conotoxins MVIIA and MVIIC. We found that these t-toxins were well expressed in cultured mammalian cells, primary cultures of hippocampal neurons, and in neurons of mice injected with lentivirus encoding t-toxins. The capability to block one or both channels was then confirmed by electrophysiological recordings of HEK293-$Ca_v2.2$ cells and rat hippocampal neurons in vitro (13). Overall, we found that AgaIVA and MVIIA provided the best blocking capabilities and in fact that these two t-toxins were as effective as the soluble toxins in fully inhibiting their respective target channels in a cell (13).

For the subsequent in vivo proof-of-function of t-toxins in mice, we have chosen the nigrostriatal pathway because of its unique behavioral phenotype after unilateral inhibition. The nigrostriatal pathway is part of the basal ganglia, which are associated with a variety of functions, including motor control and learning. They consist of two primary input structures (subthalamic nucleus (STN) and striatum), two primary output structures (substantia nigra pars reticulate (SNpr) and globus pallidus internal segment (GPi)), and two intrinsic nuclei (globus pallidus pars externa (GPe) and substantia nigra pars compacta (SNpc)) (58, 59). The SNpc consists of dopaminergic neurons, receives input from the striatum and sends most of its output back to the striatum via the medium forebrain bundle (mfb) (Fig. 6a) (58, 59). Basal ganglia disorders, like Parkinson's disease, which is caused by the death of dopaminergic (DA) neurons in SNpc, are typically characterized by an inability to correctly initiate and terminate voluntary movements, an inability to suppress involuntary movements, and an abnormal muscle tone (60). In the past, the neurotoxin 6-hydroxydopamine (6-OHDA) has been used extensively for the induction of unilateral lesion of DA neurons in the substantia nigra to induce circling behavior in rats. These studies played an essential role in the dissection of the nigrostriatal pathway and its role in motor coordination (61–64). In response to the resulting striatal DA depletion, a receptor-mediated supersensitivity, caused by increasing affinity and number of striatal D2 receptors in denervated postsynaptic neurons, develops (62). This supersensitivity on the side of the lesioned hemisphere usually causes the reversal of the rotational direction if DA agonists like apomorphine are administered (Fig. 6b) (61, 62).

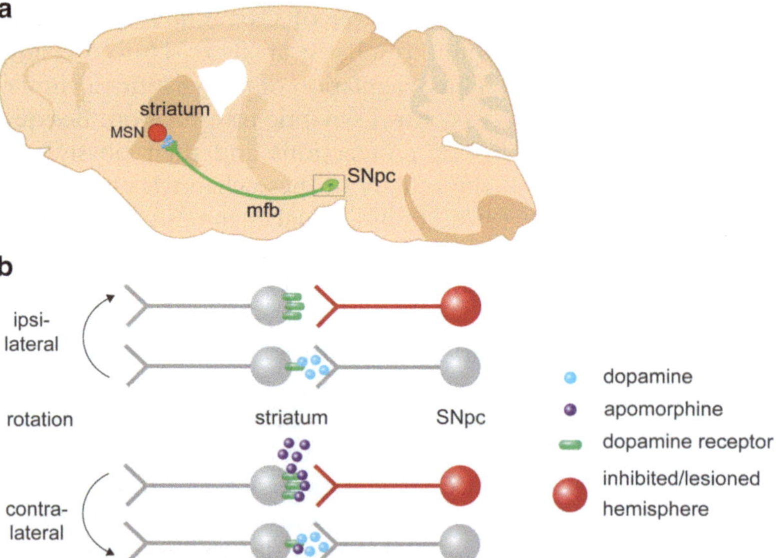

Fig. 6. Simplified representation of the nigrostriatal pathway and induced rotational behavior. (**a**) Sagittal view of the basic nigrostriatal pathway. Dopaminergic neurons of the substantia nigra pars compacta (*SNpc*) project to the striatum via the medial forebrain bundle (*mfb*) and release dopamine to medium spiny neurons (*MSN*). This input is relayed via several circuits to the neocortex, finally regulating important physical conditions, including motor coordination. (**b**) Lesioning or inhibition of dopaminergic neurons from only one hemisphere leads to an imbalance in dopaminergic signaling. Higher activity on the intact side usually results in an ipsilateral rotational phenotype of the animal. If apomorphine, a dopamine receptor agonist, is administered, the imbalance is changed to the inhibited/lesioned side, due to increased receptor affinity and number of receptors, as a physiological reaction to the missing innervation. This results in a change of the rotational behavior to contralateral circling (From Auer (26)).

Likewise, C57/Bl6 mice, which were stereotactically injected in SNpc with lentiviruses encoding for both t-toxins (MVIIA-PE and AgaIVA-VG), displayed a robust rotational phenotype and reversal upon apomorphine administration. These findings strongly suggest an imbalance in the motor coordination of t-toxin injected mice, resulting from the inhibition of the dopaminergic nigrostriatal pathway by the action of both t-toxins (13).

This first proof-of-function of the validity of using virally encoded t-toxins in vivo in the mouse demonstrates the general applicability of the t-toxin strategy as a straightforward method that can be used to block $Ca_v2.1$ and $Ca_v2.2$ calcium currents, resulting in cell-specific and cell-autonomous silencing of neurotransmission. These data also suggest that both t-toxins could be broadly applied for long-term inhibition of $Ca_v2.1$ and $Ca_v2.2$ channels individually or simultaneously, to allow the characterization of the channel contribution to physiological functions and circuit analyses in a wide variety of species. t-Toxin constructs used for these studies, including controls and Cre-dependent and DOX-inducible variants can be obtained from www.addgene.org.

5.3. Selective Manipulation of Sodium Voltage-Gated Currents in Drosophila and in Mice with t-Toxins

Given that initiation and propagation of electrical signals in excitable tissues depend on voltage-gated sodium channels (VGSC, Na_v), it is not surprising that many venom peptides target VGSC. Venom peptide toxins bind to different receptor sites on the channel protein. Few of them physically block the pore and prevent sodium conductance, while a great majority of toxins change channel gating by voltage-sensor trapping through binding to extracellular receptor sites (65). For example, μO-conotoxins partially block ionic influx by binding close to the pore of sodium channel types expressed in the heart (Na_v1.5), muscle (Na_v1.4), and peripheral nociceptive neurons (Na_v1.8) (66), while δ-atracotoxins and β-scorpion toxins inhibit inactivation of activated channels inducing tetanus-like bursts of action potentials followed by plateau potentials resulting in neuronal transmission block and paralysis of the prey. Circuit analyses have been done using tethered sodium toxins with these two opposed types of activities (Table 1). For instance, δ-atracotoxin Hv1a has been used in its tethered form to alter the rhythmicity of circadian neurons in Drosophila with the GAL4-UAS transgenic system (20), while our study employed the t-MrVIA μO-conotoxin and mouse BAC transgenesis to target nociceptive neurons and interfere with pain perception (13, 14). In both cases, the tethered toxin specifically inhibited sodium voltage-gated currents within the genetically targeted neuronal population. Functional analyses of mouse nociceptive neurons expressing t-MrVIA revealed that the t-toxin acts at the membrane, as it can be released by enzymatic cleavage of the GPI anchor (14). Importantly, the studies in transgenic mice indicated that t-MrVIA produced preferential inactivation of Na_v1.8 channels without compensation by Na_v1.7, as it occurs in Na_v1.8 knockout mice (14). This is interesting, since this channel is a target for the treatment of pain, and its relatively depolarized activation voltage dependence may allow it to continue to function when nociceptive neurons are depolarized in the cold (67). Thus, these studies using t-toxins, taken together with previous studies on cold-sensitive neurons (68, 69), indicate that Na_v1.8 is the likely channel to encode for cold perception. As research on venom peptide toxins and synthetic peptide ligands progresses, it would be interesting to identify and test other antagonists and inactivation blockers of VGSC channels expressed in central neurons for neurocircuitry studies.

6. Further Applications and Developments of t-Toxins

6.1. Neurocircuitry

Four extensions of the present studies are of immediate interest for studies of neurocircuitry and cell networks. First, although reversible expression of the tethered toxins can be achieved using established methods (70), as we have shown with the DOX-inducible

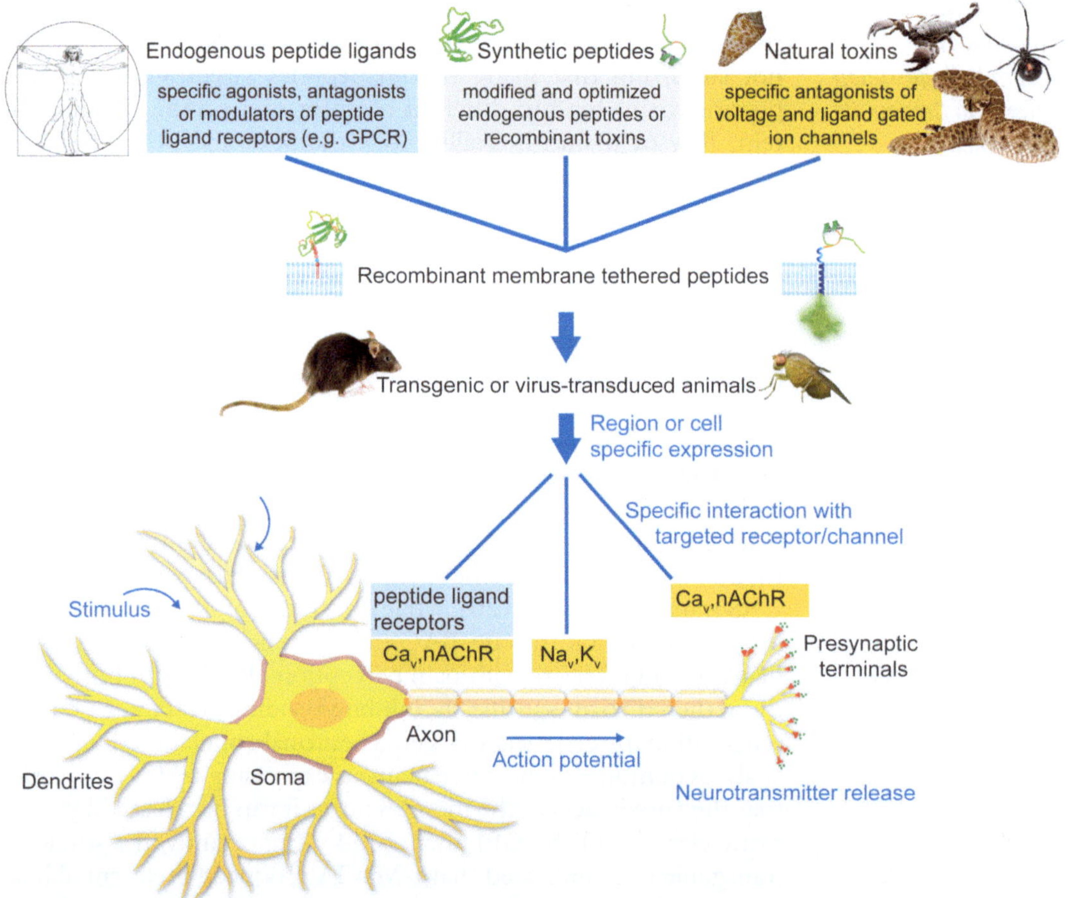

Fig. 7. Applications of the tethered-toxin/tethered-peptide strategy. Endogenous peptide ligands, natural toxins, and synthetic, modified versions of ligands or toxins can be integrated into recombinant membrane-attached fusion constructs and applied in vitro in transfected or transduced cells in cell culture, or in vivo in transgenic or virus-transduced animals. The t-peptide retains the specificity of the toxin/peptide ligand, allowing controlled manipulation of distinct subtypes of ion channels and receptors in a given neuronal circuit without affecting other channels/receptors in the cell (From Holford et al. (16)).

t-toxins, development of strategies for the rapid regulation of these activities for use in short-term experiments (i.e., light-inducible methods) remains an important goal. Second, the cell-autonomous modulatory action of t-toxins and their selectivity for specific ion channels and receptors could be further exploited by directing t-peptide molecules to subcellular compartments within the neuron (Fig. 7). For instance, t-toxins could be directed to the axon initial segment where sodium channels are concentrated (71), or to the dendritic compartment (72). Third, the application of this strategy to other toxins, such as conantokins (73) or σ-conotoxins (74), will allow cell-specific manipulation of NMDA receptor activity or serotonin receptors in vivo. Fourth, extension of this genetic approach to other (nontoxin) peptides, including hormones and neuropeptides (Fig. 7), offers interesting opportunities for the analysis of the roles of these ligands in specific cell types. Indeed, our strategy has been used to specifically activate G-protein-coupled receptors (GPCRs) (31, 32). Given the small size of most peptide toxins, we believe that the creation of novel specificities by mutagenesis will extend the use of this approach to receptors and ion channels for which natural toxins have not been identified yet. Thus, we anticipate that tethered toxins and other peptides will become critical instruments for the genetic dissection of CNS cells and circuits.

6.2. Cell-Based Therapies

The tethered-peptide strategy also represents a potential new avenue for the development of genetic therapies for chronic diseases caused by malfunction of ion channels and peptide ligand receptors. For instance, cases of severe chronic pain in humans, resistant to analgesics and opioids, are currently being treated with ω-conotoxin MVIIA (commercialized as Prialt) (75–77). However, the use of this toxin requires the implantation of intrathecal microinfusion pumps, which allow constant administration of the soluble drug to minimize the substantial side effects due to block of $Ca_v2.2$ channels present in the CNS. Genetic targeting of t-MVIIA to nociceptive neurons in transgenic mice has shown that these mice are protected from inflammatory and neuropathic pain (13). Therefore, the t-toxin could be a viable alternative therapy to avoid uncontrolled diffusion of the injected toxin and the necessity for repetitive treatments once safe viral methods for genetic intervention in humans will be implemented. Other disorders which have been traced to mutations in genes encoding ion channels or regulatory proteins, such as channelopathies (78), could also benefit from the use of specific toxins if these could be selectively targeted to the affected neuronal population. For instance, t-toxins producing partial or total block of particular ion channel subtypes could be used in disorders caused by missense mutations that result in channel hyperactivity. Examples of hyperactive disorders include familial hemiplegic migraine type 1 (FHM-1) caused by gain-of-function mutations in P/Q-type ($Ca_v2.1$) calcium channels (79, 80), or different types of epilepsy such as autosomal dominant nocturnal frontal lobe epilepsy (ADNFLE) associated to mutations in nAChRs (81, 82). It could be interesting to use t-toxins to dissect the circuitry of the disease in these or other mouse mutant models of ion channel mutations (Fig. 7).

6.3. Drug Discovery

A major problem in the application of peptide toxins for therapeutic purposes has been the scarcity of obtaining the venom product. To circumvent this, most toxins are synthesized chemically, but this too comes along with significant hurdles, for example, attaining the correct disulfide scaffold with in vitro folding. However, even when the toxin is successfully synthesized, soluble toxins have a limited time of application that makes their use in vivo problematic.

The t-toxin strategy overcomes these limitations with the ability to recombinantly synthesize the toxins or peptide ligands in the cell itself and coexpress it with the molecular target (receptor or channel) to be screened. This coexpression cell system theoretically also allows the introduction of point mutations to interconvert tethered agonists into antagonists. Several recent reports use the t-peptide technology to characterize point mutants of peptide hormones against class B1 GPCRs (12, 32). These features make the t-toxin and t-peptide genetic approaches to alternative strategies for drug discovery and development of targeted therapeutics.

7. Concluding Remarks

The physiologies of all cells depend upon the activities of ion channels and receptors expressed at the cell surface. With an ever-increasing number of isolated and characterized peptide toxins, the tethered-toxin strategy opens the door to experimental studies that will greatly expand our knowledge of molecules, cells, and circuits. The possibility to combine t-toxins with transgenic and viral methodologies, as well as inducible and reversible strategies, provides a new genetic-based method to map functional neuronal circuits and synaptic changes in vivo and therefore could greatly contribute to create integrated functional connectivity maps of the mammalian nervous system.

References

1. Inchauspe, C. G., Martini, F. J., Forsythe, I. D., and Uchitel, O. D. (2004) Functional Compensation of P/Q by N-Type Channels Blocks Short-Term Plasticity at the Calyx of Held Presynaptic Terminal, *J Neurosci* 24: 10379–10383.
2. Akopian, A. N., Souslova, V., England, S., Okuse, K., Ogata, N., Ure, J., Smith, A., Kerr, B. J., McMahon, S. B., Boyce, S., Hill, R., Stanfa, L. C., Dickenson, A. H., and Wood, J. N. (1999) The tetrodotoxin-resistant sodium channel SNS has a specialized function in pain pathways, *Nat Neurosci* 2: 541–548.
3. Drago, J., McColl, C. D., Horne, M. K., Finkelstein, D. I., and Ross, S. A. (2003) Neuronal nicotinic receptors: insights gained from gene knockout and knockin mutant mice, *Cell Mol Life Sci* 60: 1267–1280.
4. Gong, S., Yang, X. W., Li, C., and Heintz, N. (2002) Highly efficient modification of bacterial artificial chromosomes (BACs) using novel shuttle vectors containing the R6Kgamma origin of replication, *Genome Res* 12: 1992–1998.
5. Tolu, S., Avale, M. E., Nakatani, H., Pons, S., Parnaudeau, S., Tronche, F., Vogt, A., Monyer, H., Vogel, R., de Chaumont, F., Olivo-Marin, J. C., Changeux, J. P., and Maskos, U. (2010) A versatile system for the neuronal subtype specific expression of lentiviral vectors, *FASEB J* 24: 723–730.
6. Hatten, M. E., and Heintz, N. (2005) Large-scale genomic approaches to brain development and circuitry, *Annu Rev Neurosci* 28: 89–108.
7. Luo, L., Callaway, E. M., and Svoboda, K. (2008) Genetic dissection of neural circuits, *Neuron* 57: 634–660.
8. Terlau, H., and Olivera, B. M. (2004) Conus Venoms: A Rich Source of Novel Ion Channel-Targeted Peptides, *Physiol Rev* 84: 41–68.
9. Twede, V. D., Miljanich, G., Olivera, B. M., and Bulaj, G. (2009) Neuroprotective and

cardioprotective conopeptides: an emerging class of drug leads, *Curr Opin Drug Di De* 12: 231–239.

10. Phui Yee, J. S., Nanling, G., Afifiyan, F., Donghui, M., Siew Lay, P., Armugam, A., and Jeyaseelan, K. (2004) Snake postsynaptic neurotoxins: gene structure, phylogeny and applications in research and therapy, *Biochimie* 86: 137–149.
11. Dutertre, S., and Lewis, R. J. (2010) Use of venom peptides to probe ion channel structure and function, *J Biol Chem* 285: 13315–13320.
12. Ibañez-Tallon, I., Wen, H., Miwa, J. M., Xing, J., Tekinay, A. B., Ono, F., Brehm, P., and Heintz, N. (2004) Tethering naturally occurring peptide toxins for cell-autonomous modulation of ion channels and receptors in vivo, *Neuron* 43: 305–311.
13. Auer, S., Sturzebecher, A. S., Juttner, R., Santos-Torres, J., Hanack, C., Frahm, S., Liehl, B., and Ibanez-Tallon, I. (2010) Silencing neurotransmission with membrane-tethered toxins, *Nat Methods* 7: 229–236.
14. Stürzebecher, A. S., Hu, J., Smith, E. S., Frahm, S., Santos-Torres, J., Kampfrath, B., Auer, S., Lewin, G. R., and Ibanez-Tallon, I. (2010) An in vivo tethered toxin approach for the cell-autonomous inactivation of voltage-gated sodium channel currents in nociceptors, *J Physiol* 588: 1695–1707.
15. Miwa, J. M., Ibanez-Tallon, I., Crabtree, G. W., Sánchez, R., Sali, A., Role, L. W., and Heintz, N. (1999) lynx1, an endogenous toxin-like modulator of nicotinic acetylcholine receptors in the mammalian CNS, *Neuron* 23: 105–114.
16. Holford, M., Auer, S., Laqua, M., and Ibañez-Tallon, I. (2009) Manipulating neuronal circuits with endogenous and recombinant cell-surface tethered modulators, *Front Mol Neurosci* 2: 21.
17. Ibañez-Tallon, I., Miwa, J. M., Wang, H. L., Adams, N. C., Crabtree, G. W., Sine, S. M., and Heintz, N. (2002) Novel modulation of neuronal nicotinic acetylcholine receptors by association with the endogenous prototoxin lynx1, *Neuron* 33: 893–903.
18. Miwa, J., Stevens, T., King, S., Caldarone, B., Ibaneztallon, I., Xiao, C., Fitzsimonds, R., Pavlides, C., Lester, H., and Picciotto, M. (2006) The Prototoxin lynx1 Acts on Nicotinic Acetylcholine Receptors to Balance Neuronal Activity and Survival In Vivo, *Neuron* 51: 587–600.
19. Tekinay, A. B., Nong, Y., Miwa, J. M., Lieberam, I., Ibanez-Tallon, I., Greengard, P., and Heintz, N. (2009) A role for LYNX2 in anxiety-related behavior, *Proceedings of the National Academy of Sciences* 106: 4477–4482.
20. Wu, Y., Cao, G., Pavlicek, B., Luo, X., and Nitabach, M. N. (2008) Phase coupling of a circadian neuropeptide with rest/activity rhythms detected using a membrane-tethered spider toxin, *PLoS Biol* 6: e273.
21. Hruska, M., Keefe, J., Wert, D., Tekinay, A. B., Hulce, J., Ibanez-Tallon, I., and Nishi, R. (2009) Prostate stem cell antigen is an endogenous lynx1-like prototoxin that antagonizes alpha7 containing nicotinic receptors and prevents programmed cell death of parasympathetic neurons, *J. Neurosci.*
22. Olivera, B. M., McIntosh, J. M., Cruz, L. J., Luque, F. A., and Gray, W. R. (1984) Purification and sequence of a presynaptic peptide toxin from Conus geographus venom, *Biochemistry* 23: 5087–5090.
23. Ferber, M., Sporning, A., Jeserich, G., DeLaCruz, R., Watkins, M., Olivera, B. M., and Terlau, H. (2003) A Novel Conus Peptide Ligand for K+Channels, *J Biol Chem* 278: 2177–2183.
24. Shon, K. J., Olivera, B. M., Watkins, M., Jacobsen, R. B., Gray, W. R., Floresca, C. Z., Cruz, L. J., Hillyard, D. R., Brink, A., Terlau, H., and Yoshikami, D. (1998) mu-Conotoxin PIIIA, a new peptide for discriminating among tetrodotoxin-sensitive Na channel subtypes, *The Journal of Neuroscience: The Official Journal of the Society for Neuroscience* 18: 4473–4481.
25. Nicke, A., Loughnan, M. L., Millard, E. L., Alewood, P. F., Adams, D. J., Daly, N. L., Craik, D. J., and Lewis, R. J. (2003) Isolation, structure, and activity of GID, a novel alpha 4/7-conotoxin with an extended N-terminal sequence, *J Biol Chem* 278: 3137–3144.
26. Auer, S. (2010) Development of new membrane-tethered toxins as genetic tools for in vitro and in vivo silencing of ion channels, *Dissertation, Freie Universität Berlin,* http://www.diss.fu-berlin.de/diss/receive/FUDISS_thesis_000000017398. Accessed 07 February 2011.
27. Millard, E. L., Nevin, S. T., Loughnan, M. L., Nicke, A., Clark, R. J., Alewood, P. F., Lewis, R. J., Adams, D. J., Craik, D. J., and Daly, N. L. (2009) Inhibition of neuronal nicotinic acetylcholine receptor subtypes by alpha-Conotoxin GID and analogues, *J Biol Chem* 284: 4944–4951.
28. Terlau, H., Stocker, M., Shon, K. J., McIntosh, J. M., and Olivera, B. M. (1996) MicroO-conotoxin MrVIA inhibits mammalian sodium channels, but not through site I, *J Neurophysiol* 76: 1423–1429.
29. Bulaj, G., Zhang, M.-M., Green, B. R., Fiedler, B., Layer, R. T., Wei, S., Nielsen, J. S., Low, S. J., Klein, B. D., Wagstaff, J. D., Chicoine, L.,

Harty, T. P., Terlau, H., Yoshikami, D., and Olivera, B. M. (2006) Synthetic μO-Conotoxin MrVIB Blocks TTX-Resistant Sodium Channel NaV1.8 and Has a Long-Lasting Analgesic Activity†, *Biochemistry* 45: 7404–7414.

30. Hruska, M., Ibañez-Tallon, I., and Nishi, R. (2007) Cell-autonomous inhibition of alpha 7-containing nicotinic acetylcholine receptors prevents death of parasympathetic neurons during development, *J Neurosci* 27: 11501–11509.

31. Choi, C., Fortin, J. P., McCarthy, E., Oksman, L., Kopin, A. S., and Nitabach, M. N. (2009) Cellular dissection of circadian peptide signals with genetically encoded membrane-tethered ligands, *Curr Biol* 19: 1167–1175.

32. Fortin, J.-P., Zhu, Y., Choi, C., Beinborn, M., Nitabach, M. N., and Kopin, A. S. (2009) Membrane-tethered ligands are effective probes for exploring class B1 G protein-coupled receptor function, *Proc Natl Acad Sci USA* 106: 8049–8054.

33. Paulick, M. G., and Bertozzi, C. R. (2008) The glycosylphosphatidylinositol anchor: a complex membrane-anchoring structure for proteins, *Biochemistry* 47: 6991–7000.

34. Vintersten, K., Testa, G., Naumann, R., Anastassiadis, K., and Stewart, A. F. (2008) Bacterial artificial chromosome transgenesis through pronuclear injection of fertilized mouse oocytes, *Methods Mol Biol* 415: 83–100.

35. Gong, S., and Yang, X. W. (2005) Modification of bacterial artificial chromosomes (BACs) and preparation of intact BAC DNA for generation of transgenic mice, *Curr Protoc Neurosci* Chapter 5: Unit 5 21.

36. Benn, S. C., Costigan, M., Tate, S., Fitzgerald, M., and Woolf, C. J. (2001) Developmental expression of the TTX-resistant voltage-gated sodium channels Nav1.8 (SNS) and Nav1.9 (SNS2) in primary sensory neurons, *J Neurosci* 21: 6077–6085.

37. Stirling, L. C., Forlani, G., Baker, M. D., Wood, J. N., Matthews, E. A., Dickenson, A. H., and Nassar, M. A. (2005) Nociceptor-specific gene deletion using heterozygous NaV1.8-Cre recombinase mice, *Pain* 113: 27–36.

38. Cockrell, A. S., and Kafri, T. (2007) Gene delivery by lentivirus vectors, *Mol Biotechnol* 36: 184–204.

39. Wong, L. F., Goodhead, L., Prat, C., Mitrophanous, K. A., Kingsman, S. M., and Mazarakis, N. D. (2006) Lentivirus-mediated gene transfer to the central nervous system: therapeutic and research applications, *Hum Gene Ther* 17: 1–9.

40. Delenda, C. (2004) Lentiviral vectors: optimization of packaging, transduction and gene expression, *J Gene Med* 6 S125–138.

41. Naldini, L., Blomer, U., Gallay, P., Ory, D., Mulligan, R., Gage, F. H., Verma, I. M., and Trono, D. (1996) In vivo gene delivery and stable transduction of nondividing cells by a lentiviral vector, *Science* 272: 263–267.

42. Zufferey, R., Donello, J. E., Trono, D., and Hope, T. J. (1999) Woodchuck hepatitis virus posttranscriptional regulatory element enhances expression of transgenes delivered by retroviral vectors, *J Virol* 73: 2886–2892.

43. Akkina, R. K., Walton, R. M., Chen, M. L., Li, Q. X., Planelles, V., and Chen, I. S. (1996) High-efficiency gene transfer into CD34+ cells with a human immunodeficiency virus type 1-based retroviral vector pseudotyped with vesicular stomatitis virus envelope glycoprotein G, *J Virol* 70: 2581–2585.

44. Aiken, C. (1997) Pseudotyping human immunodeficiency virus type 1 (HIV-1) by the glycoprotein of vesicular stomatitis virus targets HIV-1 entry to an endocytic pathway and suppresses both the requirement for Nef and the sensitivity to cyclosporin A, *J Virol* 71: 5871–5877.

45. Lois, C., Hong, E. J., Pease, S., Brown, E. J., and Baltimore, D. (2002) Germline transmission and tissue-specific expression of transgenes delivered by lentiviral vectors, *Science* 295: 868–872.

46. Szulc, J., Wiznerowicz, M., Sauvain, M.-O., Trono, D., and Aebischer, P. (2006) A versatile tool for conditional gene expression and knockdown, *Nat Meth* 3: 109–116.

47. Nagy, A. (2000) Cre recombinase: the universal reagent for genome tailoring, *Genesis* 26: 99–109.

48. Hamilton, D. L., and Abremski, K. (1984) Site-specific recombination by the bacteriophage P1 lox-Cre system. Cre-mediated synapsis of two lox sites, *J Mol Biol* 178: 481–486.

49. Wang, X. (2009) Cre transgenic mouse lines, *Methods Mol Biol* 561: 265–273.

50. Changeux, J. P., Kasai, M., and Lee, C. Y. (1970) Use of a snake venom toxin to characterize the cholinergic receptor protein, *Proc Natl Acad Sci USA* 67: 1241–1247.

51. Lewis, R. J. (2009) Conotoxins: molecular and therapeutic targets, *Prog Mol Subcell Biol* 46: 45–65.

52. Tsetlin, V. I., and Hucho, F. (2004) Snake and snail toxins acting on nicotinic acetylcholine receptors: fundamental aspects and medical applications, *FEBS Lett* 557: 9–13.

53. Tsetlin, V. I., Kasheverov, I. E., Zhmak, M. N., Utkin, Y. N., Vulfius, C. A., Smit, A. B., and Bertrand, D. (2006) Alpha-conotoxin analogs with enhanced affinity for nicotinic receptors and acetylcholine-binding proteins, *J Mol Neurosci* 30: 77–78.

54. Auer, S., and Ibañez-Tallon, I. (2010) "The King is dead": Checkmating ion channels with tethered toxins, *Toxicon* 56: 1293–1298.
55. Chua, J. J., Kindler, S., Boyken, J., and Jahn, R. (2010) The architecture of an excitatory synapse, *J Cell Sci* 123: 819–823.
56. Catterall, W. A., and Few, A. P. (2008) Calcium channel regulation and presynaptic plasticity, *Neuron* 59: 882–901.
57. Zucker, R. S., and Regehr, W. G. (2002) Short-term synaptic plasticity, *Annu Rev Physiol* 64: 355–405.
58. Mink, J. W. (1996) The Basal Ganglia: Focused selection and inhibition of competing motor programs, *Prog Neurobiol* 50: 381–425.
59. Groenewegen, H. J. (2003) The basal ganglia and motor control, *Neural Plast* 10: 107–120.
60. Takakusaki, K., Saitoh, K., Harada, H., and Kashiwayanagi, M. (2004) Role of basal ganglia-brainstem pathways in the control of motor behaviors, *Neurosci Res* 50: 137–151.
61. Ungerstedt, U., and Arbuthnott, G. W. (1970) Quantitative recording of rotational behavior in rats after 6-hydroxy-dopamine lesions of the nigrostriatal dopamine system, *Brain Res* 24: 485–493.
62. Schwarting, R. K. W., and Huston, J. P. (1996) The unilateral 6-hydroxydopamine lesion model in behavioral brain research. Analysis of functional deficits, recovery and treatments, *Prog Neurobiol* 50: 275–331.
63. Da Cunha, C., Wietzikoski, E. C., Ferro, M. M., Martinez, G. R., Vital, M. A. B. F., Hipólide, D., Tufik, S., and Canteras, N. S. (2008) Hemiparkinsonian rats rotate toward the side with the weaker dopaminergic neurotransmission, *Behav Brain Res* 189: 364–372.
64. Deumens, R., Blokland, A., and Prickaerts, J. (2002) Modeling Parkinson's disease in rats: an evaluation of 6-OHDA lesions of the nigrostriatal pathway, *Exp Neurol* 175: 303–317.
65. Catterall, W. A., Cestele, S., Yarov-Yarovoy, V., Yu, F. H., Konoki, K., and Scheuer, T. (2007) Voltage-gated ion channels and gating modifier toxins, *Toxicon* 49: 124–141.
66. Leipold, E., DeBie, H., Zorn, S., Borges, A., Olivera, B. M., Terlau, H., and Heinemann, S. H. (2007) muO conotoxins inhibit NaV channels by interfering with their voltage sensors in domain-2, *Channels (Austin)* 1: 253–262.
67. Baker, M. D., and Wen, H. (2010) Tethered-toxin debut gets cold reception, *J Physiol* 588: 1663.
68. Carr, R. W., Pianova, S., and Brock, J. A. (2002) The effects of polarizing current on nerve terminal impulses recorded from polymodal and cold receptors in the guinea-pig cornea, *J Gen Physiol* 120: 395–405.
69. Zimmermann, K., Leffler, A., Babes, A., Cendan, C. M., Carr, R. W., Kobayashi, J., Nau, C., Wood, J. N., and Reeh, P. W. (2007) Sensory neuron sodium channel Nav1.8 is essential for pain at low temperatures, *Nature* 447: 855–858.
70. Mansuy, I. M., and Bujard, H. (2000) Tetracycline-regulated gene expression in the brain, *Curr Opin Neurobiol* 10: 593–596.
71. Garrido, J. J., Giraud, P., Carlier, E., Fernandes, F., Moussif, A., Fache, M.-P., Debanne, D., and Dargent, B. (2003) A Targeting Motif Involved in Sodium Channel Clustering at the Axonal Initial Segment, *Science* 300: 2091–2094.
72. Lewis, T. L., Mao, T., Svoboda, K., and Arnold, D. B. (2009) Myosin-dependent targeting of transmembrane proteins to neuronal dendrites, *Nat Neurosci* 12: 568–576.
73. Layer, R. T., Wagstaff, J. D., and White, H. S. (2004) Conantokins: peptide antagonists of NMDA receptors, *Curr Med Chem* 11: 3073–3084.
74. England, L. J., Imperial, J., Jacobsen, R., Craig, A. G., Gulyas, J., Akhtar, M., Rivier, J., Julius, D., and Olivera, B. M. (1998) Inactivation of a serotonin-gated ion channel by a polypeptide toxin from marine snails, *Science* 281: 575–578.
75. Zamponi, G. W., Lewis, R. J., Todorovic, S. M., Arneric, S. P., and Snutch, T. P. (2009) Role of voltage-gated calcium channels in ascending pain pathways, *Brain Res Rev* 60: 84–89.
76. Miljanich, G. P. (2004) Ziconotide: neuronal calcium channel blocker for treating severe chronic pain, *Curr Med Chem* 11: 3029–3040.
77. Staats, P. S., Yearwood, T., Charapata, S. G., Presley, R. W., Wallace, M. S., Byas-Smith, M., Fisher, R., Bryce, D. A., Mangieri, E. A., Luther, R. R., Mayo, M., McGuire, D., and Ellis, D. (2004) Intrathecal ziconotide in the treatment of refractory pain in patients with cancer or AIDS: a randomized controlled trial, *J Am Med Assoc* 291: 63–70.
78. George, A. L. (2005) Inherited disorders of voltage-gated sodium channels, *J Clin Invest* 115: 1990–1999.
79. Ophoff, R. A., Terwindt, G. M., Vergouwe, M. N., van Eijk, R., Oefner, P. J., Hoffman, S. M., Lamerdin, J. E., Mohrenweiser, H. W., Bulman, D. E., Ferrari, M., Haan, J., Lindhout, D., van Ommen, G. J., Hofker, M. H., Ferrari, M. D., and Frants, R. R. (1996) Familial hemiplegic migraine and episodic ataxia type-2 are caused by mutations in the Ca2+ channel gene CACNL1A4, *Cell* 87: 543–552.

80. Tottene, A., Conti, R., Fabbro, A., Vecchia, D., Shapovalova, M., Santello, M., van den Maagdenberg, A. M. J. M., Ferrari, M. D., and Pietrobon, D. (2009) Enhanced excitatory transmission at cortical synapses as the basis for facilitated spreading depression in Ca(v)2.1 knockin migraine mice, *Neuron* 61: 762–773.

81. Klaassen, A., Glykys, J., Maguire, J., Labarca, C., Mody, I., and Boulter, J. (2006) Seizures and enhanced cortical GABAergic inhibition in two mouse models of human autosomal dominant nocturnal frontal lobe epilepsy, *Proc Natl Acad Sci USA* 103: 19152–19157.

82. Steinlein, O. K., Mulley, J. C., Propping, P., Wallace, R. H., Phillips, H. A., Sutherland, G. R., Scheffer, I. E., and Berkovic, S. F. (1995) A missense mutation in the neuronal nicotinic acetylcholine receptor alpha 4 subunit is associated with autosomal dominant nocturnal frontal lobe epilepsy, *Nat Genet* 11: 201–203.

Chapter 9

Genetic Methods for Anatomical Analysis of Neuronal Circuits

Leah R. DeBlander, Aldis P. Weible, and Clifford G. Kentros

Abstract

An integral part of neural circuit analysis involves understanding cellular-level connectivity. With recent technical advances, this capability has finally been realized. In this chapter, we will discuss various methods of targeting specific cell types in transgenic mice, with an emphasis on the minimal promoter transgenic design. Minimal promoter transgenics is a "modular" transgenic approach in which a "driver" (i.e., a Cre or transactivator) line is crossed to different "payload" (i.e., floxed or tetO) lines to obtain region- and/or cell-specific patterns of transgenic expression. We will describe how this technique, when combined with the recombinant rabies viral tracer developed by Wickersham and colleagues (Nat Methods 4:47–49, 2007), enables the determination of monosynaptic inputs of specific cell types in the brain.

Key words: Transgenic, Rabies, Monosynaptic connectivity, tTA, tetO, Neural circuit

1. Introduction

The mammalian brain is arguably the most complex structure the planet has ever seen. The sheer number of neurons (~10^{13} in the human brain) of innumerable types, each with thousands of interconnections, truly boggles the mind. However, neither lesions nor even extracellular unit recordings are good at distinguishing cell type, so the vast majority of the studies that provide the basis for our current understanding of brain function in a behaving animal largely ignore these innumerable cell types. However, the set of technologies collectively known as "mouse molecular genetics" has the potential to revolutionize systems neuroscience, not so much because of their molecular specificity, but because of their *anatomical* specificity: they allow us to address the individual elements of neural circuits, specific neuronal cell types, rather than having to content ourselves with brain regions.

Alexei Morozov (ed.), *Controlled Genetic Manipulations*, Neuromethods, vol. 65,
DOI 10.1007/978-1-61779-533-6_9, © Springer Science+Business Media, LLC 2012

Part of the challenge is deciding how many different kinds of neurons there are. Currently, neurons are often separately defined by their electrophysiological, morphological, and anatomical properties (1) and their gene expression profiles (2). In the past, it has been difficult to bring these levels of characterization together to form a more complete definition of neuronal cell type. The beauty of the modern "modular" genetically modified mice (such as animals that express recombinases, see Chaps. 1 and 2) is that they allow for a diverse set of manipulations to be performed on the same group(s) of neurons in the mammalian brain. In this way, the wide range of current and future genetic tools can be deployed in a genetically predetermined set of neurons to reveal not only their identity (and that of their synaptic partners) but also their role in the function of an awake, behaving animal. The goals of this chapter are first to outline the tools in general that are available for studying neural circuits in mice, and then expand specifically upon using transgenic targeting of recombinant rabies virus as an anatomical tracing technique.

Electrical engineers analyze electrical circuits via a technique called nodal analysis: basically, one manipulates one node (e.g., by putting in a variable resistor) and records from other points (nodes) in the circuit. Much of the excitement generated by the development of techniques for manipulating (e.g., optogenetics) stems from the desire to do something very similar in awake, behaving brains: manipulating particular neurons in a circuit while recording from others (3). However, providing the proper context for such experiments requires identifying neuronal cell types and knowing the connectivity between them (in other words, knowing the circuit diagram). The "modular" transgenic mice are perhaps best suited toward this end, as one can use the same "driver" (i.e., the Cre or transactivator) line to express a variety of transgenes in the same subsets of neurons by crossing it to different "payload" lines (i.e., the floxed or tetO line). In this way, a variety of analyses can be performed on the same cell type, such as determining the anatomical connectivity of a transgenically targeted cell type and then determining its functional connectivity. In this chapter, we set out the basic strategy our lab has taken toward this end and then zero in on a particular method: the determination of the monosynaptic inputs of particular neurons via transgenic targeting of recombinant rabies virus.

2. Transgenic Mice for the Analysis of Neural Circuits

The first step toward characterizing a neural circuit is the ability to target neuronal cell types. Using minimal promoter, BAC (bacterial artificial chromosome) or knock-in gene transfer techniques, exogenous genes can be expressed in all or some subset of a specific cell type defined by its gene expression profile. Minimal promoter transgenic constructs, historically referred to simply as "transgenics," are made using restriction enzymes or PCR to isolate a small (~7–12 kb of 5′ sequence), highly conserved region of the promoter to drive cell type specific expression of a gene. BAC transgenic constructs are typically made by recombining a larger promoter region (~30–100 kb of 5′ sequence) with the driver gene in special strains of *E. coli*; these promoter regions are stored in BAC vectors (vectors with carrying capacities for large DNA fragments) and are part of larger genomic libraries. Both minimal promoter and BAC transgenics are injected into the nucleus of a fertilized egg at the single-cell stage and rely on random integration events for incorporation of the exogenous DNA. In contrast, knock-in transgenics use targeted recombination to incorporate exogenous genes near the promoter region of interest in an embryonic stem cell.

The two exogenous genes commonly expressed in this manner are Cre recombinase and the tetO transactivator transcription factor (tTA). These genes are part of the Cre/Lox (Chaps. 1 and 2) and tTA/tetO inducible systems and are the most widely used systems in mouse transgenics. Cre mediates rearrangement of DNA flanked by loxP sites, introducing excision or inversion events which can be used to control gene expression. The tTA mediates transcription of any gene preceded by the tetO promoter. While both systems are inducible, the tTA/tetO system holds the advantage because it can be turned on and off in a reversible manner. Induction of Cre on the other hand results in irreversible DNA rearrangement. Regardless, there are reasons why use of the Cre/Lox system may be preferable. First, a large number of cell type specific lines have been designed using this system, many of which are available from GENSAT and the Allen Brain Institute (see Sect. 4). Additionally, many effector genes employing viral delivery systems have been designed to take advantage of these lines.

Viral vectors have been utilized as effective gene delivery systems, and they offer several advantages both as an alternative to and in combination with transgenic technologies. The ease of simply injecting viral vectors into a region of interest to deliver an effector gene offers an obvious advantage over the costly and time-consuming process of designing transgenic lines to deliver the very same gene (for more information on the types of viral delivery systems available, see Chap. 2a, 2b). Additionally, complementary viral vectors are often preferred over transgenics by investigators working with cell type specific Cre lines because the time point of Cre activity is determined by the time of viral injection instead of developmental activation of the promoter driving Cre expression.

Many effector genes were designed to manipulate the activity of neurons. One of the first neuronal silencers used in mammals that has been delivered to neurons both by viral vector and transgenically was the allatostatin G-protein-coupled receptor (4). Because the allatostatin receptor is found in insects but not mammals, infusion of the allatostatin peptide results in the silencing of only those neurons expressing the receptor. Allatostatin can be a robust silencer, but its slow time course of activation and deactivation makes it less than ideal for certain experimental manipulations (5). Other peptide-induced systems like the ivermectin receptor silencer which, much like allatostatin, also prevents neuronal depolarization and the Vamp/Mist system which prevents neurotransmission (6) were developed around the same time and will likely prove to have similar limitations. Recently developed systems incorporating light-activated ion channels appear to have overcome issues of time course (Chap. 3e).

The problem with viral technologies, peptide-induced silencers, and light-activated ion channel actuators is that they limit electrophysiological manipulations to the cannulation site. A different kind of tool is needed to easily target large, multiple, or irregularly shaped structures without invasive surgeries or large volume cannulations. In order to circumvent these limitations, a process called molecular evolution was used by the Roth Laboratory to develop modified muscarinic receptors that respond to a biologically inert molecule with the ability to cross the blood–brain barrier. Several different receptor types were developed using this method, including ones that activate or silence the neurons expressing it. Both receptor types respond to the biologically inert molecule CNO (7). With the addition of these new transgenic tools, it will be much easier to perform specific lesioning experiments in awake, behaving animals, providing a cell type or structure specific behavioral correlate. Both cell type specific and structure specific activity modulators afford differing degrees of resolution for nodal analysis, and both can provide different kinds of information about functional connectivity within a circuit.

In summary, transgenics have the potential to greatly expand our capacity to analyze the circuitry of the mammalian brain by providing tools that act with the granularity of the system. First, we must identify the specific cell types in individual neuroanatomical regions that express the transgene, through analyses of cytoarchitecture and gene expression. These data then inform subsequent experiments investigating their functional connectivity. Because neurons communicate with one another through electrical activity, the functional connectivity of a cell can be assessed by driving or inhibiting its electrical activity and recording the subsequent change at other nodes within the circuit, essentially performing the nodal analysis described above. However, one cannot understand such circuit manipulations without an idea of the circuit diagram, so circuit analysis requires a detailed anatomical analysis of the transgenic cell type, including its anatomical connectivity.

3. Neuron-Specific Connectivity via Transgenic Targeting of Recombinant Rabies Virus

Anatomical analysis of neural circuits requires not only the determination of cell type, which is difficult enough, but determining the patterns of connectivity between different cells, a much harder proposition. Although their many successes have provided the basis for our current understanding of the connectivity of the CNS, traditional techniques for the analysis of neural connectivity all share the same key flaw; they do not distinguish between cell types. Typically, with these tracers, a bolus is taken up by all the cells near the injection site, revealing much about connectivity between brain regions but little about local connections. In other words, though these traditional techniques give us a rough idea of the transfer of information through the brain, they leave us almost entirely ignorant of how information is processed at each way station. Transgenic targeting of anatomical tracers has the specificity to provide just this kind of connectivity information.

Historically, a variety of molecules have been used for anatomical tracing studies in the mammalian CNS. Some tracers such as wheat germ agglutinin (WGA) and the herpesvirus move in both anterograde and retrograde directions (8–10). Other tracers like the rabies virus move in an exclusively retrograde direction (8), while certain types of biotinylated dextran amine (BDA) move only in the anterograde direction (11). All of these tracing methods have drawbacks. Viral tracers and lectins such as WGA move multiple synapses, often in an activity-dependent manner. This can be partially mitigated by controlling the timing postinjection, but it remains easy to confuse second, third, and higher-order neurons, especially when strong connections are overrepresented. Herpesvirus and WGA move both anterograde and retrograde, confusing the direction of information flow. None of the tracers are cell specific unless injected into a single neuron, and, with the exception of the BDA, they do not fill a cell sufficiently to allow morphological analysis. Additionally, many viral tracers (e.g., (12)) kill the infected cells after only a few days of infection. There are also a few examples of transgenically targeted tracing studies, all having the advantage of more or less cell-specific tracer expression. One approach is to express protein tracers with transgenic mice, such as the case of DsRed-coupled wheat germ agglutinin (13) or the heavy chain of tetanus toxin (14). An alternative approach is to use transgenic mice to target recombinase-dependent viruses. However, while these methods represent a great improvement on traditional techniques, the tracer also continues to jump synapses, and so do not resolve the confusion of multiple synaptic steps.

A new method that has vastly improved upon previous anatomical tracing methods is the transgenically targeted rabies viral (RV) tracer. In this technique, a recombinant RV that has its

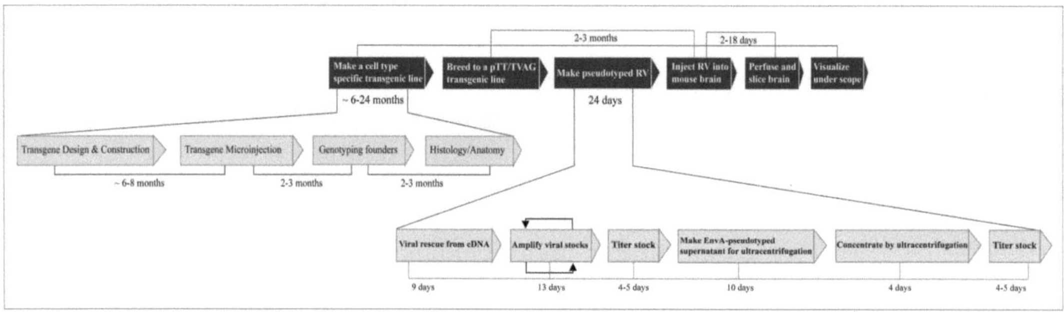

Fig. 1. Schematics and timelines for the steps involved in using pseudotyped rabies virus to investigate anatomical connectivity in cell type specific transgenic mouse lines.

glycoprotein gene replaced by a fluorophore is pseudotyped with the avian viral coat protein, EnvA. The resulting pseudotyped rabies can only infect cells expressing the avian receptor, TVA. When the TVA receptor and rabies glycoprotein are supplied *in trans* to a mammalian neuron, that neuron and its presynaptic partners are fluorescently labeled (15). Recent techniques allow single-cell transfection in vivo of the genes encoding the rabies glycoprotein and TVA receptor, allowing for the characterization of unicellular monosynaptic retrograde connectivity (16). These genes can also be expressed in a transgenic mouse line using the tTA/tetO system in order to achieve cell type or structure specific anatomical analysis. Unlike previous anatomical studies using wild-type rabies, providing the rabies glycoprotein in trans limits the spread of the pseudotyped rabies to the transgenic neurons' monosynaptic partners (17). Subsequent results are then far easier to interpret. In this way, the combination of virus and transgenics enables the anatomical properties of a genetically defined cell type to be investigated and compared across brain structures in an interpretable manner, while the unicellular rabies infection provides an elegant approach to studying microcircuits (Fig. 1).

4. Materials

4.1. Minimal Promoter Transgenics

NucleoBond BAC extraction kit (Clontech: Midi kit, cat. no. 740 579; Maxi kit, cat. no. 740 414.10).

BAC DNA handling: BAC DNAs are incredibly large (usually hundreds of kilobases) circular pieces of DNA. In order to prevent sheering, use wide-bore pipette tips to transfer BACs and do not freeze the DNA. Store BACs at 4°C in TE buffer.

TBE gels, TBE buffer: TBE buffer works better than TAE to achieve the resolution necessary to excise a fragment of digested BAC

DNA from a gel containing hundreds of other fragments of varying size. Use a 0.8% agarose gel for the best resolution of the larger bands. Multiply the length of the gel in centimeters by 8 V to determine the appropriate voltage at which to run the gel.

Ultracompetent cells (Inoue method): Ligating large pieces of DNA together is energetically unfeasible and therefore a statistically rare ligation event. For this reason, cells used for constructing minimal promoter transgenics should have a competence level of 10^9. We have achieved the highest competency levels using the Inoue method.

Gene Clean kit (MP Biomedicals, cat. no. 1001–600): We use the Gene Clean kit with some minor modifications to purify minimal promoter constructs for microinjection. The modifications are as follows:

Step 5. Incubation of DNA/NaI with GLASSMILK at room temperature. We incubate for 20–30 min to maximize DNA binding to the silica beads.

Step 10. We repeat the wash step twice for a total of three washes in NEW wash.

Step 12. Resuspend in 0.2 μm filtered low TE for microinjection. Spin the GLASSMILK down and transfer the DNA/low TE to a new tube. Spin down again and carefully remove the resuspended DNA to avoid any residual GLASSMILK contamination. Repeat one last time to ensure all the GLASSMILK is gone for a total of three transfers and three spins.

We also use the Gene Clean kit to purify BAC promoter fragments from TBE gels; for this, follow the instructions in the Gene Clean booklet for gel extraction from TBE gels.

GENSAT: http://www.gensat.org/index.html

Allen Brain Atlas: http://mouse.brain-map.org/welcome.do;jsessionid=8BBB582EC534AD1050D22D926F1F4972

BAC Distributor: Children's Hospital Oakland Research Institute http://bacpac.chori.org/libraries.php

UCSC Genome Browser: http://genome.ucsc.edu/

Mouse Microinjection Facility: http://www.neuro.uoregon.edu/ionmain/TMF/index.html

4.2. Anatomical Tracing Using Rabies Virus in Transgenic Mice

4.2.1. Viral Rescue and Amplification (From Wickersham's Nature Protocols (18))

Cell lines: HEK-293T/17 cells (ATCC, cat. no. CRL-11268)

The 293T-TVA800 (ref. 32) cell line can be requested from Dr. John Young (jyoung@salk.edu).

The BHK-B19G2 and BHK-EnvARGCD2 lines can be requested from the authors.

Plasmids (if rescuing virus from cDNA): genome vectors, such as cSPBN-4GFP (GenBank, accession number GU299211), pTIT-N (GenBank, accession number GU299212), pTIT-P (GenBank, accession number GU299213), pTIT-G (GenBank, accession number GU299214), pTIT-L (GenBank, accession number GU299215), and pCAGGS-T7 (GenBank, accession number GU299216).

At present, these plasmids are still only available by request from individual labs. The rabies virus protein expression vectors pTIT-N, pTIT-P and pTIT-L26, and pTIT-G27 and the G-deleted genome vector pSADΔG-Ppu14 can be requested from Dr. Karl-Klaus Conzelmann (conzelma@lmb.uni-muenchen.de). The genome vector cSPBN33—with convenient restriction sites flanking the glycoprotein gene, but not itself a glycoprotein deletion mutant—can be requested from Dr. Matthias Schnell (matthias.schnell@jefferson.edu). The G-deleted version cSPBN-4GFP, derived from cSPBN by replacement of the glycoprotein gene with that of the enhanced green fluorescent protein 34, can be requested from the authors. The T7 RNA polymerase expression vector pCAGGS-T7 (ref. (35)) can be requested from Dr. Robert Lamb (ralamb@ northwestern.edu), or the nearly identical pC-T7 can be requested from Matthias Schnell.

Reagents: Dulbecco's phosphate-buffered saline (DPBS) (Gibco, cat. no. 14190)

10× Dulbecco's phosphate-buffered saline (DPBS) (Gibco, cat. no. 70011)

Trypsin 0.05% (Gibco, cat. no. 25300)

Dulbecco's modified Eagle's medium (DMEM), high glucose (Gibco, cat. no. 11995)

Fetal bovine serum (FBS) (Hyclone, cat. no. SH30071.02)

100× Antibiotic–antimycotic (Gibco, cat. no. 15240)

95% Ethanol (e.g., Pharmco-AAPER, cat. no. 111000190)

Poly-L-lysine 0.01% (Sigma, cat. no. P4832)

Lipofectamine 2000 (Invitrogen, cat. no. 11668–027) if rescuing virus from cDNA

Opti-MEM I (Gibco, cat. no. 11058) dilution medium for Lipofectamine 2000

Dimethyl sulfoxide (DMSO) (EMD, cat. no. B10323-74)

16% Paraformaldehyde (PFA), 10-ml ampoules (Ted Pella, cat. no. 18505)

Sucrose (Sigma, cat. no. S9378)

Normal cell culture medium: 1,000 ml DMEM + 100 ml FBS + 10 ml antibiotic–antimycotic. Where not otherwise specified, "medium" refers to normal cell culture medium throughout the protocol. This can be stored at 4°C for several months.

Cell freezing medium: 10% DMSO + 90% normal medium (see above). As it is used in small quantities, this is most efficiently prepared immediately before use in the quantity desired.

Plasmid DNA (if generating virus from cDNA): Prepare using standard molecular biological techniques in the quantities listed below in Step 2, with results verified by restriction digestion and gel electrophoresis. This can be stored indefinitely at −20°C.

1:6 Poly-L-lysine: 100 ml poly-L-lysine + 500 ml DPBS. This can be stored at 4°C for several months.

70% Ethanol: For 1 l of 70% ethanol, 734 ml 95% ethanol + 263 ml deionized water. This can be stored in sealed glass bottles indefinitely.

20% Sucrose in DPBS (30 ml needed per prep.): For 50 ml, 10 g sucrose + DPBS to a total of 50 ml. Filter-sterilize with 0.45-μm filter (e.g., Steriflip) before use or storage. After filter sterilization, the solution can be stored at 4°C indefinitely.

2% Paraformaldehyde in PBS: To prepare, for example, 80 ml of the solution, mix 10 ml of 16% PFA, 8 ml 10× DPBS, and 62 ml distilled water. Store at 4°C for up to 1 month.

4.2.2. Surgical Setup

Stereotaxic Kopf Instruments: Model 900 small animal stereotaxic instrument with Model 921 nonrupture ear bars

Surgical anesthetics: ketamine HCL 100 mg/kg and dexmedetomidine HCL 40 μg/kg

Postoperative analgesia: buprenorphine HCL 60 μg/kg

Injection syringe: Model 7000.5 500-nl Hamilton syringe, part number 86259

Grip cement: Dentsply, Milford Delaware, powder (cat. no. 675571) and liquid (cat. no. 675572)

Surgical drill: Ideal Micro-DrillTM, Roboz Quality Surgical Instruments part number RS-6300, with a 0.8-mm-diameter carbide bur part number RS-6280C-1

4.2.3. Imaging

Olympus BX61 (microscope), DP72 (camera), BX-UCB (control box)

Prior ProScan III (motorized stage), Lumen 200Pro (light source)

Stitch software Metamorph Premier

5. Methods

5.1. Designing Cell Type Specific Transgenic Mice

An important part of dissecting a neural circuit using inducible systems is the targeted expression of the tTA or Cre driver line. There are several ways to go about genetic targeting of a cell type. The three approaches most commonly used are BAC (19), knock-in, and minimal promoter transgenics (20). A fourth orthogonal approach, called subtractive transgenics, employs both the Cre/Lox and the tTA/tetO inducible systems and could, in theory, provide an even finer degree of anatomical resolution (Note 1). Although BAC and knock-in transgenics are very popular for targeting a neuronal subtype, we have found as users of these types of lines that their utility as driver lines can be unpredictable (Note 2) (21). In contrast, minimal promoter transgenics has been used to great effect by a number of groups including our own (20). For this reason, this section will focus mainly on the design, construction, and evaluation of minimal promoter constructs.

All four approaches mentioned rely on cis-acting sequences to convey the entirety or some subset of native expression to an exogenous gene. The collection of cis-acting sequences that control a gene's expression are called promoters. Although promoters are generally thought to be located 5′ to a gene's coding region, there are also cis-acting sequences 3′ to the coding region and within the intronic or exonic region of the gene that contribute to its expression pattern (22). BAC and minimal promoter transgenics are essentially equivalent in this respect as both types of transgenics likely include only a subset of the cis-acting sequences that are important for native gene expression. In these two types of lines, one would expect to sometimes see only a subset of the cells that express the native gene in wild-type animals. In theory, BACs should exhibit a larger proportion of expressing cells compared with minimal promoter transgenics. However, in our experience, there is still no way to predict the expression pattern a BAC or minimal promoter transgenic will ultimately yield. Because the mysteries surrounding a gene's specific expression profile, especially in the CNS, are still being worked out, establishing a BAC or minimal promoter transgenic line will require some trial and error (Notes 3, 4).

5.2. Choosing the Minimal Promoter

Transgenes targeted to a neuronal type or a specific structure can be designed using minimal promoters to drive exogenous gene expression. The advantage of using a minimal promoter based approach is that transgenic lines may exhibit a subset of native expression, allowing structure-specific anatomical, physiological, and behavioral analysis. In addition, minimal promoters can be designed to target neurons with specific gene expression profiles. Often in minimal promoter transgenics and likely to a lesser degree in BAC transgenics, a large number of transgenic copies incorporate into the same insertion site (12). It is likely that multiple-copy insertion helps amplify expression of the gene, leading to more effective driver and effector lines, though in minimal promoter transgenics, multiple insertions may also lead to greater instability. Finally, minimal promoter design and construction is far easier than either BAC or knock-in transgenics.

A useful place to begin design of a minimal promoter construct is at the Allen Brain Institute (see Sect. 4). Allen Brain has amassed an extensive database containing RNA in situ data for thousands of genes expressed in the mouse brain. This database is searchable by structure, by gene, and by characteristics of gene expression like RNA intensity levels and is an invaluable resource for finding promoters to target specific structures. If the structure of interest cannot be located within the search parameters, the database must be searched by hand. Look through the in situ data to find a very strong expressing gene that is as specific as possible to the target structure. Also make sure to choose several candidate genes as ease of strategy and degree of sequence conservation will favor specific promoters.

If the goal of the minimal promoter construct is to target a genetically distinct population of neurons, it is essential to determine the degree of conservation across the 5′ untranslated region and the 3′-most promoter region as well as the degree of cell-type specificity. One must also consider sequence conservation while designing a minimal promoter construct to target a structure in the CNS. Performing the sequence analysis will help inform what and how much sequence is included in the construct. Furthermore, the degree of conservation will determine which genes to move forward with and where and how much promoter is included in the construct.

Some of this work may have already been done for you. Before performing any conserved sequence analysis, query GENSAT (see Sect. 4) or the Allen Brain Institute for a transgenic line already using that gene's promoter. Next, visit UCSC's genome browser (see Sect. 4) to locate a BAC construct containing the sequence you want to work with. The selected BAC's accession number should begin with RP23 or RP24. Many BACs are available from CHORI (see Sect. 4), one of the main BAC repositories. If your BAC is available from CHORI, use the genome browser to examine the degree of sequence conservation immediately upstream of the translational start site.

Choosing a promoter region to move forward with is a compromise between ease of strategy and degree of sequence conservation. Devise a strategy to isolate a region of promoter near the translational start site. The promoter bite should be approximately 7–12 kb to ensure that standard cloning methods remain applicable. Next, amplify and extract the BAC DNA. Nucleobond's BAC extraction (see Sect. 4) kits, in our experience, give the best yields of BAC DNA. Use either PCR or restriction enzymes to isolate the chosen promoter fragment from the rest of the BAC. If there is a digestion strategy to excise the promoter fragment where one of the enzymes used for excision cuts the BAC many times and the other enzyme cuts it only a few times, use a shotgun cloning strategy. In the shotgun cloning approach, a large quantity of BAC DNA is digested with the chosen enzymes, and all of the fragments are included in the ligation reaction. If neither enzyme is a frequent cutter, digest a massive amount of the BAC and gel extract the promoter fragment. We recommend using the Gene Clean (see Sect. 4) kit for promoter fragments larger than 10 kb.

5.3. Constructing the Optimal Driver Gene

A driver gene like the tTA should be expressed at the highest levels possible in order to achieve an effective dose of the gene following the tetO promoter. Toward this end, there are several elements that work through complementary mechanisms to increase transgene expression. It is widely known that the addition of an intron to a transgenic construct will increase expression levels in eukaryotic cells. Many viruses employ this trick to increase their own transcript levels in their host. The presence of an intron is thought to "fool" eukaryotic cells into treating the transcript as one of their own. Intron A elements are borrowed from a number of organisms and are used to increase transcription levels of a transgene. A second element, the PRE, is often used in combination with intron A and is thought to facilitate transport from the cell's nucleus. Like intron A, the PRE is another element found in viral genes. Studies have suggested that the PRE interacts with a nuclear export pathway to facilitate increased viral mRNA transport from the cell's nucleus. Because of these complementary effects, the combination of a PRE and an intron can significantly increase expression levels (23).

Typically, the ease with which the transgenic neurons can be visualized is an important consideration during construct design. The ability to visualize the presence of the driver or effector gene often is essential for subsequent experiments, especially those involving slice electrophysiology or anatomical analysis. To do this, it is necessary to express multiple genes from the same RNA. There are several ways to design a multicistronic construct that work with differing degrees of success. Fusion proteins are ideal because the presence of the marker protein always indicates the presence of the effector protein. However, fusing two proteins together can alter one or both of the proteins' functionality. Alternatively, the IRES (internal ribosome entry site) sequence, which allows ribosomes to bind and initiate translation in the middle of the RNA transcript, poses no risk of altering the functionality of the protein of interest. However, we and others have observed variation between protein expression from an IRES and protein expression from the upstream 5′ UTR (24). Often we see the IRES-driven protein expressed in places the upstream protein is not, or we see it only in a subset of the anatomical locations that the upstream protein is observed. For this reason, we would recommend using the 2A element over the IRES and as an alternative to a fusion protein if altering protein functionality is a concern. The 2A is a sequence found in the genome of some types of viruses that separates the coding regions of two genes. The 2A sequence encodes a short peptide whose sequence facilitates cotranslational cleavage between its N-terminal glycine and proline. If the peptide is placed between two protein-coding regions, these two protein regions will separate during translation (25).

5.4. Assessing the Founder Lines

After completion, the transgene is linearized and injected into mouse oocytes. The microinjection and subsequent implantation procedure are then performed by a facility that specializes in transgenic microinjection, BAC injections, and knock-ins (see Sect. 4). Approximately 21 days postimplantation, the F1 generation will be born. Although some laboratories use southern blots to genotype their founding generation because they are more sensitive, we would suggest genotyping the F1 generation by PCR because of the relative ease of the technique. Before the animals reach weaning age at postnatal day 21, design and test primers for single-copy sensitivity. This can be done by diluting the transgene in genomic DNA to simulate a single-copy insertion event. Ensure that several primer pairs work and are single-copy sensitive. Each animal that tests positive for the presence of the transgene represents a unique insertion event within the genome.

Both single- and multiple-copy insertion events can occur after microinjection. Occasionally, multiple copies will insert at different sites within the genome. If this occurs, interpreting the subsequent anatomical results can be quite difficult. For example, one might observe genotypically positive animals expressing tTA at a very high level and other genotypically positive animals from the same parent that appear phenotypically negative for tTA. Our laboratory has also observed transgenic founder animals that fail to transmit the transgene to the next generation. This may be due to a transgenic insertion that subsequently disrupted a gene that was important for germ cell viability. A transgene's site of insertion in the genome almost certainly influences its anatomical expression pattern, and in some cases, the effect of that insertion site may supersede the minimal promoter's effect. Additionally, if the transgene has interrupted a gene's coding region, this may affect future experiments. For this reason, transgenic single- and double-positive controls should be included in any experiment.

Knowing the exact location of a transgene insertion within the genome could prove to be very important for several reasons. Often it is necessary to breed transgenic animals to homozygosity in order to achieve high enough Ns. However, as the exact insertion site is not known in minimal promoter and BAC transgenics, breeding to homozygosity presents a problem. The transgene insertion site may also have interrupted a gene that another laboratory is interested in. The insertion event effectively creates a knockout that could be used in further experiments. Several techniques have been developed to determine the transgene insertion site. One such technique involves digesting with restriction enzymes and ligating such that a portion of the transgene and the flanking genomic DNA are contained in the same circularized piece of DNA. By using primers targeted to the transgene portion, a product can be obtained that also contains a portion of the flanking genomic sequence. Sequencing the product will reveal the location of the insertion site (26). While nontrivial to execute, the potential benefits of the addition of this step are clear. It may help improve transgenic design, may explain variations in transgene expression, will allow genotyping of homozygous animals, and may result in a useful knockout line as well.

The final selection process is screening for tTA lines that can effectively drive gene expression at the tetO promoter. To accomplish this, genotypically positive members of the F0 generation are bred to a tetO effector line and the functionality of the resulting cross is assessed. Ideally, the tetO effector line used for screening would be one whose effector gene expression most often mirrors the expression pattern of the driver line (Note 4) while also amplifying expression levels. RNA in situ hybridization is usually the simplest way to visualize the expression pattern of new lines due to the relative ease of designing an appropriate probe. Alternatively, coexpression of a fluorophore in the tetO line will allow confirmation of the expression pattern with fluorescence microscopy immediately after sectioning. Either method will reveal whether and to what extent the tTA line drove transgene expression in the tetO effector line. This is particularly important as tTA expression patterns alone do not reliably predict the levels of RNA expressed by the effector line in the presence of tTA. Additionally, our laboratory has seen discrepancies between tTA expression patterns and the resulting expression of a transgene driven by the tetO promoter. In other words, do not expect that the tTA expression will be predictive of the effector gene's expression (Note 3).

5.5. Anatomical Analysis Using Recombinant Rabies Virus in Transgenic Mice

Below we have included a summary of the steps involved in making high-titer rabies virus. For a more detailed account of the steps involved, please refer to Wickersham's Nature Protocols (2010) (18).

To understand the utility of the rabies virus (RV) as a transsynaptic tracer, it is important to understand its life cycle and molecular makeup (27, 28). Rabies virus is a negative-sense-strand RNA virus with a 12-kb genome. Unlike many viruses, the rabies genome remains RNA based throughout its life cycle, making it difficult to customize using traditional cloning methods. The first rescue of RV from cDNA was performed by Schnell and colleagues (29), paving the way for in depth study of the virus. The rabies genome contains five genes: N, P, L, M, and G. The neurotropic nature of the rabies virus is conferred through the glycoprotein coat G; deleting G results in a replication incompetent virus. G interacts with receptors on the host cell leading to adsorption and penetration. Upon entering the cell, the rabies virus uncoats and the polymerase L begins transcribing a leader RNA and 5 capped mRNAs coding the five genes. The switch from transcription to replication is regulated by the intracellular ratio of the nucleoprotein N to leader RNA. Viral replication begins with the synthesis of the full-length positive-sense strand of the rabies genome which then serves as a template for the negative-sense-strand RNA. N, the phosphoprotein P, and L form the NPL complex which organizes around the negative-sense genomic RNA to form the ribonucleoprotein core of the virus. In the next step, the matrix protein assembles around the ribonucleoprotein core, and finally, the protein coat is assembled, allowing the virus to bud off from the host cell.

There are several factors that make RV an incredibly useful tool for analyses of neuroanatomical connectivity. Transport, primarily at synaptic junctions, is exclusively retrograde, infecting cells upstream from the host cell (30–34). Rabies also preferentially infects neurons, exhibiting a very low affinity for glia (35). Initial infection and retrograde transport is rapid, occurring in a matter of days (34). Finally, amplification of the signal through replication results in unequivocal labeling of first-order (i.e., initially infected) and upstream neurons (15, 17, 34, 36).

The detailed understanding of RV's structure makes possible the precise genetic targeting of the pathways being studied. As an envelope virus, RV's glycoprotein coat or envelope can be replaced to change its tropism. The process of replacing a virus' coat protein with that of another is referred to as pseudotyping and can be used to change the viral tropism without any alteration to the viral genome. Technical advances by Wickersham and colleagues (15, 36) allow cell type specific infection and monosynaptic spread of the virus from the initial infection site. In their system, the envelope protein gene G is removed from the rabies' genome and replaced with a fluorophore. The recombinant rabies is then repackaged in an avian sarcoma virus protein coat, EnvA. Using the system, Wickersham et al. demonstrated in vitro selective infection of transgenic neurons expressing the EnvA receptor TVA. Recently, this system has taken a step forward (17) by incorporating transgenic mice and an inducible tTA/tetO system to get cell type specific infection in vivo (37). Transgenic mice containing both components of the tTA/tetO system can deliver the avian receptor TVA and the rabies glycoprotein to certain types of neurons, therefore enabling directed infection of recombinant rabies. Because the virus is pseudotyped with the EnvA coat protein, only those neurons expressing the TVA receptor will be infected. Furthermore, because the RV glycoprotein is provided *in trans* by the same cells, new viral particles assembled in the cell can transport transsynaptically and infect upstream neurons. However, because the recombinant RV genome lacks the coding for the native glycoprotein, transport is restricted to a single synapse. In this way, primarily infected neurons and their presynaptic partners can be labeled and characterized (Fig. 2).

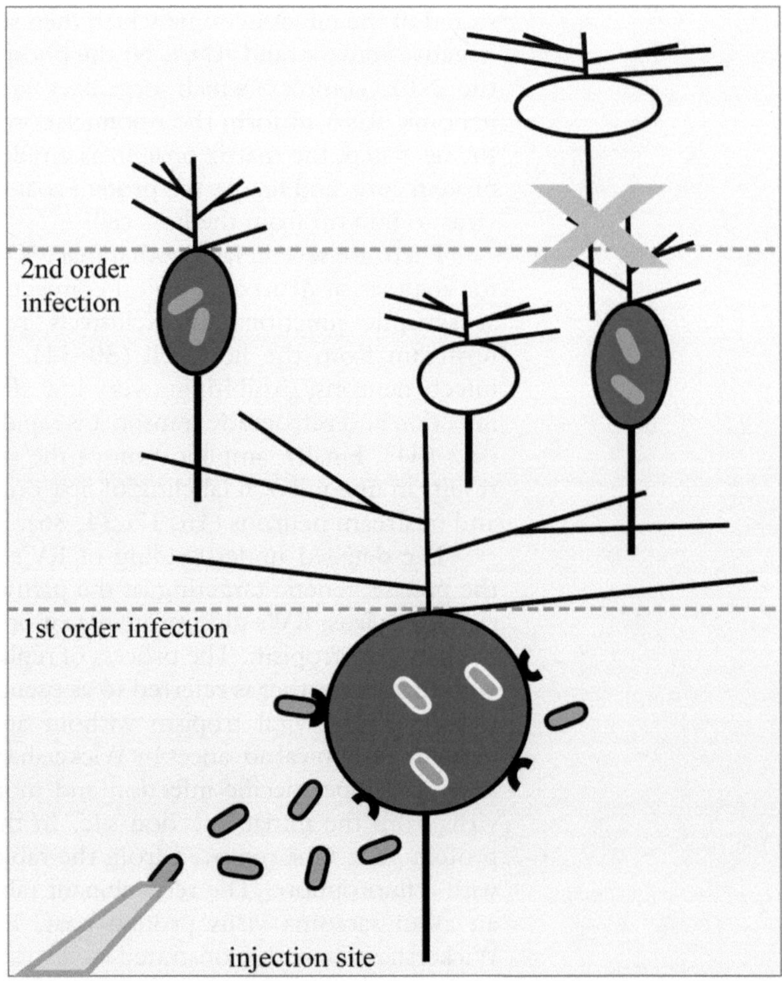

Fig. 2. Pseudotyped RV (*black* and *gray bullets*) is injected into a transgenic mouse brain. Because the virions bear the avian viral coat protein, they can only infect neurons expressing the TVA receptor (*black semicircles*) which is not native to mammals. The virions also lack the gene coding for the RV glycoprotein. Instead, the RV glycoprotein is supplied by the transgenic neurons, so virions assembled inside the primarily infected neurons have the native RV's glycoprotein (*gray* and *white bullets*). These phenotypically wild-type RV particles can move retrograde to infect neurons that synapse directly on the transgenic neurons. Because the secondarily infected neurons do not contain the RV glycoprotein, they cannot pass on the infection.

5.6. Rabies Viral Rescue and Amplification

Making and modifying rabies virus has traditionally been difficult. However, recent changes to the viral amplification step described by Wickersham and colleagues (18) have improved the likelihood of generating a high-titer stock. The first step to making high-titer EnvA-pseudotyped rabies is the RV cDNA rescue. If you already have starter virus, skip the rescue portion of the protocol. The cDNA rescue of the G-deleted RV is accomplished using the full-length cDNA clone of the viral genome where the gene coding for the glycoprotein is replaced by a fluorophore. The cDNA clone also contains a T7 promoter at the 3′ end of the genome so the T7 DNA-dependent RNA polymerase can generate sense strand copies of the RV genome. The T7 promoter is separated from the genome by a self-cleaving RNA sequence called the hammerhead ribozyme (38). The inclusion of the ribozyme allows recovery of the precise 3′ end of the RV genome. Individual expression vectors for T7 polymerase driven by the chicken actin promoter with a CMV enhancer and the five RV genes driven by the T7 promoter are also used in the rescue.

For viral rescue, HEK-293T cells are grown overnight in a poly-L-lysine (PLL, see Reagents)-coated flask or plate in normal cell culture medium. On the following day, a mixture of genome vectors, pTIT-N, pTIT-R, pTIT-G, pTIT-L, pCAGGS-T7, or pC-T7, combined with Opti-MEM and Lipofectamine 2000 is applied to the cells. The cells and DNA mixture are incubated for 6 h to allow transfection to occur. The cells are then washed and grown for 7 days in normal cell culture medium. Change the medium daily for the first 3 days, discarding supernatant. On days 4–7, change the medium daily but collect the supernatant instead of discarding it. Starting on the fourth day, fluorescing cells should become visible, indicating successful viral assembly. Filter-sterilize all collected supernatants. These can be stored at 4°C but will experience a drop in titer of 4–5% each day.

The modifications to the RV protocol at the amplification and ultracentrifugation steps described by Wickersham and colleagues (18) have increased the viral titer compared to previous protocols (18). After viral rescue, the G-deleted virus is pseudotyped during the amplification step. The amplification of the pseudotyped RV allows for more efficient infection of the transgenic neurons expressing the TVA receptor and RV G protein in the complementing mouse line. The RV is pseudotyped by taking the G-deleted virus in its native envelope from the previous steps and infecting cells that express the EnvA viral coat protein. BHK-EnvA2 cells are thawed and expanded for plating. Expand enough cells to grow on 16 plates. The plate number ensures there is a high enough volume of supernatant for ultracentrifugation (150 ml). Do not use PLL with EnvA2 cells as it slightly decreases the titer. These cells are grown overnight in medium and infected with the viral supernatant collected from the previous step. Count cells on one plate and infect the other 15 at an MOI of 2 in 22.5 ml. Approximately 4.5×10^8 IU of unpseudotyped virus is needed (15 plates × ~1.5×10^7 cells per plate × 2 IU per cell). The MOI at this step is kept low to minimize contamination of the resulting pseudotyped stock by the original virus coated with RV G protein. Incubate this mixture overnight in medium and discard the supernatant as it will be contaminated with the infecting stock. Wash twice with DPBS to remove any remaining contaminating virus. Following a second overnight incubation in medium, collect supernatant from the 15 plates for concentration by ultracentrifugation. Plate cells for titering the supernatant. Reserve 0.5 ml of the supernatant for titering.

The ultracentrifugation step concentrates the viral particles. In this step, 25 ml of filtered supernatant is layered on top of 5 ml of 20% sucrose in DPBS. Place the tubes in the ultracentrifuge and spin at 50,000 g for 2 h at 4°C. Aspirate off the supernatant and replace with 100 µl DPBS to prevent desiccation of the virus. Decontaminate all surfaces thoroughly. Pool resuspended virus, vortex, and spin down. Aliquot into 0.5-ml cryotubes in 20-µl aliquots. Label and store at –80°C.

5.7. Anatomical Analysis in Cell-Type Specific Transgenic Mice

Following amplification of the pseudotyped RV, the virions are aliquoted in appropriate volumes to minimize the number of freeze–thaw cycles and any resulting reduction in titer. The RV aliquots are then flash frozen in liquid nitrogen and stored at –80°C until needed. After thawing an RV aliquot on ice, a small volume of virions can be injected into the mouse brain through a simple surgery using a Hamilton syringe attached to a stereotax. The transgenic lines that provide the TVA receptor and rabies glycoprotein *in trans* will be referred to as pTT/TVAG lines throughout the text. These animals contain a transgene consisting of the tetO promoter preceding the TVA receptor and the rabies glycoprotein genes. These genes are separated from one another by a sequence that codes for F2A cotranslational self-cleaving peptide.

Several preliminary controls should be done using both pTT/TVAG single-positive animals and double-positive animals bred to a tTA driver line. These types of controls should be done for every single-expressing pTT/TVAG line and are designed to identify transgenic leak. First, a single positive from each pTT/TVAG line should be tested for leak. In our lab, we found one line exhibiting leak from the tetO promoter independent of tTA expression. In order to test for leak, RV is injected into an accessible area of the mouse brain, usually somewhere in cortex, and then the sliced brain is scanned for viral infection around the injection site. In leaky lines, usually labeled cells will be seen around the injection site without any transport to other areas of the cortex. A second necessary control is designed to detect leak in double-positive animals. If a double-positive animal shows infection in a brain region that does not show detectable levels of transgene expression, then it cannot be used for subsequent experiments. Although we have never seen infection in this type of control, the fact remains that even one TVA receptor could result in rabies viral infection. For this reason, levels of RNA or protein that are undetectable by antibody or riboprobe label could still cause infection.

Following investigation of each pTT/TVAG founding line for the presence of transgenic leak, the nonleaky lines can be used in experiments. At this point prior to surgery, it is important to determine where, at what level, and at what cell density the transgene is expected to express in the line. Do not rely on tTA expression to predict the expression of the TVA and glycoprotein genes (Note 4). An anatomical survey of a double positive from each founding line should be done to determine the transgene expression profile. These results will then influence the injection volume and the location of the injection site. Volumes typically range from 50 to 150 nl with a larger volume being used for areas with low levels of transgene expression or low transgenic cell counts. In transgenic lines with low transgenic cell counts, the probability of injecting near the area expressing the transgene is significantly lowered. The larger volume of RV is meant to counteract this problem. Conversely, in an area with dense transgene expression, injecting a smaller bolus of RV may make interpreting local connectivity much easier.

5.8. Virus Injection

Elements of the surgical methodology detailed below have been described previously (17). Some of the specifics are matters of preference (e.g., the choice of anesthesia or the type of ear bars used). Others, however, are of particular importance, given the nature of the viral tracer, and will be emphasized accordingly.

In our previous study, mice were anesthetized with a cocktail of ketamine (100 mg/kg) and dexmedetomidine hydrochloride (40 μg/kg). Dexamethasone (0.1 mg/kg) and atropine (0.03 mg/kg) were administered presurgically to ameliorate possible inflammation and respiratory irregularities, respectively. Eyes were kept moist with a thin layer of antibacterial ophthalmic ointment. Mice were positioned in a stereotaxic frame, and the head held in position with atraumatic ear bars. A portion of the scalp was excised, and the surface of the skull cleaned thoroughly with 3% H_2O_2. After drilling a small hole drilled in the skull overlying the injection site, the head was leveled in the X, Y, and Z planes. After injection, the wound was covered with sterile petroleum jelly, and the skull was capped with grip cement (Dentsply, Milford DE). Mice were administered with buprenorphine (0.06 mg/kg) for postoperative analgesia and individually housed following surgery.

Because of the amplification of label inherent in the viral tracing technique, particular care must be taken during the injection process, especially when using transgenic lines with nonspecific expression patterns. We perform our injections manually with a

500-nl 33-ga. Hamilton syringe attached to the stereotaxic frame (Kopf Instruments). After filling the syringe, carefully wipe down the sides and blot the tip with cotton applicators. Lower the syringe to the dura at the target coordinates to obtain a starting depth. The syringe may then be passed through a small incision in the dura and lowered at 500-μm increments, resting 2 min between increments, to a resting depth at the target. Maintain the syringe at depth for 5 min to allow the tissue to settle and then proceed with the injection in 25 nl steps, waiting 1 min between steps. To avoid back pressure, maintain the syringe at depth for a minimum of 5 min after the full volume is injected, then raise the syringe 200 μm and allow it to set for at least an additional 2 min before removal.

Depending on the location of the injection site and the desired spread and density of label, some additional ideas should be considered. For extremely superficial injections (e.g., into sensory cortices), a pulled glass pipette may be preferred for the injection as the 33-ga. Hamilton may cause some dimpling of the cortical surface. If injections are being made in lines with nonspecific expression patterns, accuracy of injection is extremely important and the total volume should be kept quite small (e.g., for a titer of 9.9×10^8/ml, this would mean a total volume of no more than 50 nl). Conversely, if the expression pattern is specific, larger injection volumes (50–200 nl) may be used while avoiding the complication of initial infection in adjacent structures. In these instances, it may be desirable to lower the titer of the virus as high concentrations, when injected into specific, densely expressing lines can result in an overwhelmingly high amount of initial and secondary infection at the injection site (Fig. 3).

After surgery, the animal is left to recover for approximately 7 days. The 7-day recovery period was chosen somewhat arbitrarily, and it may be worth comparing longer and shorter RV incubation

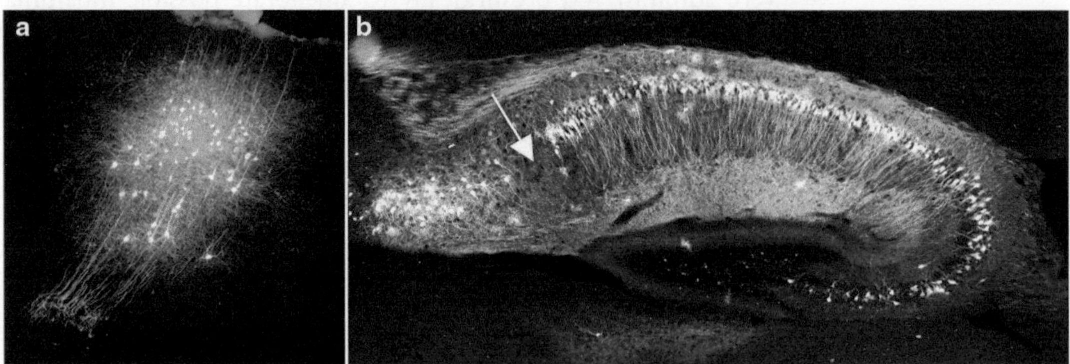

Fig. 3. (**a**) Rabies viral infection after injection into M1 cortex in a transgenic mouse. (**b**) Rabies viral infection and transport in the hippocampus. The *arrow* indicates the injection site, centered in the subicular region of the hippocampus. A large viral bolus was given resulting in primary infection of transgenic neurons in both the CA1 and subiculum. In both pictures, the pial surface is toward the top.

periods. There are several reasons to vary the incubation time of the RV. A shorter incubation time could be used to visualize primarily infected neurons in order to do morphological analysis to help determine cell type. We have only tried a shorter incubation time once. After a 2-day infection, we observed faintly labeled neurons around the injection site, but they were so faint that it was difficult to see distinct morphologies. Processing the infected tissue with an RV glycoprotein antibody might solve this problem by amplifying the label enough to visualize morphology. Although we have not yet tried longer incubation times, there are several reasons to do so. First, we think it is possible that if secondarily infected neurons also express the transgene, they could allow the RV to spread retrograde to third-order neurons. After a 7-day incubation period, we have not seen any evidence of transcomplementation, but a longer period of incubation might allow this to occur. Investigation into longer and shorter incubation periods should be performed on double-positive transgenics that express the transgene in well-characterized circuits. Our lab is currently using the hippocampal circuit to investigate transcomplementation and shorter incubation times to visualize the primarily infected cells' morphology.

On the seventh day postinfection, the animal is perfused and the brains are fixed in 2% paraformaldehyde followed by a 6% sucrose wash. Brains are then frozen and cut into 30-μm sections using a cryostat. The resulting infection can be viewed immediately after slicing. It is advisable to use a high-quality microscope and automatic stitching software as poor quality optics will make analysis nearly impossible and manual postimage processing will significantly lower the throughput of the experiments. With the right microscope and stitching software, the rabies viral tracer technique can become a high-throughput tool for anatomical analysis (see Sect. 4).

The utility of this anatomical tracing technique comes from its ability to trace the monosynaptic connections of a cell type in order to characterize and quantify its presynaptic partners. Two requirements must be satisfied in order to categorize and quantify the infected cells. The primarily and secondarily infected cells must be distinguishable from one another, and the antibody label must be interpretable enough to quantify. Our lab is currently investigating strategies for meeting both of these requirements.

Following successful infection, primarily infected and secondarily infected neurons can be characterized by their morphology and the genes they express. Although our lab has the most experience using RNA in situ, it cannot be used to examine gene expression profiles in RV-infected cells because the virus shuts down host transcription (see Note 4). For this reason, we are currently learning to work with antibodies in order to characterize the cell types. Finding antibodies that strongly and specifically label neurons can be difficult. Therefore, it is advisable to assess the primarily infected cell's morphology and connectivity in order to provide some clues about cell type and then choose several versions of the appropriate primary antibody to test for effectiveness. Our lab is currently using fixed brain slices for antibody stain, and this makes the primary antibody requirements more stringent. One way to fix this might be to an adaptor antibody to amplify the signal of the secondary or use a brighter secondary (e.g., quantum-dot-conjugated secondaries). It might also be worth trying floating sections to increase the penetrance and the amount of tissue exposed to the antibody.

In order to distinguish between primary and secondarily infected neurons, the injection site must be labeled with a marker stable enough to remain in the animal's tissue for at least 7 days postinjection. Then the primarily infected neurons can be inferred by the number of cells in the labeled area that express the TVA receptor. We are currently looking into quantum dots as well as various dyes and cell tracers to help distinguish between primary and secondary infection as knowing the number of primarily infected neurons is absolutely crucial to the utility of the technique.

The rabies viral tracer technique especially in combination with transgenics is still in its early stages. Our laboratory is currently building up a suite of antibody labels to characterize the primarily and secondarily infected neurons using antibody staining. For this to be a high through put system, the antibodies we use must label cells in a strong and specific manner so the quantification of cell types can be automated. Concurrent with this exigency is the ability to distinguish between primary and secondary infection by labeling the injection site. Once we are able to do this, we can begin quantifying the local connectivity between the transgenic neurons and other cell types identified by antibody label. This system offers a new and improved way to visualize, characterize, and quantify transgenic neurons and their presynaptic partners; however, we and others have found that the ratio of connectivity between the primarily infected cells and their presynaptic partners in well-characterized circuits is usually several orders of magnitude lower than that seen in the literature (16, 17). Even in single-cell infection in vivo, the number of presynaptic partners do not reflect the numbers in the literature (16). For this reason, although the system can provide an accurate ratio of cell type specific connections, it cannot be considered strictly quantitative. This, of course, is true of all other systems of connectivity analysis as well: this technique is still an unambiguous improvement on prior art.

6. Notes

1. The subtractive approach to transgenic targeting of cell types uses both the Cre/Lox system and the tTA/tetO system to achieve even finer-grained anatomical analysis and is like the

intersectional approach used to analyze neural circuits in *Drosophila* (Chap. 1b). Because Cre recombinase can excise any DNA region flanked by loxP sites, if the tTA gene is floxed, then its expression can be turned off wherever Cre is expressed. In this case, the tTA is expressed wherever Cre is not expressed, and this allows for targeting of nonoverlapping subsets of cells. A second related strategy is to flank a gene like neomycin with loxP sites and place it in front of the tTA. In this case, the tTA will be expressed wherever Cre is expressed because the presence of the interfering sequence is removed. If these strategies are employed using cell type specific Cre and tTA transgenic lines, the cell type specific targeting is determined by the combination of two cell type specific lines instead of one.

There are several surmountable problems associated with this technique. First, this approach should not be used in combination with minimal promoter floxed neomycin tTA or floxed tTA transgenic lines (unpublished data). The reason for this is that often during an insertion event, multiple copies of the transgene are integrated into the genome (12). The orientation of loxP sites to one another as well as their 5′ to 3′ directionality is important, and if there are more than two loxP sites in the same vicinity, then the system becomes unpredictable. Second, in order to take advantage of this very specific targeting of the tTA, the Cre, the tTA, and the tetO lines must be bred to obtain triple-positive animals, and this requires an enormous number of breedings and time-consuming genotyping. Third, Cre-recombinase-mediated DNA excision is not reversible, so it is also important to determine when during development Cre recombinase is performing the excision because all the daughter cells will inherit the altered DNA. If expression at the promoter driving Cre happens too early in development in a double-positive floxed neomycin tTA/Cre line before cell types are completely differentiated, then the predicted target cell type may only be a subset of the cell types that are actually targeted. Conversely, in the floxed tTA strategy, the cell type to be targeted may not get targeted at all if the excision event happens too early.

The first and second problems can be resolved if the floxed neomycin tTA and floxed tTA genes are always knocked in to the locus of the gene of interest. If even just the tTA is a knock-in, then those animals can be genotyped for homozygosity. The homozygous animals can then be bred to tetO lines thereby increasing the chance of getting triple positives for further experiments. To overcome the dosage issues that often come up in knock-in lines, Mayford's tetO/tTA amplifier construct could be included in the knock-in design to amplify expression of the tTA (39). The third problem requires choosing the right combination of Cre and tTA promoter lines. It may be that certain Cre/tTA combinations will never work because of their time points of expression during development.

2. In BAC transgenesis, a large chunk of native promoter drives expression of an exogenous driver gene in order to phenocopy the expression of a native gene. BAC transgene construction is a difficult and time-consuming technique to master, and once made, they are not guaranteed to express as predicted. Much of this can be attributed to the size of the BAC construct which is hundreds of kilobases long. BACs are so large they must be recombined with the driver gene inside special strains of *E. coli*. DNA sheering can occur when injecting such a large molecule of DNA through a microinjection needle, and this may help explain why BAC transgenics do not always phenocopy expression. Additionally, unlike knock-ins, the site of genomic insertion is not predetermined, and this may also contribute to the occasional unpredictable expression pattern. Knock-ins are created by targeted recombination of a driver gene to the end of a native gene's exon or into one of the gene's introns (40). The context provided by the native gene results in nearly identical expression of the exogenous gene. The recombination event occurs in embryonic stem (ES) cells from a mouse strain bearing a certain coat color. The transgenic ES cells are then injected into the blastocyst of a developing mouse of the same strain but different coat color. The resulting animal is a chimera bearing patches of fur of both colors. Once the chimera is made, the transgenic chromosome must incorporate into the gametes before the knock-in animal can be used, a process requiring several generations of breeding.

   Even with this very direct approach to targeting a cell type, the expression of the exogenous gene does not always appear to recapitulate native expression (unpublished data). This may be due to a gene dose that is undetectable using our methods or some more exotic problem (e.g., (12)). One possible explanation is that the dose of the native gene in some knock-in lines does not translate into an effective dose of the driver gene. As Chandler and colleagues have shown, at least in BAC transgenics, transgene copy number correlates the levels of gene expression. Furthermore, we have observed in our laboratory that driver lines with weak expression generally cannot produce detectable levels of expression from the tetO promoter. Another explanation for the altered expression we have observed is that the mutant copy of the gene becomes silenced resulting in reduced expression of the transgene (41). In either case, the solution may be to breed these problem lines to homozygosity or amplify expression by breeding to the tetO/tTA transgenic line.

The issues outlined above should be considered before undertaking the time-consuming process of designing BAC or knock-in transgenics. Clearly, the current strategies for knock-in design are not yet optimized. For future knock-in designers, it is worth considering several tools that might provide solutions to the two main problems often encountered while working with knock-ins. To alleviate the gene dose issue, an amplifier cassette could be incorporated into the tTA targeting construct. One such amplifier is the tetO/tTA construct designed by Mayford and colleagues (39). This construct allows greater tTA expression through induction of a second tTA gene at the tetO promoter by the cell type specific tTA. This would be an ideal addition to the construction of a knock-in line because it amplifies expression of the transgene without altering expression levels of the native gene. We have also seen knock-in lines whose expression pattern appeared to be only a subset of the native gene's expression pattern. The only way to ensure this does not happen is to knock in the tTA gene in frame with the last exon of the native gene's coding region and separate these two coding regions with the 2A self-cleaving peptide (25).

3. Our lab has observed variability in transgenic expression between individuals of the same line as well as variability from generation to generation. The term "epigenetics" (42) is often used to explain this phenomenon, though this is due largely to a lack of understanding of the underlying mechanisms involved in the observed drift in expression patterns (41).
4. Our lab has also observed variation in gene expression between tTA and tetO lines in double-positive animals. In other words, the tTA expression pattern is not always predictive of the tetO line's expression pattern. We have attributed this to a combination of insertional effects and epigenetic mechanisms operating at the two sites of transgenic insertion.
5. Although the main function of the rabies genes appears obvious from their names, many of the genes serve multiple functions. The rabies matrix protein M is just one example of a multipurpose gene in the rabies virus. The matrix protein, in addition to encapsulating the RNP complex, is also thought to play a role in the switch from transcription to replication (43), and in order to promote self-replication, it interferes with the host's translational machinery. Specifically, the matrix protein has been shown to bind host translational factors like eIf3, preventing translation at conventional eukaryotic 5′ UTRs, while translation at viral IRESs remains unaffected (44). The functional implications of this are several. First, rabies virus should not be used as a gene delivery system if part of the subsequent experiment is to perform functional analyses on the infected cell. Additionally, it may be difficult to do further analysis to characterize the infected cells. RNA in situ cannot be used to

characterize cells after 7-day incubation with rabies, and antibody staining may or may not be effective after a 7-day incubation period depending on the half-life of the antigen personal observation.

References

1. Bernard, A., Sorensen, S. A., and Lein, E. S. (2009) Shifting the paradigm: new approaches for characterizing and classifying neurons, *Current Opinion in Neurobiology 19*, 530–536.
2. Sugino, K., Hempel, C. M., Miller, M. N., Hattox, A. M., Shapiro, P., Wu, C., Huang, Z. J., and Nelson, S. B. (2006) Molecular taxonomy of major neuronal classes in the adult mouse forebrain, *Nat Neurosci 9*, 99–107.
3. Cardin, J. A., Carlen, M., Meletis, K., Knoblich, U., Zhang, F., Deisseroth, K., Tsai, L. H., and Moore, C. I. (2009) Driving fast-spiking cells induces gamma rhythm and controls sensory responses, *Nature 459*, 663–667.
4. Tan, W., Janczewski, W. A., Yang, P., Shao, X. M., Callaway, E. M., and Feldman, J. L. (2008) Silencing preBotzinger Complex somatostatin-expressing neurons induces persistent apnea in awake rat, *Nat Neurosci 11*, 538–540.
5. Wehr, M., Hostick, U., Kyweriga, M., Tan, A., Weible, A. P., Wu, H., Wu, W., Callaway, E. M., and Kentros, C. (2009) Transgenic silencing of neurons in the mammalian brain by expression of the allatostatin receptor (AlstR), *J Neurophysiol 102*, 2554–2562.
6. Tervo, D., and Karpova, A. Y. (2007) Rapidly inducible, genetically targeted inactivation of neural and synaptic activity in vivo, *Curr Opin Neurobiol 17*, 581–586.
7. Armbruster, B. N., Li, X., Pausch, M. H., Herlitze, S., and Roth, B. L. (2007) Evolving the lock to fit the key to create a family of G protein-coupled receptors potently activated by an inert ligand, *Proc Natl Acad Sci USA 104*, 5163–5168.
8. Callaway, E. M. (2008) Transneuronal circuit tracing with neurotropic viruses, *Curr Opin Neurobiol 18*, 617–623.
9. Harrison, P. J., Hultborn, H., Jankowska, E., Katz, R., Storai, B., and Zytnicki, D. (1984) Labelling of interneurones by retrograde trans-synaptic transport of horseradish peroxidase from motoneurones in rats and cats, *Neuroscience Letters 45*, 15–19.
10. Itaya, S. K., Van Hoesen, G. W., and Barnes, C. L. (1986) Anterograde transsynaptic transport of WGA-HRP in the limbic system of rat and monkey, *Brain Research 398*, 397–402.
11. Reiner, A., Veenman, C. L., Medina, L., Jiao, Y., Del Mar, N., and Honig, M. G. (2000) Pathway tracing using biotinylated dextran amines, *Journal of Neuroscience Methods 103*, 23–37.
12. Lin, T., Yasumoto, H., and Tsai, R. Y. (2007) BAC transgenic expression efficiency: bicistronic versus ATG-fusion strategies, *Genesis 45*, 647–652.
13. Sugita, M., and Shiba, Y. (2005) Genetic tracing shows segregation of taste neuronal circuitries for bitter and sweet, *Science 309*, 781–785.
14. Maskos, U., Kissa, K., St Cloment, C., and Brulet, P. (2002) Retrograde trans-synaptic transfer of green fluorescent protein allows the genetic mapping of neuronal circuits in transgenic mice, *Proc Natl Acad Sci USA 99*, 10120–10125.
15. Wickersham, I. R., Lyon, D. C., Barnard, R. J., Mori, T., Finke, S., Conzelmann, K. K., Young, J. A., and Callaway, E. M. (2007) Monosynaptic restriction of transsynaptic tracing from single, genetically targeted neurons, *Neuron 53*, 639–647.
16. Marshel, J. H., Mori, T., Nielsen, K. J., and Callaway, E. M. (2010) Targeting single neuronal networks for gene expression and cell labeling in vivo, *Neuron 67*, 562–574.
17. Weible, A. P., Schwarcz, L., Wickersham, I. R., DeBlander, L., Wu, H., Callaway, E. M., Seung, H. S., and Kentros, C. G. (in press) Transgenic targeting of recombinant rabies virus reveals monosynaptic connectivity of specific neurons., *The Journal of Neuroscience*.
18. Wickersham, I. R., Sullivan, H. A., and Seung, H. S. (2010) Production of glycoprotein-deleted rabies viruses for monosynaptic tracing and high-level gene expression in neurons, *Nat Protoc 5*, 595–606.
19. Yang, X. W., Model, P., and Heintz, N. (1997) Homologous recombination based modification in Escherichia coli and germline transmission in transgenic mice of a bacterial artificial chromosome, *Nat Biotechnol 15*, 859–865.
20. Mayford, M., Abel, T., and Kandel, E. R. (1995) Transgenic approaches to cognition, *Current Opinion in Neurobiology 5*, 141–148.

21. Jensen, P., Farago, A. F., Awatramani, R. B., and Dymecki, S. M. (2008) Redefining the central serotonergic system based on genetic lineage, *Fund Clin Pharmacol 22*, 115–115.
22. Visel, A., Rubin, E. M., and Pennacchio, L. A. (2009) Genomic views of distant-acting enhancers, *Nature 461*, 199–205.
23. Sun, J., Li, D. F., Hao, Y. L., Zhang, Y. W., Fan, W. L., Fu, J. J., Hu, Y. Z., Liu, Y., and Shao, Y. M. (2009) Posttranscriptional Regulatory Elements Enhance Antigen Expression and DNA Vaccine Efficacy, *DNA Cell Biol 28*, 233–240.
24. Bochkov, Y. A., and Palmenberg, A. C. (2006) Translational efficiency of EMCV IRES in bicistronic vectors is dependent upon IRES sequence and gene location., *Biotechniques 41*, 283–290.
25. Szymczak, A. L., Workman, C. J., Wang, Y., Vignali, K. M., Dilioglou, S., Vanin, E. F., and Vignali, D. A. (2004) Correction of multi-gene deficiency in vivo using a single 'self-cleaving' 2A peptide-based retroviral vector, *Nat Biotechnol 22*, 589–594.
26. Liang, Z., Breman, A. M., Grimes, B. R., and Rosen, E. D. (2008) Identifying and genotyping transgene integration loci, *Transgenic Res 17*, 979–983.
27. Finke, S., and Conzelmann, K. K. (2005) Replication strategies of rabies virus, *Virus Res 111*, 120–131.
28. Albertini, A., Schoehn, G., Weissenhorn, W., and Ruigrok, R. (2008) Structural aspects of rabies virus replication, *Cellular and Molecular Life Sciences 65*, 282–294.
29. Schnell, M. J., T. Mebatsion, and Karl-Klaus Conzelmann (1994). "Infectious rabies viruses from cloned cDNA." EMBO J 13(18): 4195-4203.
30. Iwasaki, Y., and Clark, H. F. (1975) Cell to cell transmission of virus in the central nervous system. II. Experimental rabies in mouse, *Lab Invest 33*, 391–399.
31. Charlton, K. M., and Casey, G. A. (1979) Experimental oral and nasal transmission of rabies virus in mice, *Can J Comp Med 43*, 10–15.
32. Ugolini, G. (1995) Specificity of Rabies Virus as a Transneuronal Tracer of Motor Networks - Transfer from Hypoglossal Motoneurons to Connected 2nd-Order and Higher-Order Central-Nervous-System Cell Groups, *J Comp Neurol 356*, 457–480.
33. Tang, Y., Rampin, O., Giuliano, F., and Ugolini, G. (1999) Spinal and brain circuits to motoneurons of the bulbospongiosus muscle: Retrograde transneuronal tracing with rabies virus, *The Journal of Comparative Neurology 414*, 167–192.
34. Kelly, R. M., and Strick, P. L. (2000) Rabies as a transneuronal tracer of circuits in the central nervous system, *Journal of Neuroscience Methods 103*, 63–71.
35. Tsiang, H., Koulakoff, A., Bizzini, B., and Berwald-Netter, Y. (1983) Neurotropism of rabies virus. An in vitro study, *J Neuropathol Exp Neurol 42*, 439–452.
36. Wickersham, I. R., Finke, S., Conzelmann, K. K., and Callaway, E. M. (2007) Retrograde neuronal tracing with a deletion-mutant rabies virus, *Nat Methods 4*, 47–49.
37. Mansuy, I. M., and Bujard, H. (2000) Tetracycline-regulated gene expression in the brain, *Current Opinion in Neurobiology 10*, 593–596.
38. Hean, J., and Weinberg, M. S. (2008) The Hammerhead Ribozyme Revisited: New Biological Insights for the Development of Therapeutic Agents and for Reverse Genomics Applications In *RNA and the Regulation of Gene Expression: A Hidden Layer of Complexity* (Morris, K. V., Ed.), Caister Academic Press.
39. Reijmers, L. G., Perkins, B. L., Matsuo, N., and Mayford, M. (2007) Localization of a stable neural correlate of associative memory, *Science 317*, 1230–1233.
40. Roebroek, A. J. M., Wu, X., and Bram, R. J. (2002) Knockin Approaches, In *Transgenic Mouse* (Hofker, M. H., and Deurson, J. v., Eds.), pp 187–200.
41. Whitelaw, E., Sutherland, H., Kearns, M., Morgan, H., Weaving, L., and Garrick, D. (2001) Epigenetic Effects on Transgene Expression, In *Gene Knockout Protocols* (Tymms, M. J., and Kola, I., Eds.), pp 351–368, Humana Press Inc., Totowa.
42. Bird, A. (2007) Perceptions of epigenetics, *Nature 447*, 396–398.
43. Finke, S., Mueller-Waldeck, R., and Conzelmann, K. K. (2003) Rabies virus matrix protein regulates the balance of virus transcription and replication, *J Gen Virol 84*, 1613–1621.
44. Komarova, A. V., Real, E., Borman, A. M., Brocard, M., England, P., Tordo, N., Hershey, J. W., Kean, K. M., and Jacob, Y. (2007) Rabies virus matrix protein interplay with eIF3, new insights into rabies virus pathogenesis, *Nucleic Acids Res 35*, 1522–1532.

Chapter 10

Functional Circuitry Analysis in Rodents Using Neurotoxins/Immunotoxins

Kazuto Kobayashi, Kana Okada, and Nobuyuki Kai

Abstract

Immunotoxin cell targeting is a transgenic animal technology used to eliminate specific cell types from a complex neural circuitry by using cytotoxic activity of recombinant immunotoxins, which are composed of an antibody variable region fused to bacterial toxin fragments. This technology provides a useful approach for studying the neural circuitry mechanism underlying a variety of brain functions. The present chapter provides a detailed experimental strategy for immunotoxin cell targeting and its application to neural circuitry analysis, in particular focusing on the basal ganglia circuitry, which is implicated in the control of motor functions.

Key words: Transgenic animal, Recombinant immunotoxin, Genetic cell ablation, Neural circuitry, Basal ganglia, Motor control

1. Introduction

An understanding of neural mechanisms underlying brain functions has advanced along with the development of techniques that permit manipulation of the functions of neuronal types having particular identities. One useful approach is a selective elimination of the neuronal types from the complex neural circuitry. When elimination of a subpopulation of neurons leads to physiological and behavioral changes, it is possible to characterize the role of a specific neuronal type in various brain functions. The standard procedure for elimination of small areas or specified nuclei in the brain is electric lesion, in which a steel wire is introduced into the target region, and the passage of an electric current through it disrupts the cells surrounding the tip of the electrode (1). Another procedure is the infusion of excitatory amino acids, which produces

Alexei Morozov (ed.), *Controlled Genetic Manipulations*, Neuromethods, vol. 65,
DOI 10.1007/978-1-61779-533-6_10, © Springer Science+Business Media, LLC 2012

disruption of neuronal cell bodies in the vicinity of the infusion micropipette (1). In addition, some neurotoxic compounds are also used to remove selective neurons based on their pharmacological specificities, such as 1-methyl-4-phenyl-1,2,3,6-tetrahydropyridine (MPTP) for dopaminergic cell groups (2) and N-(2-chloroethyl)-N-ethyl-2-bromobenzylamine (DSP-4) for noradrenergic cell groups (3). However, the use of drugs that are based on pharmacological properties has been limited to a small number of cell groups in the brain.

The diversity of gene expression that distinguishes neuronal types in the brain provides us the technology that enables the selective targeting of those neuronal types. Immunotoxin cell targeting is a transgenic technique for cell-type-specific and inducible ablation of neurons (4–6) (see Fig. 1). In this technique, we generate transgenic animals (mice and rats) that are engineered to express the target molecule of a recombinant immunotoxin, e.g., the human interleukin-2 receptor α-subunit (IL-2Rα) under the control of a cell-type-specific gene promoter/enhancer. Subsequently, these animals are treated with the corresponding immunotoxin at the desired time by using appropriate injection methods. In particular, local injection of the immunotoxin with stereotaxic surgery permits localized elimination of neuronal types within the specified brain regions. Moreover, we can control the extent of cell elimination by adjusting the injection doses of the immunotoxin or injection volumes of the solution.

In the present chapter, we describe a detailed experimental strategy for immunotoxin cell targeting and its application to the study of neural circuitry analysis, in particular focusing on the basal ganglia circuitry, which mediates the control of motor functions.

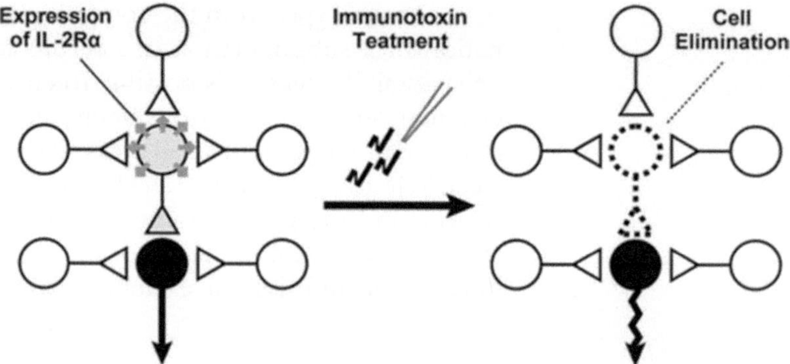

Fig. 1. Strategy of immunotoxin cell targeting. Transgenic animals are generated that express target molecules of recombinant immunotoxin (human IL-2Rα) in specific cell types in the neural circuitry. Injection of the immunotoxins induces elimination of specific neuronal types bearing IL-2Rα.

2. Materials

2.1. Recombinant Immunotoxins

Immunotoxins are conjugates of monoclonal antibodies and toxins that possess cytotoxic activity against mammalian cells. Anti-Tac(Fv)-based immunotoxins, including anti-Tac(Fv)-PE40, anti-Tac(Fv)-PE38, and anti-Tac(Fv)-PE38-KDEL (7–9), are used for immunotoxin cell targeting. These proteins are composed of the variable heavy and light chains of anti-Tac antibody, a monoclonal antibody specific for human IL-2Rα fused to a truncated form of *Pseudomonas* exotoxin (PE) (Fig. 2a). The immunotoxins selectively

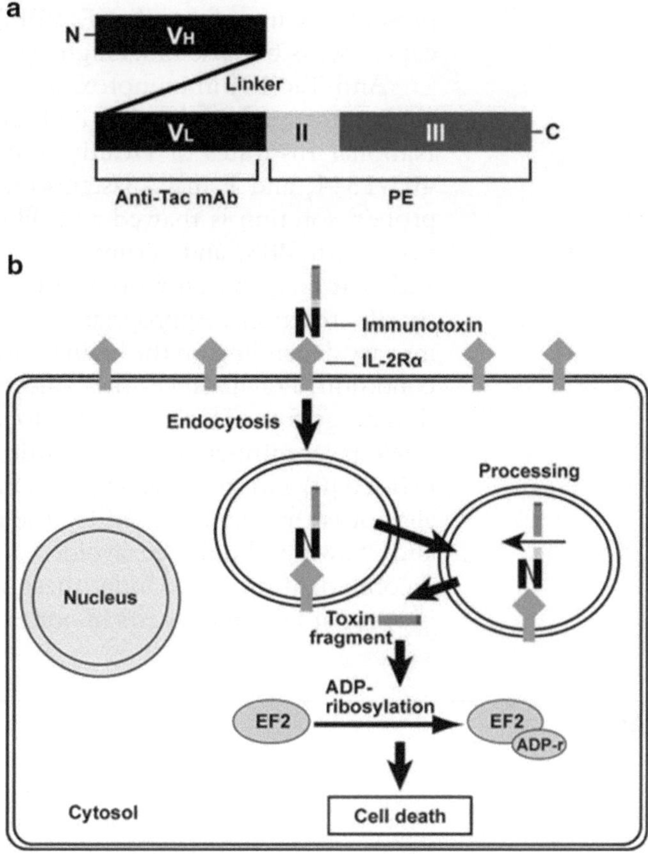

Fig. 2. Recombinant immunotoxins for cell targeting. (**a**) Structure of anti-Tac(Fv)-based immunotoxins. The immunotoxins consist of the variable regions (V_H and V_L) of anti-Tac monoclonal antibody (mAb) for human IL-2Rα and a truncated form of *Pseudomonas* exotoxin (*PE*) containing domains II (translocation domain) and III (catalytic domain). In the anti-Tac(Fv)-PE38-KDEL immunotoxin, the KDEL sequence is included in its C-terminus. (**b**) Reaction of anti-Tac(Fv)-based immunotoxins. The immunotoxins bind human IL-2Rα and are incorporated into the cells. They are processed in the endocytotic vesicles, after which their C-terminal fragment containing the PE catalytic domain is translocated into the cytosol. The fragment catalyzes ADP-ribosylation of elongation factor-2 (EF-2), inhibiting *de novo* protein synthesis and thereby resulting in apoptosis of the cells bearing the receptor.

bind human IL-2Rα on the cell surface and are then incorporated into the cells through endocytosis. They are processed in the vesicles and their C-terminal fragment containing the catalytic domain of PE is translocated into the cytosol. This fragment catalyzes the ADP-ribosylation of elongation factor-2, thus inhibiting protein synthesis and finally resulting in disruption of the cells expressing the receptor (see Fig. 2b). Anti-Tac(Fv)-PE38 is missing PE amino acids 365–380 of anti-Tac(Fv)-PE40, and anti-Tac(Fv)-PE38-KDEL contains the same deletion and the C-terminal KDEL sequence (9). Anti-Tac(Fv)-PE38 and anti-Tac(Fv)-PE38-KDEL possess greater inhibitory activity for in vitro protein synthesis than anti-Tac(Fv)-PE40, and especially anti-Tac(Fv)-PE38-KDEL shows the most significant cytotoxic activity against target cells. At present, the use of anti-Tac(Fv)-PE38-KDEL is recommended for experiments because of its high cytotoxic activity.

Anti-Tac(Fv)-immunotoxins are provided by Dr. Ira Pastan (Laboratory of Molecular Biology, National Cancer Institute, National Institutes of Health; Tel: (301) 496-4797; Fax: (301) 402-1344; and E-mail: pastani@mail.nih.gov). A frozen stock of protein solution is thawed and diluted to appropriate concentrations with PBS, and serum albumin (Sigma, St. Louis, MO) is added at final concentrations of 0.25–1 mg/ml for blocking non-specific reactions. Appropriate concentrations of the immunotoxins vary depending on the brain region for injection or experimental condition. For instance, the concentrations for intracranial injection range from 10 μg/ml (the dorsal striatum) to 60 μg/ml (the nucleus accumbens). The solution is separated into aliquots (10–20 μl) and stored at −80°C. On the day of the experiment, an aliquot is thawed and used for the injection. Repetition of freeze and thawing should be avoided because of aggregation of the recombinant protein. Under these storage conditions, the protein is active at least for 2 years in our experience.

3. Methods

3.1. Generation of Transgenic Animals

The specificity and level of human IL-2Rα transgene expression are key factors for remarkable achievement of immunotoxin cell targeting. Transgenic mice or rats are generated that express IL-2Rα under the control of a cell-type-specific gene promoter functioning in the neuronal types of interest. The use of a short DNA fragment containing the promoter occasionally triggers transgene expression in ectopic sites where the gene promoter is not normally activated, and the pattern of ectopic expression varies among the transgenic lines. If these ectopic sites are localized within or near the targeted brain regions, these cells will respond to immunotoxin treatment. To avoid this issue, we need to select the transgenic lines appropriate for specific cell ablation. In contrast, a knock-in approach by homologous recombination in mouse embryonic stem cells is expected to restrict transgene expression to the target cell types, but this approach disrupts the gene locus used for the knock-in. This difficulty can be overcome by the use of the internal ribosomal entry site which promotes cap-independent translation from an internal initiation codon in the mRNAs (10), or 2A peptide derived from the Picornavirus family, which facilitates cleavage of the precursor composed of multiple proteins (11, 12). Moreover, the application of bacterial artificial chromosomes (BACs) provides a powerful system to confer cell-type-specific gene expression (13, 14). Even in the BAC transgenic animals, we need to choose the lines that show appropriate gene expression.

For the detection of transgene expression, in situ, hybridization for IL-2Rα mRNA (15) and immunohistochemistry with anti-IL-2Rα antibody (1:500 dilution, Sigma) are available. The fusion protein composed of IL-2Rα and GFP (or YFP) is sensitive to immunotoxin treatment (16, 17), and the protein tagged with FLAG peptide also possesses immunotoxin responsiveness (18). These fused or tagged proteins provide advantages for immunohistochemical detection of transgene products and observation of endogenous fluorescence.

3.2. Intracranial Injection

Animals are subjected to stereotaxic surgery for intracranial injection of recombinant immunotoxins, as schematically illustrated in Fig. 3. An animal is anesthetized and placed in the apparatus (Model 900/Mouse Adaptor Model 51625, David Kopf Instruments, Tujunga, CA), and the skull is maintained by a head holder. The capillary holder is moved in measured distances along the three axes, i.e., anterior-posterior, dorsal-ventral, and lateral-medial, by manipulating the adjusting knobs. After cutting the scalp open, a hole is drilled through the skull in an appropriate position from the bregma. A glass micropipette is introduced into a target brain region by using the coordinates according to a brain atlas for the mouse (19) or the rat (20). The solution is injected through the micropipette connected to a 10-μl Hamilton syringe, which is attached to a microinfusion pump (ESP-32, EICOM, Kyoto, Japan) by Teflon tubing.

For preparation of micropipettes, a series of glass capillaries are made from glass tubing (inner diameter = 0.75 mm, outer diameter = 1.0 mm, Sutter Instrument, Navato, CA) by using a puller (P-97, Sutter Instrument). The tapered part of the micropipette is gently scraped with the cut edge of the diamond cutter disk at the appropriate position on the taper (at the position 40–60 μm in diameter).

Before immunotoxin injection, dye injection is carried out to check the coordinates of the micropipette tip and to investigate the reach of the injected solution. The dye, pontamine sky blue

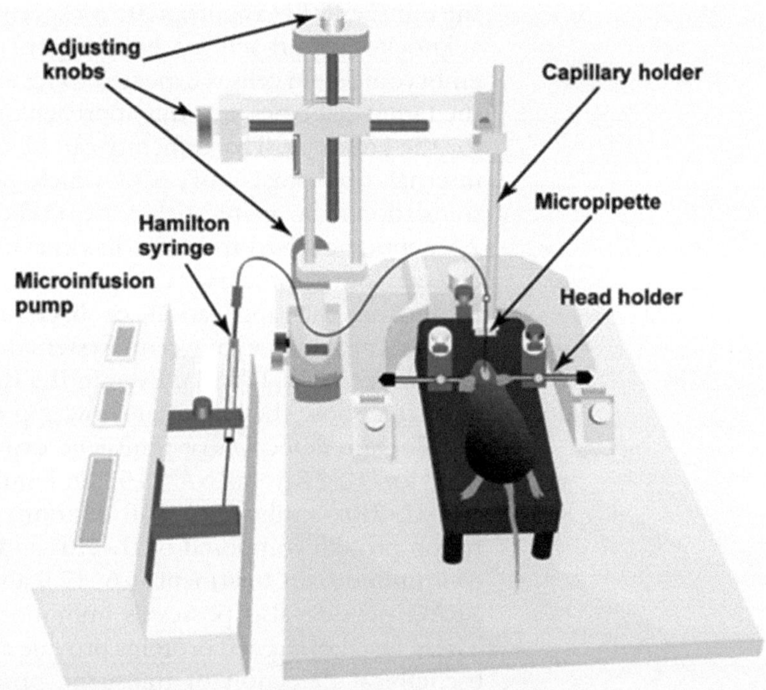

Fig. 3. Schematic illustration of stereotaxic surgery equipment for intracranial injection of recombinant immunotoxins. An animal is anesthetized and placed in the apparatus, and the skull is maintained by a head holder. The capillary holder moves in measured distances along the three axes by manual adjustment of knobs. After cutting the scalp open, a hole is drilled through the skull in an appropriate position from bregma. A glass micropipette is introduced into the targeted brain region by using the coordinates according to the brain atlas. Solution is injected through the micropipette connected to a Hamilton syringe, which is connected via Teflon tubing to a microinfusion pump.

(AVOCADO Research Chemicals, Morecambe, UK), is diluted to a final concentration of 3.6% in PBS containing serum albumin (0.25–1.0 mg/ml). The capillary and syringe are filled with Fluorinert FC77 (3M, Saint Paul, MN) followed by aspiration of the dye into the middle of the micropipette. Injection is performed at a constant flow rate of 0.1 μl/min. The volume of the solution varies depending on the targeted brain areas, but we generally inject 0.3–0.5 μl of the solution/site. After the injection, the micropipette is left in place for 3 min and then removed. Immediately after dye injection, the brains are taken, and their sections are stained with neutral red. The position of the micropipette tip and the reach of the injected dye are checked under a microscope.

Appropriate doses of recombinant immunotoxins are injected using the aforementioned procedure, in which dye is exchanged for immunotoxin solution. For selective ablation of the target neurons, it is important to adjust the dose of immunotoxin. We generally inject 5–20 ng of immunotoxin/site, and the dose is dependent on the neuronal type, brain region, and transgenic strain (transgene expression level). The number of injection sites varies dependent on the target regions. In general, cell ablation begins 3–4 days after immunotoxin injection, and we can check the ablation histologically by 7 days after the injection. We need to adjust the experimental conditions for the intracranial injection to accomplish appropriate results of cell ablation.

3.3. Analysis of Neural Circuit and Behavior

Immunotoxin cell targeting is applicable for the research on the neural circuitry that mediates different brain functions (cf., (15–17, 21–24)). Here, we describe representative applications of this technology for research on the basal ganglia circuitry. The basal ganglia mediate a variety of motor functions (25–27). The striatum is the major structure of the basal ganglia that receives convergent excitatory inputs from many cortical and thalamic areas and projects to the output nuclei, including the substantia nigra pars reticulata (SNr) and entopeducular nucleus (EPN) (see Fig. 4a). Striatal projections are composed of two subpopulations of GABAergic medium spiny neurons that constitute the direct and indirect pathways. The striatonigral spiny neurons in the direct pathway provide monosynaptic inhibition to the output nuclei, whereas the striatopallidal spiny neurons in the indirect pathway inhibit the globus pallidus (GP). The GP neurons send inhibitory projections to the output nuclei and the subthalamic nucleus (STN), and the STN neurons provide glutamatergic projections to the output nuclei and the GP. In addition, dopaminergic neurons located in the substantia nigra pars compacta (SNc) modulate different neuronal populations expressing dopamine receptor subtypes in the striatum (28–30). Immunotoxin targeting has been used for analysis of the neural circuit mechanism that mediates dopamine-related motor behavior by eliminating specific neuronal types that constitute the basal ganglia.

To study the roles of the striatopallidal neurons in dopamine-related motor behavior, ablation of these neurons was induced by immunotoxin cell targeting (15). Because the striatopallidal neurons are known to express dopamine D_2 receptor (D_2R), mutant mice were generated that express the human IL-2Rα gene cassette under the control of the mouse D_2R gene (Fig. 4b). A solution of immunotoxin (10 μg/ml) was injected into the dorsal striatum using the stereotaxic approach (six sites, 0.5 μl/site) (Fig. 4c). The immunotoxin injection eliminated the majority of the striatal neurons expressing D_2R in the mutants, whereas it did not influence the striatal neurons expressing dopamine D_1 receptor (D_1R) (Fig. 4d). In addition to the elimination of the striatopallidal neurons in the injected D_2R mutants, a decrease in the number of striatal cholinergic interneurons was also observed because of expression of D_2R in some cholinergic interneurons. The behavioral and physiological changes derived from cholinergic ablation

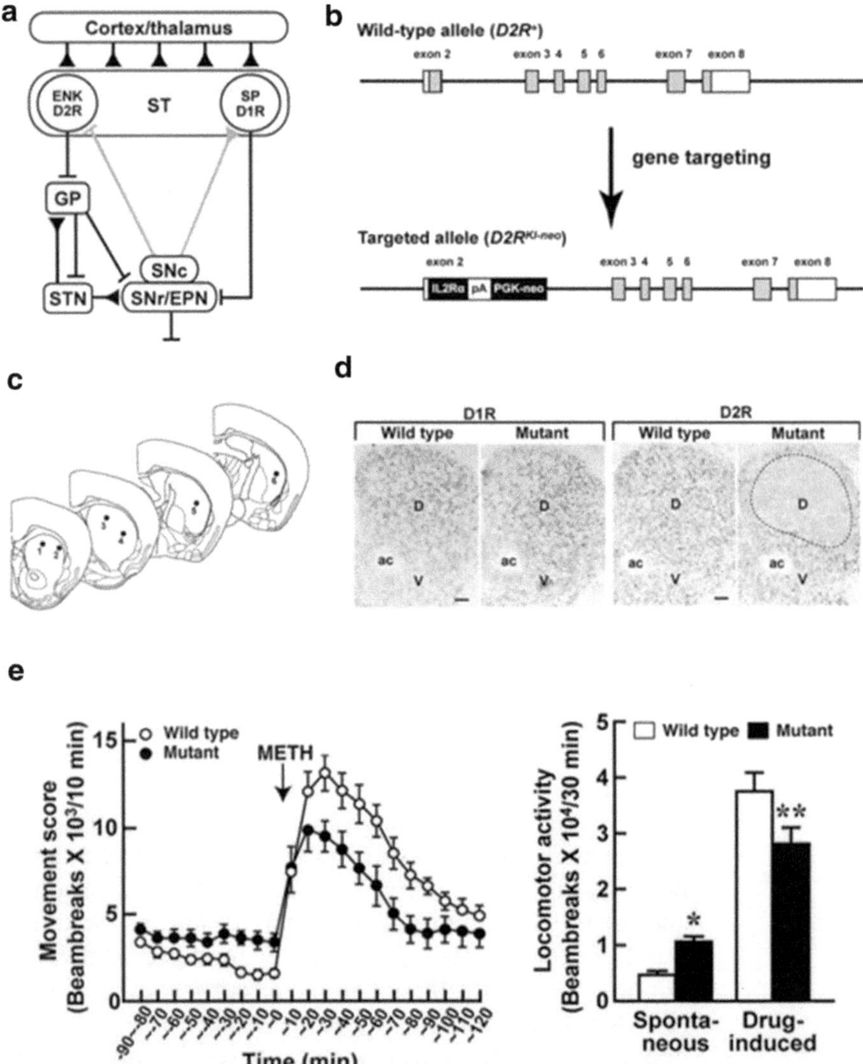

Fig. 4. Immunotoxin targeting of D_2R-containing striatopallidal neurons in the dorsal striatum. (**a**) Anatomical structure of basal ganglia circuitry. The striatonigral neurons containing D_1R and substance P (SP) project to the output nuclei including the SNr and the EPN, whereas the striatopallidal neurons containing D_2R and enkephalin (ENK) project to the GP. The glutamatergic excitatory pathway is shown by the *black line* and *triangle*, and the GABAergic inhibitory pathway is shown by the *black line* and *bar*. The dopaminergic pathway is indicated by the *gray line* and *triangle* (facilitation) or *bar* (inhibition). See the text for detailed explanation of the circuitry structure. (**b**) Strategy for knock-in mutagenesis of the human IL-2Rα gene cassette into the mouse D_2R locus. The IL-2Rα gene cassette was introduced into the initiation codon in exon 2 by gene targeting. pA, polyadenylation signal; and PGK-neo, phosphoglycerate kinase promoter-neo resistance gene. (**c**) Coordinates for intrastriatal injection of immunotoxins. The immunotoxin solution was injected into six sites in the dorsal striatum. (**d**) Elimination of striatal neuronal types containing D_2R in the mutant mice by immunotoxin injection. For in situ hybridization, sections through the mouse striatum were prepared and hybridized with riboprobe for D_2R or D_1R sequence. The dorsal (*D*) and ventral (*V*) regions of the striatum are indicated. ac, anterior commissure. *Scale bar*, 200 μm. (**e**) Impaired motor behavior by ablation of striatopallidal neurons. On day 7 after immunotoxin injection, mice were tested for locomotor activity in an open field. After monitoring spontaneous locomotion for 1.5 h, the mice were administered methamphetamine (METH, 2.0 mg/kg, s.c.), and the movement score for a 10-min period was measured. The time course of the movement score is indicated in left panel. The total number for the movement score in a 30-min test period was calculated to evaluate spontaneous locomotor activity during the pretreatment (−30 to 0 min) and drug-induced locomotor activity after METH treatment (10–40 min). The injected mutant mice showed an increase in spontaneous locomotor activity and reduced drug-induced motor activation (Two-way ANOVA, Tukey-HSD test, $^*p<0.05$, $^{**}p<0.05$ versus the wild-type mice) (Modified from ref. (15)).

with a chemical neurotoxin were compared with those changes in the injected D_2R mutants to evaluate the changes derived from the elimination of the striatopallidal neurons. The striatopallidal elimination caused an increase in spontaneous movement and a reduction in motor activation dependent on systemic treatment with methamphetamine, which stimulates dopamine release (Fig. 4e). For characterization of changes in the neural circuitry, the expression level of markers that reflect spontaneous neural activity of some basal ganglia nuclei was studied. These data suggested that the striatopallidal elimination induced hyperactivity of the GP neurons and hypoactivity of the output neurons, resulting in the increased spontaneous movement. Normally, the striatopallidal neurons appear to inhibit the GP activity and maintain the output neuron activity, causing suppression of spontaneous movement. In addition, systemic dopamine treatment inhibits the striatopallidal activity through D_2R signaling and induces the activation of the striatonigral neurons as well as the GP and STN neurons. Another marker study with dopamine-induced c-*fos* expression suggested that the striatopallidal elimination attenuated dopamine-induced activation of the striatonigral neurons with no apparent influence on the activation of the GP and STN neurons, leading to impaired motor activation. In response to dopamine stimulation, the striatopallidal neurons appear to remove inhibition of the GP neurons and to influence the striatonigral activation for the full suppression of output neuron activity, thereby resulting in the expression of dopamine-induced motor behavior.

For study of the roles of the STN neurons in dopamine-related motor behavior, elimination of these neurons was induced by immunotoxin cell targeting (17). Transgenic mice were generated that express human IL-2Rα fused to GFP in the STN neurons under the control of the mouse neuropsin gene promoter (Fig. 5a). A solution of immunotoxin (10 μg/ml) was injected into the STN using the stereotaxic approach (one site, 0.3 μl) (Fig. 5b). The immunotoxin injection removed the majority of the STN neurons indicated by IL-2Rα/GFP in the transgenic mice (Fig. 5c). Ablation of these neurons increased spontaneous movement and reduced hyperactivity in response to dopamine stimulation (Fig. 5d). Single-unit recording of some basal ganglia nuclei indicated that the STN ablation reduced spontaneous firing activity of the output neurons with no change in spontaneous GP firing, resulting in the increased spontaneous movement. These data suggested that the STN neurons facilitate the output activity through the projection to the output neurons, causing the suppression of spontaneous movement. In addition, dopamine stimulation normally activates the STN activity and causes upregulation of the GP activity and downregulation of the output activity. Single-unit recording after dopamine stimulation indicated that the STN ablation blocked dopamine-induced elevation of the GP firing activity

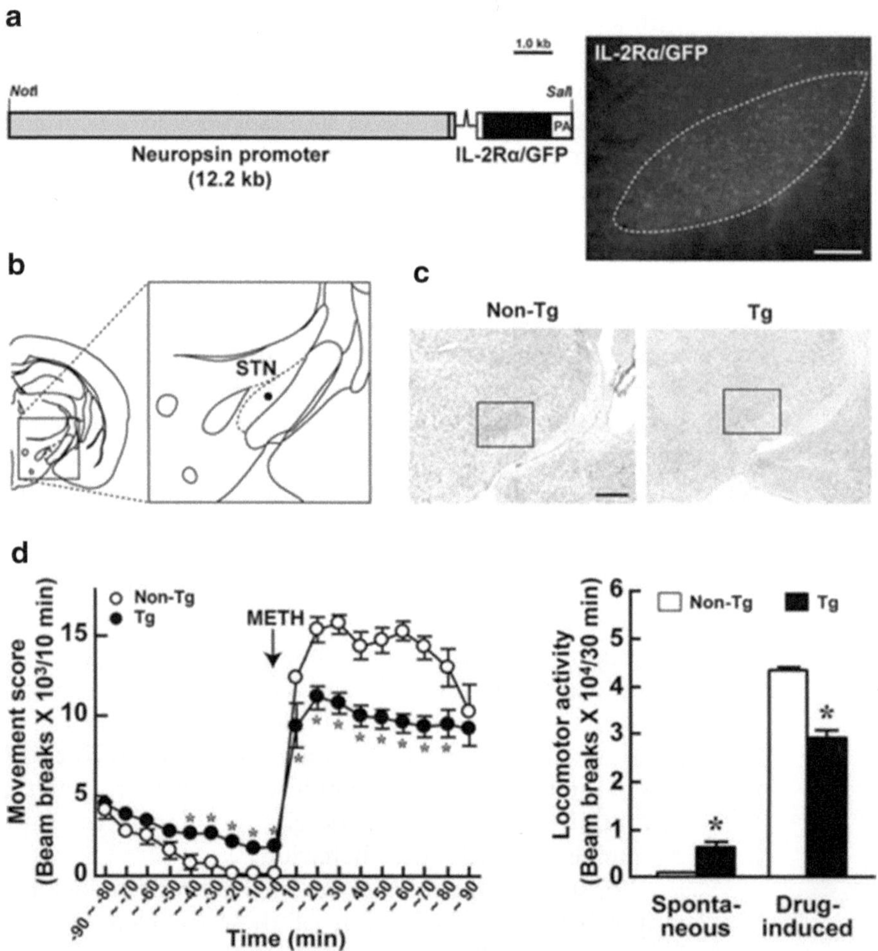

Fig. 5. Immunotoxin targeting of the STN-related pathways. (**a**) Generation of transgenic mice for STN targeting. The left panel depicts the transgene construct containing a 12.2-kb DNA fragment of the neuropsin gene promoter connected to a human IL-2Rα/GFP fusion gene. The *right panel* shows a fluorescent microscopic image of a section through the STN, indicating the expression of the IL-2Rα/GFP fusion gene in the STN neurons of the transgenic mice. *Scale bar*, 100 μm. (**b**) Coordinate for intrasubthalamic injection of recombinant immunotoxin. (**c**) Elimination of the STN neurons in the transgenic (*Tg*) mice. Neutral red staining of sections through the STN prepared from the non-Tg and Tg mice injected with the immunotoxin. *Scale bar*, 400 μm. (**d**) Disturbance in motor behavior by the STN ablation. On day 7 after immunotoxin injection, mice were tested for locomotor activity in the open field, as described in the legend of Fig. 4. The time course for the movement score is indicated in the *left panel*. The total number for the movement score in a 30-min test period was calculated to evaluate spontaneous locomotor activity during the pretreatment (–30 to 0 min) and drug-induced locomotor activity after METH treatment (10–40 min). The injected transgenic mice showed an increase in spontaneous locomotor activity and attenuated drug-induced motor activation (Two-way ANOVA, Tukey test, $*p < 0.05$ versus the nontransgenic mice) (Modified from ref. (17)).

and attenuated dopamine-induced suppression of the output firing activity, leading to the reduction in dopamine-induced hyperactivity. The data suggested that in response to dopamine stimulation the STN activation acts to suppress the output neuron activity predominantly through the GP-mediated pathways (the subthalamo-pallido-nigral pathways), thereby leading to the full expression of dopamine-induced hyperactivity.

4. Note

In the present chapter, we described two applications of immunotoxin cell targeting for research on the basal ganglia circuitry that mediates dopamine-induced motor activation. This technology has been also used for studying the functions of other neuronal types in the ventral striatum and the cerebellum (16, 23, 24). The technology is further applicable for the study of cellular function in the peripheral tissues (18, 21). By devising the method of immunotoxin treatment, we can target different types of neurons in the central and peripheral nervous systems. In addition, the application of Cre-*loxP*-mediated site-specific recombination (31, 32) and tetracycline-regulated gene expression (33, 34) confers advantages for cell-type-specific and high-level expression of the IL-2Rα transgene. The existence of established transgenic lines for Cre recombinase or tetracycline transactivator will facilitate generation of transgenic animals for specific cell ablation. The production of the reporter transgenic animals is now in progress in our laboratory. Moreover, the use of highly efficient retrograde transport viral vectors permits the induction of transgene expression in neuronal cell bodies innervating the brain region that receives the vector injection (35, 36, 37). Immunotoxin treatment of specific brain regions containing IL-2Rα-expressing neurons will selectively eliminate the desired neural pathways. This system provides a new approach to investigate the behavioral and physiological roles of specific neural pathways in brain functions. Future extension of immunotoxin cell targeting, together with further development of these new experimental tools, will promote a more detailed understanding of the neural circuitry mechanisms underlying a wide range of brain functions.

References

1. Carlson NR (2009) Methods and strategies of research. In Physiology of Behavior, 10th edn., Chapter 5. Allyn and Bacon, Boston, pp. 134–168.
2. Langston JW, Ballard P, Tetrud JW, Irwin I (1983) Chronic parkinsonism in humans due to a product of meperidine-analog synthesis. Science 219:979–980.
3. Jonsson G, Hallman H, Ponzio F, Ross S (1981) DSP4 (N-(2-chloroethyl)- N-ethyl-2-bromobenzylamine): a useful denervation tool for central and peripheral noradrenaline neurons. Eur J Pharmacol 72: 173–188.
4. Kobayashi K, Morita S, Sawada H, Mizuguchi T, Yamada K, Nagatsu I, Fujita K, Kreitman RJ, Pastan I, Nagatsu T (1995) Immunotoxin-mediated conditional disruption of specific neurons in transgenic mice. Proc Natl Acad Sci USA 92:1132–1136.
5. Kobayashi K, Pastan I, Nagatsu T (1997) Controlled genetic ablation by immunotoxin-mediated cell targeting. In: Houdebine LM (ed) Transgenic Animals: Generation and Use. Harwood Academic Publishers, Amsterdam, pp. 331–336.
6. Kobayashi K (2007) Controlled cell targeting system to study the brain neural circuitry. Neurosci Res 58:118–123.
7. Chaudhary VK, Queen C, Junghans RP, Waldmann TA, FitzGerald DJ, Pastan I (1989) A recombinant immunotoxin consisting of two antibody variable domains fused to *Pseudomonas* exotoxin. Nature 339:394–397.

8. Batra JK, FitzGerald D, Gately M, Chaudhary VK, Pastan I (1990) Anti-Tac(Fv)-PE40, a single chain antibody *Pseudomonas* fusion protein directed at interleukin 2 receptor bearing cells. J Biol Chem 265:15198–15202.
9. Kreitman RJ, Bailon P, Chaudhary VK, FitzGerald DJ, Pastan I (1994) Recombinant immunotoxins containing anti-Tac(Fv) and derivatives of Pseudomonas exotoxin produce complete regression in mice of an interleukin-2 receptor-bearing human carcinoma. Blood 83:426–434.
10. Kobayashi T, Kida Y, Kaneko T, Pastan I, Kobayashi K (2001) Efficient ablation by immunotoxin-mediated cell targeting of the cell types that express human interleukin-2 receptor depending on the internal ribosome entry site. J Gene Med 3:505–510.
11. Furler S, Paterna J-C, Weibel M, Büeler H (2001) Recombinant AAV vectors containing the foot and mouth disease virus 2A sequence confer efficient bicistronic gene expression in cultured cells and rat substantia nigra neurons. Gene Ther 8:864–873.
12. Szymczak AL, Workman CJ, Wang Y, Vignali KM, Dilioglou S, Vanin EF, Vignali DAA (2004) Correction of multi-gene deficiency in vivo using a single 'self-cleaving' 2A peptide-based retroviral vector. Nat Biotech 22:589–594.
13. Heintz N (2001) Bac to the future: The use of bac transgenic mice for neuroscience research. Nat Rev Neurosci 2:861–870.
14. Yang XW, Gong S (2005) An overview on the generation of BAC transgenic mice for neuroscience research. Curr Protoc Neurosci Chapter 5:Unit 5.20.
15. Sano H, Yasoshima Y, Matsushita N, Kaneko T, Kohno K, Pastan I, Kobayashi K (2003) Conditional ablation of striatal neuronal types containing dopamine D2 receptor disturbs coordination of basal ganglia function. J Neurosci 23:9078–9088.
16. Watanabe D, Inokawa H, Hashimoto K, Suzuki N, Kano M, Shigemoto R, Hirano T, Toyama K, Kaneko S, Yokoi M, Moriyoshi K, Suzuki M, Kobayashi K, Nagatsu T, Pastan I, Nakanishi S (1998) Ablation of cerebellar Golgi cells disrupts synaptic integration involving GABA inhibition and NMDA receptor activation in motor coordination. Cell 95:17–27.
17. Yasoshima Y, Kai N, Yoshida S, Shiosaka S, Koyama Y, Kayama Y, Kobayashi K (2005) Subthalamic neurons coordinate basal ganglia function through differential neural pathways. J Neurosci 25:7743–7753.
18. Matsusaka T, Xin J, Niwa S, Kobayashi K, Akatsuka A, Hashizume H, Wang Q, Pastan I, Fogo BA, Ichikawa I (2005) Genetic engineering of glomerular sclerosis in the mouse via control of onset and severity of podocyte-specific injury. J Am Soc Nephrol 16:1013–1023.
19. Franklin KBJ, Paxinos G (2007) The Mouse Brain in Stereotaxic Coordinates, 3rd ed. Elsevier Academic Press, San Diego. http://www.elsevierdirect.com/product.jsp?isbn=9780123694607.
20. Paxinos G, Watson C (2007) The Rat Brain in Stereotaxic Coordinates, 6th ed. Elsevier Academic Press, San Diego. http://www.elsevierdirect.com/product.jsp?isbn=9780125476126&dmnum=CWS1 -->
21. Sawada H, Nishii K, Suzuki T, Hasegawa K, Hata T, Nagatsu I, Kreitman RJ, Pastan I, Nagatsu T, Kobayashi K (1998) Autonomic neuropathy in transgenic mice caused by immunotoxin targeting of the peripheral nervous system. J Neurosci Res 51:162–173.
22. Kaneko S, Hikida T, Watanabe D, Ichinose H, Nagatsu T, Kreitman RJ, Pastan I, Nakanishi S (2000) Synaptic integration mediated by striatal cholinergic interneurons in basal ganglia function. Science 289:633–637.
23. Hikida T, Kaneko S, Isobe T, Kitabatake Y, Watanabe D, Pastan I, Nakanishi S (2001) Increased sensitivity to cocaine by cholinergic cell ablation in nucleus accumbens. Proc. Natl. Acad. Sci. USA 98:13351–13354.
24. Hikida T, Kitabatake Y, Pastan I, Nakanishi S (2003) Acetylcholine enhancement in the nucleus accumbens prevents addictive behaviors of cocaine and morphine. Proc Natl Acad Sci USA 100:6169–6173.
25. Alexander GE, Crutcher MD (1990) Functional architecture of basal ganglia circuits: neural substrates of parallel processing. Trends Neurosci 13:266–271.
26. DeLong MR (1990) Primate models of movement disorders of basal ganglia origin. Trends Neurosci 13:281–285.
27. Gerfen CR, Wilson CJ (1996) The basal ganglia. In Swanson LW, Björklund A, Hökfelt T (eds) Handbook of Chemical Anatomy, Vol. 12, Elsevier, Amsterdam, pp 37–468.
28. Gerfen CR, Engber TM, Mahan LC, Susei Z, Chase TN, Monsma FJ Jr, Sibley DR (1990) D_1 and D_2 dopamine receptor-regulated gene expression of striatonigral and striatopallidal neurons. Science 250:1429–1432.
29. Kawaguchi Y, Wilson CJ, Augood SJ, Emson PC (1995) Striatal interneurones: chemical, physiological and morphological characterization. Trends Neurosci 18:527–535.
30. Kawaguchi Y (1997) Neostriatal cell subtypes and their functional roles. Neurosci Res 27:1–8.
31. Gu H, Zou YR, Rajewsky K (1993) Independent control of immunoglobulin switch recombination at individual switch regions evidenced

through Cre-loxP-mediated gene targeting. Cell 73:1155–1164.

32. Gu H, Marth JD, Orban PC, Mossomann H, Rajewsky K (1994) Deletion of a DNA polymerase β gene segment in T cells using cell type-specific gene targeting. Science 265:103–106.
33. Gossen M, Bujard H (1992) Tight control of gene expression in mammalian cells by tetracycline responsive promoter. Proc. Natl. Acad. Sci. USA 89:5547–5551.
34. Gossen M, Freundlieb S, Bender G, Müller G, Hillen W, Bujard H (1995) Transcriptional activation by tetracyclines in mammalian cells. Science 268: 1766–1769.
35. Kato S, Inoue K, Kobayashi K, Yasoshima Y, Miyachi S, Inoue S, Hanawa H, Shimada T, Takada M, Kobayashi K (2007) Efficient gene transfer via retrograde transport in rodent and primate brains using a human immunodeficiency virus type 1-based vector pseudotyped with rabies virus glycoprotein. Hum Gene Ther 18:1141–1152.
36. Kato S, Kobayashi K, Inoue K, Kuramochi M, Okada T, Yaginuma H, Morimoto K, Shimada T, Takada M, Kobayashi K (2011) A lentiviral strategy for highly efficient retrograde gene transfer by pseudotyping with fusion envelope glycoprotein. Hum Gene Ther 22: 197–206.
37. Kato S, Kuramochi M, Kobayashi K, Fukabori R, Okada K, Uchigashima M, Watanabe M, Tsutsui Y, Kobayashi K (2011) Selective neural pathway targeting reveals key roles of thalamostriatal projection in the control of visual discrimination. J Neurosci: in press.

Chapter 11

Analysis of Neuronal Circuits with Optogenetics

Feng Zhang, Le Cong, Garret D. Stuber, Antoine Adamantidis, and Karl Deisseroth

Abstract

Optogenetic tools enable precise control of molecularly defined cell types in the mammalian brain in vivo. By combining genetic targeting with an optical fiber system, neurons expressing light-sensitive microbial opsins in the mammalian brain can be directly modulated in vivo, and thereby allowing investigators to probe the causal role of specific cell populations and activity patterns in behavioral tests. Here we compare the current set of microbial opsins and suggest considerations for choosing the most optimal opsin based on experimental needs. We also provide a detailed protocol for the construction of the optical fiber system for conducting optogenetics experiments in vivo.

Key words: Optogenetics, Channelrhodopsin, Halorhodopsin, Optical stimulation, Neural circuits

1. Introduction

The mammalian brain consists of diverse circuits formed by distinct types of excitatory, inhibitory, and modulatory cells that exhibit great temporal and physical complexity from molecular, cellular, to the behavioral level. One of the major goals of systems neuroscience has been to decipher the functional organization of distinct cell types in brain circuits and behavior. We and our collaborators have developed a technology called optogenetics, leveraging molecular genetics and naturally occurring or engineered light-sensitive proteins, to enable precise temporal control of genetically defined neurons in the mammalian brain in vivo, using a light delivery system. This chapter serves as a guide for choosing the right opsin, the light-controllable membrane-actuating component of optogenetics, based on the experimental requirements, overviews strategies for targeting opsin expression into specific types of cells in the brain, and provides further details on procedure for

Alexei Morozov (ed.), *Controlled Genetic Manipulations*, Neuromethods, vol. 65,
DOI 10.1007/978-1-61779-533-6_11,

adapting optical fibers for delivering light into the brain of free-moving mammals during behavioral tests.

2. Related Information

Previous reports have described detailed characterizations, especially their respective biophysical properties, of all opsins presented in this chapter (1–12). Likewise, the application approaches suitable for freely moving mammals, including the optical fiber methodologies, have also been detailed elsewhere (13). We and others have developed and applied this technology to control specific neural cell populations? in rodents during a variety of behavioral tests such as sleep-wake cycle architecture (14), conditioned place preference (9, 15), motor control (16, 17), and rodent disease models (18–21).

3. Opsins

We have engineered a number of naturally occurring and synthetic opsins to enable optical control of electrical signaling in neurons (Fig. 1a). Since all of these opsins originate from nonmammalian sources and require retinal-based cofactors (e.g., all-*trans* retinal for the microbial opsins) to fold properly and respond to light, a major concern was their ability to function in vivo without the additional exogenous cofactors. Nonetheless, there are sufficient levels of retinoids present in the mammalian brain so that these opsins can be expressed and rendered fully functional in the intact mammalian brain without the needs to supplement cofactors (2, 4). Detailed comparison of the properties and experimental applications of various opsins can be found in Table 1.

Several algal light-sensitive cation-permeable channelrhodopsins (ChRs) (5, 22, 23) have been characterized in nonneural systems, as have many other light-activated proteins and ion conductance regulators. So far, two ChRs (Fig. 1a) have been most successfully developed for applications in neuroscience: ChR2 from *Chlamydomonas reinhardtii* (1, 2) and VChR1 from *Volvox carteri* (5). Both ChRs have fast kinetics for the open-close cycling of channel permeability so that trains of action potentials can be triggered using brief pulses (typically 1–5 ms duration) of light flashes at a range of frequencies relevant to physiological neural signaling. While both ChR2 and VChR1 are sensitive to blue light, the activation spectra of VChR1 is sufficiently red-shifted from that of ChR2 so that neurons expressing VChR1 can be activated independent of ChR2 using orange light (around 580 nm). Compared to ChR2, the wild-type VChR1 is slower to deactivate after light-off, with a time constant of ~120 ms in contrast to the ~12 ms of ChR2.

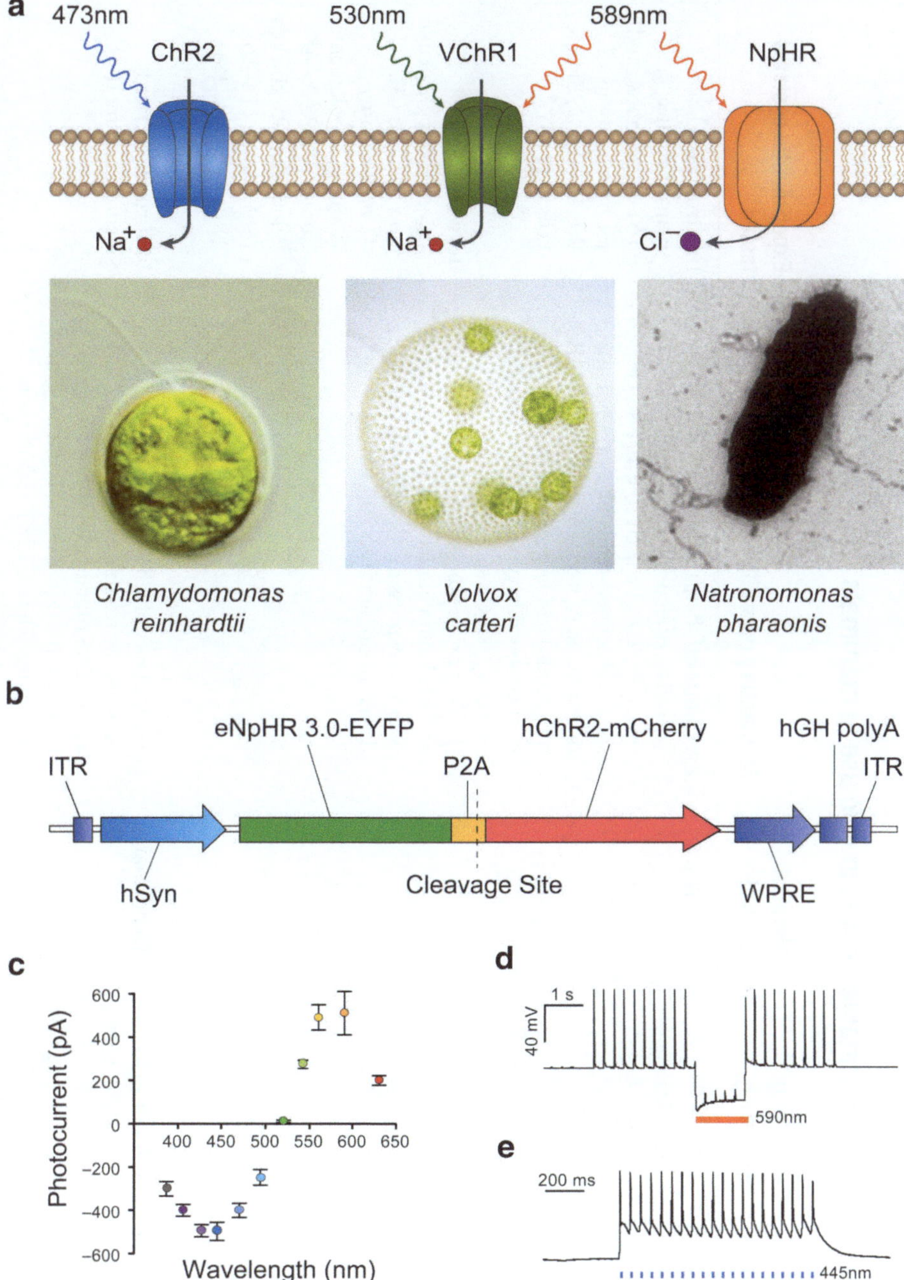

Fig. 1. Microbial opsins as optical regulators of neural activity. (**a**) Three microbial opsins enable optical activation or inhibition of neural activity. Channelrhodopsin-2 from *Chlamydomonas reinhardtii* and VChR1 from *Volvox carteri* mediate conductance of sodium ions upon exposure to blue (473 nm) and green (530 nm) light respectively. NpHR from *Natronomonas pharaonis* allows neurons to be inhibited upon illumination with yellow light (589 nm). (**b**) ChR2 and NpHR can be coexpressed in the same neuron using the 2A self-cleavage peptide. (**c**) Photocurrent spectra of neurons coexpressing ChR2 and NpHR. (**d**, **e**) *Yellow*- and *blue*-light-mediated inhibition and excitation in neurons coexpressing ChR2 and NpHR (Part **a** was modified with permission from *Nature* (13). Parts **b–e** were modified with permission from *Elsevier* (12)).

Table 1
Comparison table of optogenetic tools suitable for fast in vivo use in mammals

Opsin	Host organism	Wavelength sensitivity	Mode of control	Modulatory capabilities	Experimental systems tested
ChR2, ChR2 (H134R), ChETA	*Chlamydomonas reinhardtii*	470 nm (maximum activation)	Depolarizing	Rapid on/off, best used for precise activation of neurons on the millisecond timescale. Can be used to evoke single spikes or defined trains of action potentials over a range of frequencies. The H134R mutations yield larger photocurrents relative to wild-type ChR2, but with slower k_{off} kinetics ChETA improves the kinetic response of ChR2 and eliminates spike doublets. Neurons expressing ChETA can be driven to fire at up to 200 Hz using light	In vitro: dissociated neuron culture(1, 2, 16), acute mouse and rat brain sections (2, 29, 41–45), 293HEK cells (11, 23) In vivo: *C. elegans* (requires supplementation of ATR) (3), *D. melanogaster* (requires supplementation of ATR) (46–48), zebrafish (49), chicken (50), mouse (14–16, 18, 19, 21, 24, 28), rat (16), primate (51)
Step-function opsins (SFOs): ChR2(C128A), ChR2(C128S), and ChR2(C128T)	*Chlamydomonas reinhardtii*	470 nm (switching on), 542 nm (switching off for C128A and C128S mutants)	Depolarizing	Point mutants of ChR2 with slow or optically switchable inactivation. C128A and C128S mutants show the most prolonged activation and the highest light sensitivity, while C128T retains more temporal precision of activation. SFOs can be switched on and off with blue and green light pulses respectively	In vitro: dissociated neuron culture (8)

VChR1	*Volvox carteri*	535 nm (maximum activation), 589 nm (completely separable activation from ChR2)	Depolarizing	Red-shifted action spectra relative to ChR2. Similar to ChR2, VChR1 can be used to drive reliable action potential firing over a range of frequencies. With 589 nm light, VChR1 can be activated independently of ChR2	In vitro: dissociated neuron culture (5)
NpHR, eNpHR3.0	*Natronomonas pharaonis*	589 nm (maximum activation)	Hyper polarizing	Light-activated chloride pump. Can be used to hyperpolarize neurons with high temporal precision, capable of inhibiting single action potentials within high-frequency spike trains (up to 30 Hz). Also can be used to mediate sustained inhibition of neurons over many minutes	In vivo: *C. elegans* (requires supplementation of ATR) (16), mouse (7, 10)
				eNpHR3.0 improves the membrane trafficking of NpHR via the addition of a mammalian membrane targeting sequence from the inward-rectifying potassium channel 1.2. eNpHR is better tolerated in neurons and mediates significantly higher photocurrent	In vitro: dissociated neuron culture (7, 10, 12, 16, 51), acute mouse brain slice (10, 16)

Of note, different opsins have varying activation and inactivation kinetics, and specific mutations within the opsin core can be used to tune the kinetics of activation. Several key mutations in cysteine 128 of ChR2 (8) have been found to allow ChRs to remain open after light has been turned off. These mutant ChRs were named as step-function opsins (SFOs) because they are able to convert a brief pulse of activating light into a stable step in membrane potential by virtue of their approximately four orders of magnitude slower deactivation process compared to the wild-type ChR2. This stable activation can nevertheless be precisely terminated with a distinct, wavelength-shifted pulse of light. This allows for experimental designs that involve both short-term and long-term modulation, both of which could be precisely delivered and terminated, of the membrane potential in a defined population of cells in the mammalian brain. Specifically, ChR2 and VChR1 are ideal for testing the importance of temporal precision in spike firing and neural coding, such as probing frequency dependence of firing of a select population of neurons on behavior. SFOs on the other hand can be used to modulate the basal activity level of a neuronal population in a chronic fashion while minimizing light exposure and can allow for native asynchronous or sparse neural codes to propagate through the targeted cell type more effectively (8).

Complementary to the C128 mutations in SFOs, the E123 position in wild-type ChR2 can be altered to drastically enhance the temporal precision and responsiveness of ChR2-evoked action potentials. The new ChR2 E123T/H134R mutant, known as ChETA, has similar photocurrent amplitudes as the wild-type ChR2 but can be activated four times faster. Cells expressing ChETA can be driven to fire at up to 200 Hz (11).

Light-sensitive chloride-pumping halorhodopsins play pivotal roles in archaea for maintaining ionic homeostasis and have been known for decades to serve as light-activated single-component voltage regulators in reduced systems. We and our colleagues have found that the *Natronomonas pharaonis* variant of halorhodopsin (NpHR) can be used to inhibit neural activity through suppressing action potentials using orange light in intact tissue and behaving animals (4). Similar to ChRs, NpHR also can be activated within milliseconds of illumination, allowing inhibition of individual action potentials within high-frequency spike trains using brief light flashes. NpHR also maintains steady pump activity during chronic illumination, and therefore can be used to optically mimic reversible lesions in targeted brain structures, or loss of defined cell types. A key limitation that needs to be addressed for the NpHR-based optogenetic system is that high levels of wild-type NpHR expression in neurons can sometimes lead to toxicity resulting from poor membrane trafficking and accumulation in the endoplasmic reticulum. To improve the usability of NpHR, we have engineered an enhanced NpHR (eNpHR3.0) (12) with appropriate targeting signals to enhance membrane trafficking and proper folding in mammalian neurons. The new eNpHR3.0 achieves sufficiently high levels of HR expression in the membrane to enable optical control using even near-infrared light, a wavelength red-shifted from the typical activation maxima (12). In addition, we have further expanded the repertoire of optical control tools beyond ChRs and HRs. For alternate wavelengths of control, bacterial rhodopsins from *Halobacterium salinarum* (HsBR) and *Guillardia theta* (GtR3) can be used to inhibit neural activities using green and blue light respectively, making possible multiplexed neuromodulation using distinct bands of the full visible light spectrum to control distinct populations of cells in the brain (12).

Additionally, ChR2 and NpHR can be coexpressed in the same neuron to facilitate bidirectional optical control. In order to maintain optimal ratio of the two opsins, a 2A self-cleavage peptide can be used to ensure that equal ratio and sufficient amount of both opsins are expressed in the same cell (Fig. 1b). By shifting the color of illumination from blue to red, ChR2- and NpHR-coexpressing cells can be activated and inhibited respectively to probe the necessity as well as sufficiency of a given neural population and its signals in specific neural circuits or behaviors (Fig. 1c–e).

4. Gene Expression System

Genes encoding opsins can be selectively expressed in subsets of neurons in the intact brain using a number of well-established genetic techniques (9, 14–16, 18, 24–30). For rapid testing, high levels of opsin expression are important for achieving robust optical activation and inhibition, and can be readily achieved using viral gene delivery. Both lentiviral and adeno-associated viral (AAV) vectors (Fig. 2a) can be readily produced at high functional titer (10^9 pfu/mL for lentivirus and 10^{12} pfu/mL for AAV) and stereotactically injected into target brain regions to mediate functional levels of opsin expression. Protocols detailing the production of high titer virus have been published elsewhere (31). High titer lentiviral and AAV vectors can also be ordered from commercial facilities (e.g., University of North Carolina and University of Pennsylvania Viral Vector Cores).

5. Cell Type–Specific Opsin Targeting

Cell type-specific promoters can be used to target opsin expression to a specific population of neurons. Lentiviral and AAV vectors can package up to 5 and 2 kb of promoter sequence in addition to the

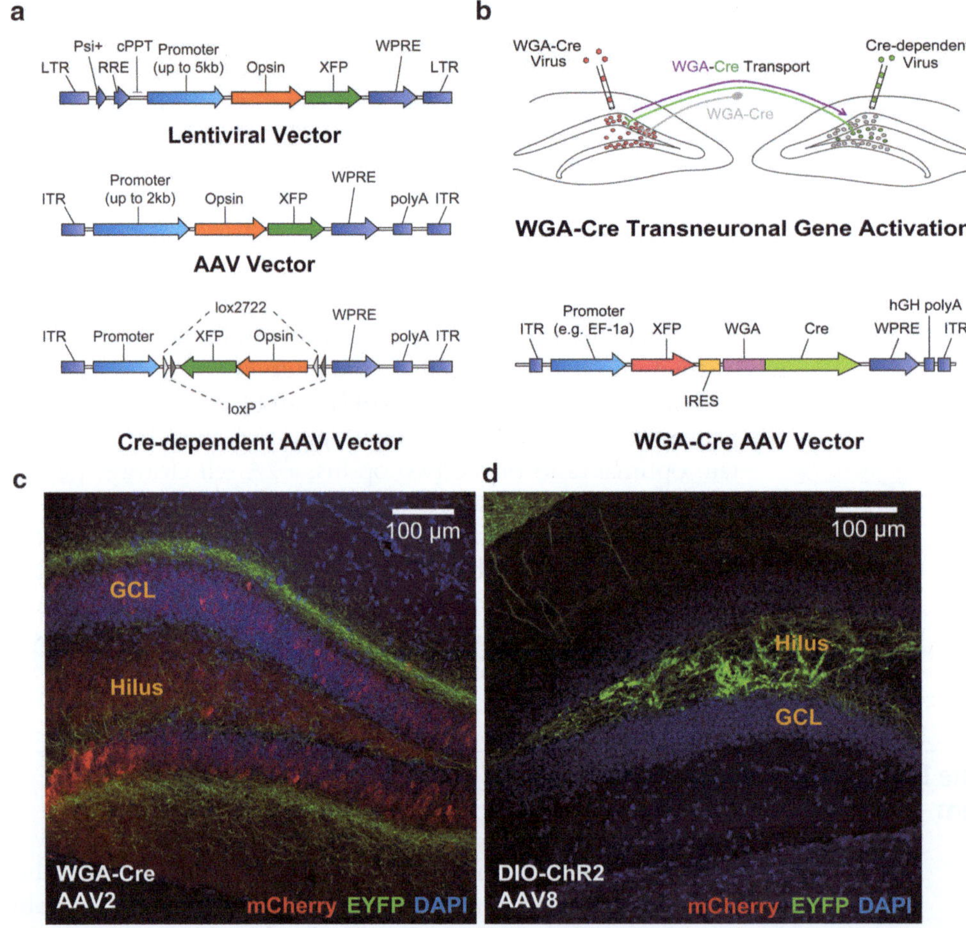

Fig. 2. Cell type-specific opsin expression systems. (**a**) Schematic of lentiviral and adeno-associated viral (AAV) vectors. Lentiviral and AAV vectors can package up to 5 and 2 kb of promoter sequence respectively. (**b**) Schematic of Cre-dependent AAV expression vector. The opsin gene is initially in the antisense orientation and flipped into the sense orientation upon exposure to Cre recombinase. (**c**) Transcellular gene activation by wheat germ agglutinin (WGA)-Cre fusion. The schematic (*top*) depicts two injection sites (one with WGA-Cre fusion gene (*bottom*) and another with Cre-dependent opsin virus) and long-range projections; Cre can be transcellularly delivered from transduced cells (*red*) to activate distant gene expression only in connected neurons that have received the Cre-dependent virus (*green*), not in others (*gray*). (**d**) Injection of WGA-Cre virus on one side of the hippocampal dentate gyrus led to activation of Cre-dependent AAV virus injected into the contralateral hippocampal dentate gyrus (Parts **a** and **b** were modified with permission from *Nature* (13). Parts **c** and **d** was modified with permission from *Elsevier* (12)).

opsin and fluorescent protein marker genes (Fig. 2a). However, many cell-specific promoters have weak transcriptional activity and cannot drive sufficient levels of opsin expression to yield robust photocurrents or sufficient level of photosuppression upon illumination. Therefore, we and others have designed Cre-recombinase-dependent AAV vectors (15, 25, 27, 30, 32) to enable cell type-specific opsin expression using transgenic mice expressing Cre under the control of native cell-specific promoter/enhancer regions (Fig. 2b). While maintaining cell type-specificity, this approach amplifies the transcriptional strength of the cell-specific promoter by allowing opsins to be expressed under strong promoters such as EF-1a in Cre-expressing neurons. The opsin gene is converted into a transcriptional-permissible form via Cre-mediated recombination. Given the increasing availability of a wide array of cell type-specific Cre-transgenic lines (33), researchers now have the ability to selectively probe the activity of a number of important cell types optogenetically, in versatile and generalizable fashion.

Subtypes of neurons defined by a given genetic marker can still be quite diverse, either receiving innervations from or sending axonal projections to distinct brain regions. Hence, the systematic study of neural circuitry depends on the successful targeting of optogenetic probes to functionally defined circuits. Since opsins are naturally membrane-localized, heterologous expression of microbial and engineered opsins render the axonal processes light sensitive. Therefore, it is possible to selectively control a homogeneous neural afferent pathway through focal injection of viral vectors and photostimulation or photosuppression of axon terminals in the target downstream brain structure. For example, to study the medial prefrontal cortical afferents in the amygdala, it is possible to inject a viral vector carrying ChR2-EYFP into the medial prefrontal cortex and implanting the stimulation optical fiber into the amygdala. It is even possible to combinatorially interrogate the role of multiple subsets of afferent fiber bundles using multiple opsins with distinct spectral sensitivities (e.g., using yellow light to activate VChR1-expressing afferents independent of ChR2-expressing afferents within the same neural tissue, which is unresponsive to yellow light).

A number of plant and microbial proteins and several viral vectors with unique anterograde- or retrograde-transporting properties (34–36) may be engineered with recombinases to activate gene expression in subpopulations of neurons with cell type and circuit specificity. For example, expression of fusion proteins containing Cre and wheat germ agglutinin (WGA) in the cell bodies of one brain region will allow the recombinase to be transsynaptically delivered only to synaptically connected neurons (Fig. 2c, d). Thus, only the neurons within functionally connected circuit nodes would be able to respond to light illumination. Similarly, retrograde-transporting viral vectors (37, 38) such as rabies virus (34) or herpes simplex virus 1 (39) can be used to deliver recombinases or transgene cassettes in a retrograde fashion. When combined with conditional expression systems, either Cre-dependent transgenic mice or viral vectors, this strategy may allow circuit-specific gene expression in a variety of mammalian animal models not limited to mice. Finally, for all viral expression systems, it is essential to characterize opsin-XFP expression intensity as a function of time post virus injection. For experiments using AAVs, it is routine to perform experiments 14–21 d postsurgery if targeting cell bodies with light and >28 d for experiments targeting presynaptic terminals.

6. In Vivo Optical Stimulation

Selective expression of opsins in genetically defined neurons makes it possible to control a subset of neurons without affecting nearby cells and processes in the intact brain, and study the behavioral consequences of their activation (ChRs) or inhibition (HR). However, light must still be delivered to the target brain structure without compromising the integrity and organization of the circuit under investigation. There are several important factors for achieving effective optical stimulation in vivo. First, it is important to choose a light source with sufficient power, as ChR2 and NpHR require 1–10 mW/mm^2 of light for reliable activation (although SFOs require approximately two orders of magnitude less light power density). For most in vivo applications, solid-state diode lasers or high-power light-emitting diodes (LEDs) can be used to provide sufficient levels of light. These diode lasers can be combined with optical fibers to deliver light into deep brain structures, such as the hypothalamus or the brain stem, with minimum perturbation to the intact brain structure, as demonstrated in rodent (Fig. 3a, b). For superficial structures such as the cortex, LEDs may be directly mounted over the brain surface. In mice, the LED may even be directly mounted over thinned skull, without performing a craniotomy. Both virus injection and delivery of the multimode fiber (50–400 μm diameter) can be guided via the same stereotactically implanted cannula, ensuring the coregistration of light illumination and opsin-expressing cells. Alternatively, optical fibers can be chronically implanted directly in brain tissue and interfaced via a fiber-coupled laser via a ferrule mounted on the skull. In the following sections, we provide a detailed protocol for setting up the fiber optic-based optical neural interface (ONI) (14, 16) for in vivo optogenetic control. Depending on the experimental design, either bilateral or unilateral stimulation may be performed. The procedures for preparing bilateral and unilateral fibers are similar and are described in detail in the following section. For behavioral experiments involving excessive rotations, a fiber commutator (Doric Lenses) can be used to relieve any tension in the optical fiber.

7. Functional Validation of Optical Control

Functional opsin expression and efficiency of optical stimulation protocols should be verified using electrophysiological and electrochemical readout methods. Depending on the targeted circuit component, intra- or extracellular recordings can be used to identify the optimal stimulation pattern (e.g., light intensity, pulse

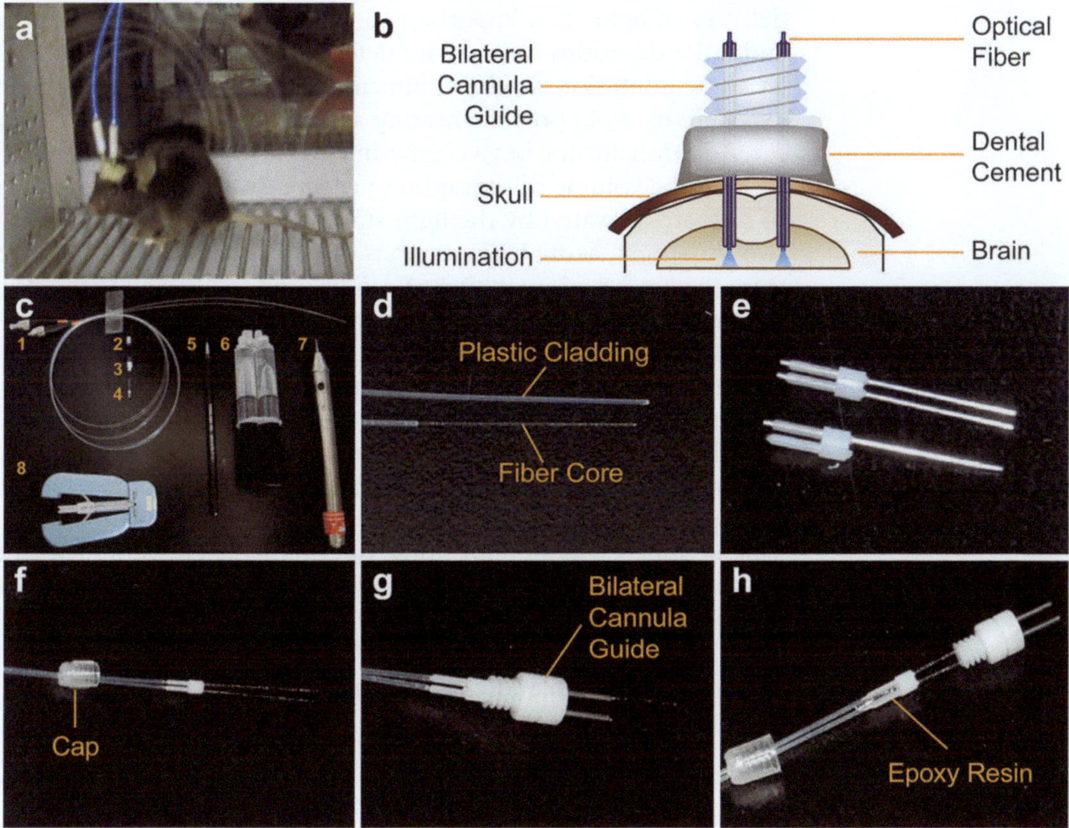

Fig. 3. Preparation of bilateral optical neural interface *(ONI)*. (**a**) Image of a free-behaving mouse with bilateral fiber implantation. (**b**) Diagram of the optical neural interface. The bilaterial cannula guide is stereotactically implanted over the target brain region and secured using dental cement. Two optical fibers are inserted through the cannula guide to illuminate the target brain structure. (**c**) All parts needed to make the fiber: optical fibers (1), dummy cannula (2), cannula guide (3), internal cannula (4), diamond scribe (5), epoxy (6), dental drill (7), fiber stripper (8). (**d**) To expose the fiber, use a fiber stripper to remove the plastic cladding. (**e**) Remove the steel cannula from the internal cannula and use the plastic pedestal to fix the optical fibers. (**f**) Thread the stripped end of the optical fibers through a dummy cannula cap (modified by drilling a hole through the *top*) and the plastic pedestal. (**g**) Insert the fiber through a bilateral cannula guide. Note the length of fiber projecting from the tip of the cannula guide and use a diamond scribe to trim the fiber so that there is only 0.5 mm of projection from the tip of the cannula guide. (**h**) Use epoxy to glue the fiber to the plastic pedestal so that the fibers are securely attached. After the epoxy hardens, make sure the cap will fit over the epoxy so that the fiber can be screwed onto the bilateral cannula guide.

frequency, and pulse duration). For in vitro verification of opsin activity, whole-cell patch-clamp electrophysiological experiments can be performed. This can either be conducted by recording light-induced activity from cell bodies that are expressing opsin variants as indentified by live fluorescent imaging or by recording from cells that are in close proximity to opsin-XFP-positive presynaptic fibers. Light can be delivered to brain slices via a LED or laser source coupled to the epifluorescence port of most microscopes. Alternatively, a small diameter optical fiber, held by a micromanipulator, can be positioned in close proximity to neurons for delivery of light. It is important to test a range of light intensities and pulse durations to find parameters that achieve reliable stimulation or inhibition, while minimizing the light intensity and light exposure to avoid possible heating and photodamage.

Additionally, for in vivo experiments, it is also helpful to characterize the volume of activation to ensure that a sufficient number of cells are activated by the light stimulation. This can be done in two ways: first, acute brain sections of different thickness prepared from the brain region of interest can be used to determine the laser output necessary to maintain a minimum power level of 1 mW/ mm^2 throughout the target brain region (14, 16); second, immunohistochemical characterization of immediate early gene (IEG) activation such as c-fos can also be used to determine the volume of activation (18).

Following experimental manipulations, it is also important to histologically verify the location of the optical fiber in the brain. This is done by examining the tissue immediately ventral to the cannula location. Take care to ensure that there is fluorescence indicative of opsin-XFP expression within 1 mm of the fiber location. This should be verified for every animal tested due to variability in expression and viral delivery.

8. Materials

1. Diode-pumped solid-state laser (473 nm for ChR2 activation, TTL modulated, 5% tolerance, FC/PC coupled at laser output; LaserGlow, cat. no. LRS-0473-PFM-00080-05; or OEM Laser Systems, BL-473-00150-CWM-SD-xx-LED-0).
2. Optical fiber (Thorlabs, cat. no. BFL37-200 for 200 μm fiber; cat. no. BFL37-300 for 300 μm fiber; fibers need to have an FC end to connect to the PC adapter on the output of the laser; length can be specified depending on the experimental setup; recommended to have at least 1 m of slack fiber from the head of the animal to the first fiber fixture). Fibers can also be constructed in lab using standard procedures, downloadable instructions available from Thorlabs.com.
3. Short wavelength fiber optic splitter for performing bilateral stimulation using two optical fibers (Fiber Optic Network Technology Canada). Also can use fused couplers (200 μm) OZ optics. Alternatively, two lasers with independent control can be used for bilateral stimulation.
4. Fiber stripping tool (300 μm fiber: Thorlabs, cat. no. T18S31; 200 μm fiber: Thorlabs, cat. No. T12S25).
5. Diamond scribe for cutting optical fiber (S90W).

6. Arbitrary waveform generator (Agilent, cat. No. 33220A).
7. Bilateral cannula guide (Plastics One, cat. no. C232G), internal/injector cannula (Plastics One, cat. no. C232I), and dummy cannula (Plastics One, cat. no.C232DC). The length of the cannula guide and internal cannula can be customized for desired target depth (recommended length for the cannula guide is 0.5 mm above the target region; internal cannula should have a projection length of 0.5 mm beyond the cannula guide tip; dummy cannula should be flush with the tip of the cannula guide). For unilateral stimulation, use the C313 series of cannula parts.
8. A custom aluminum rotating optical commutator (Doric Lenses, Québec, Canada).
9. Light power meter (Thorlabs, sensor, cat. no. S130A, and digital console, cat. no. PM100D).
10. Chronically implantable optical fibers (200 μm diameter) can be purchased from Doric Lenses. Alternatively, these can be constructed using ferrules obtained from Precision Fiber Products (MM-FER2002–2500) and 200-μm-diameter fiber from Thorlabs.

9. Method: Construction of Acute Bilateral Stimulation Fibers

Note: The optical neural interface (ONI) is prepared by modifying the small animal cannula system from Plastics One. The entire system consists of the following components: a bilateral cannula guide that has been stereotactically implanted over the target brain region, a screw cap for securing the optical fiber to the animal's head during light delivery process, a fiber guard modified from the internal cannula adapter, and a bare fiber whose length is customized based on the depth of the target region inside the brain (Fig. 3b, c). Detailed protocols for stereotactic surgery have been described previously (40). For unilateral stimulation, fibers can be prepared using unilateral cannula parts from Plastics One.

1. Use a fiber stripper to remove ~40 mm of plastic coating from the bare end of two optical fibers (Fig. 3d).
2. Remove the metal tubings from the plastic pedestal of an internal cannula. Thread the fiber through the plastic pedestal (Fig. 3e).
3. Prepare the screwing cap by drilling an ~2 mm-diameter hole centrally on the top of a dust cap to allow the optical fiber to pass through. Thread the fiber through the cap with the cap opening facing away from the FC connector on the fiber (Fig. 3f).
4. Insert the ONI fiber through a cannula guide (same dimensions as the ones implanted on the experimental animals);

ensure that they snap tightly. Use a diamond scribe to cut the fiber so that it projects 0.5 mm from the tip of the cannula guide (Fig. 3g).

Caution: It is important to use a diamond scribe so the fiber tip is flat to insure consistent illumination. After cutting, check the fiber tip under a stereomicroscope.

Slowly insert the optical fibers into the cannula guide since bare fibers without plastic cladding are very brittle.

5. Apply a very thin layer of epoxy to the interface of coated/uncoated fiber and slide the plastic pedestal onto the fiber so it is tightly against the plastic coating on the fiber. Allow the glue to dry for 10 min (Fig. 3h).

   Caution: Too much epoxy will prevent the cap from fitting over the internal cannula and screwing onto the cannula guide.

6. Connect the FC end of the fiber to the PC connector on the output port of the laser. Connect the laser control box to the waveform generator.

7. Program the waveform generator with the desired stimulation protocol (i.e., pulse frequency, duration, and intensity).

   Critical Step: Use a power meter to measure the light output at the end of the fiber before and after behavioral testing.

   Also check the fiber for uniform light output by coupling it to a laser at an intensity of 1 mW/cm^2. When light from the fiber illuminates a white surface perpendicular to the fiber, uniform tight concentric circles of light should be observed.

8. To conduct optical stimulation in free-moving animals, insert the ONI fiber through the cannula guide on the animal's head and tighten the screw cap to secure the fiber.

   Caution: Extra care should be taken during fiber insertion to prevent breakage. It is important to ensure the screw cap has been securely tightened to prevent the fiber from wiggling inside of the cannula guide.

   Critical Step: For most behaviors, it is necessary to habituate the animals to the experimental setup by practicing fiber insertion/removal several times. The animals should be placed in the behavioral testing room for at least 3 h prior to behavioral testing.

   Troubleshooting: Fiber optical commutators may be used for unilateral or bilateral stimulation experiments to alleviate excessive torsion in the fiber.

9. To begin light stimulation, turn on the output on the waveform generator to trigger the laser.

10. Cannula placement and opsin expression can be verified after behavioral testing through standard histological analysis.

10. Conclusion

Optogenetics holds promise as a versatile tool for probing the causal role of defined neurons and circuits in freely moving mammals. While additional engineering advances will be made, the basic panel of opsins, targeting methodologies, and optical hardware described here form the core technology required for in vitro or in vivo optogenetic circuit manipulation. Targeted, temporally precise control of specific neuronal subtypes may lead to insights into the cellular basis of systems-level neural circuit function as well as clinically relevant neuromodulation with fewer side effects and more robust therapeutic efficacy.

References

1. Boyden, E. S., Zhang, F., Bamberg, E., Nagel, G., and Deisseroth, K. (2005) Millisecond-timescale, genetically targeted optical control of neural activity, *Nat Neurosci 8*, 1263–1268.
2. Zhang, F., Wang, L. P., Boyden, E. S., and Deisseroth, K. (2006) Channelrhodopsin-2 and optical control of excitable cells, *Nat Methods 3*, 785–792.
3. Nagel, G., Brauner, M., Liewald, J. F., Adeishvili, N., Bamberg, E., and Gottschalk, A. (2005) Light activation of channelrhodopsin-2 in excitable cells of Caenorhabditis elegans triggers rapid behavioral responses, *Curr Biol 15*, 2279–2284.
4. Zhang, F., Wang, L. P., Brauner, M., Liewald, J. F., Kay, K., Watzke, N., Wood, P. G., Bamberg, E., Nagel, G., Gottschalk, A., and Deisseroth, K. (2007) Multimodal fast optical interrogation of neural circuitry, *Nature 446*, 633–639.
5. Zhang, F., Prigge, M., Beyriere, F., Tsunoda, S. P., Mattis, J., Yizhar, O., Hegemann, P., and Deisseroth, K. (2008) Red-shifted optogenetic excitation: a tool for fast neural control derived from Volvox carteri, *Nat Neurosci 11*, 631–633.
6. Han, X., and Boyden, E. S. (2007) Multiple-color optical activation, silencing, and desynchronization of neural activity, with single-spike temporal resolution, *PLoS ONE 2*, e299.
7. Gradinaru, V., Thompson, K. R., and Deisseroth, K. (2008) eNpHR: a Natronomonas halorhodopsin enhanced for optogenetic applications, *Brain Cell Biol 36*, 129–139.
8. Berndt, A., Yizhar, O., Gunaydin, L. A., Hegemann, P., and Deisseroth, K. (2009) Bi-stable neural state switches, *Nat Neurosci 12*, 229–234.
9. Airan, R. D., Thompson, K. R., Fenno, L. E., Bernstein, H., and Deisseroth, K. (2009) Temporally precise in vivo control of intracellular signalling, *Nature 458*, 1025–1029.
10. Zhao, S., Cunha, C., Zhang, F., Liu, Q., Gloss, B., Deisseroth, K., Augustine, G. J., and Feng, G. (2008) Improved expression of halorhodopsin for light-induced silencing of neuronal activity, *Brain Cell Biol 36*, 141–154.
11. Gunaydin, L. A., Yizhar, O., Berndt, A., Sohal, V. S., Deisseroth, K., and Hegemann, P. Ultrafast optogenetic control, *Nat Neurosci 13*, 387–392.
12. Gradinaru, V., Zhang, F., Ramakrishnan, C., Mattis, J., Prakash, R., Diester, I., Goshen, I., Thompson, K. R., and Deisseroth, K. Molecular and cellular approaches for diversifying and extending optogenetics, *Cell 141*, 154–165.
13. Zhang, F., Gradinaru, V., Adamantidis, A. R., Durand, R., Airan, R. D., de Lecea, L., and Deisseroth, K. Optogenetic interrogation of neural circuits: technology for probing mammalian brain structures, *Nat Protoc 5*, 439–456.
14. Adamantidis, A. R., Zhang, F., Aravanis, A. M., Deisseroth, K., and de Lecea, L. (2007) Neural substrates of awakening probed with optogenetic control of hypocretin neurons, *Nature 450*, 420–424.
15. Tsai, H. C., Zhang, F., Adamantidis, A., Stuber, G. D., Bonci, A., de Lecea, L., and Deisseroth, K. (2009) Phasic firing in dopaminergic neurons is sufficient for behavioral conditioning, *Science 324*, 1080–1084.
16. Aravanis, A. M., Wang, L. P., Zhang, F., Meltzer, L. A., Mogri, M. Z., Schneider, M. B., and Deisseroth, K. (2007) An optical neural interface: in vivo control of rodent motor cortex wivth integrated fiberoptic and optogenetic technology, *J Neural Eng 4*, S143–156.

17. Llewellyn, M. E., Thompson, K. R., Deisseroth, K., and Delp, S. L. (2010) Orderly recruitment of motor units under optical control in vivo, *Nat Med 16*, 1161–1165.
18. Gradinaru, V., Mogri, M., Thompson, K. R., Henderson, J. M., and Deisseroth, K. (2009) Optical deconstruction of parkinsonian neural circuitry, *Science 324*, 354–359.
19. Bi, A., Cui, J., Ma, Y. P., Olshevskaya, E., Pu, M., Dizhoor, A. M., and Pan, Z. H. (2006) Ectopic expression of a microbial-type rhodopsin restores visual responses in mice with photoreceptor degeneration, *Neuron 50*, 23–33.
20. Abbott, S. B., Stornetta, R. L., Fortuna, M. G., Depuy, S. D., West, G. H., Harris, T. E., and Guyenet, P. G. (2009) Photostimulation of retrotrapezoid nucleus phox2b-expressing neurons in vivo produces long-lasting activation of breathing in rats, *J Neurosci 29*, 5806–5819.
21. Alilain, W. J., Li, X., Horn, K. P., Dhingra, R., Dick, T. E., Herlitze, S., and Silver, J. (2008) Light-induced rescue of breathing after spinal cord injury, *J Neurosci 28*, 11862–11870.
22. Nagel, G., Ollig, D., Fuhrmann, M., Kateriya, S., Musti, A. M., Bamberg, E., and Hegemann, P. (2002) Channelrhodopsin-1: a light-gated proton channel in green algae, *Science 296*, 2395–2398.
23. Nagel, G., Szellas, T., Huhn, W., Kateriya, S., Adeishvili, N., Berthold, P., Ollig, D., Hegemann, P., and Bamberg, E. (2003) Channelrhodopsin-2, a directly light-gated cation-selective membrane channel, *Proc Natl Acad Sci USA 100*, 13940–13945.
24. Arenkiel, B. R., Peca, J., Davison, I. G., Feliciano, C., Deisseroth, K., Augustine, G. J., Ehlers, M. D., and Feng, G. (2007) In vivo light-induced activation of neural circuitry in transgenic mice expressing channelrhodopsin-2, *Neuron 54*, 205–218.
25. Zhang, F. (2008) *in Larry M. Katz Memorial Lecture, Cold Spring Harbor Laboratory Meeting on Neuronal Circuits*
26. Luo, L., Callaway, E. M., and Svoboda, K. (2008) Genetic dissection of neural circuits, *Neuron 57*, 634–660.
27. Sohal, V. S., Zhang, F., Yizhar, O., and Deisseroth, K. (2009) Parvalbumin neurons and gamma rhythms enhance cortical circuit performance, *Nature 459*, 698–702.
28. Huber, D., Petreanu, L., Ghitani, N., Ranade, S., Hromadka, T., Mainen, Z., and Svoboda, K. (2008) Sparse optical microstimulation in barrel cortex drives learned behaviour in freely moving mice, *Nature 451*, 61–64.
29. Petreanu, L., Huber, D., Sobczyk, A., and Svoboda, K. (2007) Channelrhodopsin-2-assisted circuit mapping of long-range callosal projections, *Nat Neurosci 10*, 663–668.
30. Kuhlman, S. J., and Huang, Z. J. (2008) High-resolution labeling and functional manipulation of specific neuron types in mouse brain by Cre-activated viral gene expression, *PLoS ONE 3*, e2005.
31. Kutner, R. H., Zhang, X. Y., and Reiser, J. (2009) Production, concentration and titration of pseudotyped HIV-1-based lentiviral vectors, *Nat Protoc 4*, 495–505.
32. Atasoy, D., Aponte, Y., Su, H. H., and Sternson, S. M. (2008) A FLEX switch targets Channelrhodopsin-2 to multiple cell types for imaging and long-range circuit mapping, *J Neurosci 28*, 7025–7030.
33. Gong, S., Doughty, M., Harbaugh, C. R., Cummins, A., Hatten, M. E., Heintz, N., and Gerfen, C. R. (2007) Targeting Cre recombinase to specific neuron populations with bacterial artificial chromosome constructs, *J Neurosci 27*, 9817–9823.
34. Wickersham, I. R., Lyon, D. C., Barnard, R. J., Mori, T., Finke, S., Conzelmann, K. K., Young, J. A., and Callaway, E. M. (2007) Monosynaptic restriction of transsynaptic tracing from single, genetically targeted neurons, *Neuron 53*, 639–647.
35. Maskos, U., Kissa, K., St Cloment, C., and Brulet, P. (2002) Retrograde trans-synaptic transfer of green fluorescent protein allows the genetic mapping of neuronal circuits in transgenic mice, *Proc Natl Acad Sci USA 99*, 10120–10125.
36. Sugita, M., and Shiba, Y. (2005) Genetic tracing shows segregation of taste neuronal circuitries for bitter and sweet, *Science 309*, 781–785.
37. Callaway, E. M. (2008) Transneuronal circuit tracing with neurotropic viruses, *Curr Opin Neurobiol 18*, 617–623.
38. Boldogkoi, Z., Balint, K., Awatramani, G. B., Balya, D., Busskamp, V., Viney, T. J., Lagali, P. S., Duebel, J., Pasti, E., Tombacz, D., Toth, J. S., Takacs, I. F., Scherf, B. G., and Roska, B. (2009) Genetically timed, activity-sensor and rainbow transsynaptic viral tools, *Nat Methods 6*, 127–130.
39. Lima, S. Q., Hromadka, T., Znamenskiy, P., and Zador, A. M. (2009) PINP: a new method of tagging neuronal populations for identification during in vivo electrophysiological recording, *PLoS ONE 4*, e6099.
40. Cetin, A., Komai, S., Eliava, M., Seeburg, P. H., and Osten, P. (2006) Stereotaxic gene delivery in the rodent brain, *Nat Protoc 1*, 3166–3173.
41. Zhang, Y. P., and Oertner, T. G. (2007) Optical induction of synaptic plasticity using a light-sensitive channel, *Nat Methods 4*, 139–141.

42. Petreanu, L., Mao, T., Sternson, S. M., and Svoboda, K. (2009) The subcellular organization of neocortical excitatory connections, *Nature 457*, 1142–1145.

43. Ishizuka, T., Kakuda, M., Araki, R., and Yawo, H. (2006) Kinetic evaluation of photosensitivity in genetically engineered neurons expressing green algae light-gated channels, *Neurosci Res 54*, 85–94.

44. Wang, H., Peca, J., Matsuzaki, M., Matsuzaki, K., Noguchi, J., Qiu, L., Wang, D., Zhang, F., Boyden, E., Deisseroth, K., Kasai, H., Hall, W. C., Feng, G., and Augustine, G. J. (2007) High-speed mapping of synaptic connectivity using photostimulation in Channelrhodopsin-2 transgenic mice, *Proc Natl Acad Sci USA 104*, 8143–8148.

45. Zhang, Y. P., Holbro, N., and Oertner, T. G. (2008) Optical induction of plasticity at single synapses reveals input-specific accumulation of alphaCaMKII, *Proc Natl Acad Sci USA 105*, 12039–12044.

46. Hwang, R. Y., Zhong, L., Xu, Y., Johnson, T., Zhang, F., Deisseroth, K., and Tracey, W. D. (2007) Nociceptive neurons protect Drosophila larvae from parasitoid wasps, *Curr Biol 17*, 2105–2116.

47. Pulver, S. R., Pashkovski, S. L., Hornstein, N. J., Garrity, P. A., and Griffith, L. C. (2009) Temporal dynamics of neuronal activation by Channelrhodopsin-2 and TRPA1 determine behavioral output in Drosophila larvae, *J Neurophysiol 101*, 3075–3088.

48. Schroll, C., Riemensperger, T., Bucher, D., Ehmer, J., Voller, T., Erbguth, K., Gerber, B., Hendel, T., Nagel, G., Buchner, E., and Fiala, A. (2006) Light-induced activation of distinct modulatory neurons triggers appetitive or aversive learning in Drosophila larvae, *Curr Biol 16*, 1741–1747.

49. Douglass, A. D., Kraves, S., Deisseroth, K., Schier, A. F., and Engert, F. (2008) Escape behavior elicited by single, channelrhodopsin-2-evoked spikes in zebrafish somatosensory neurons, *Curr Biol 18*, 1133–1137.

50. Li, X., Gutierrez, D. V., Hanson, M. G., Han, J., Mark, M. D., Chiel, H., Hegemann, P., Landmesser, L. T., and Herlitze, S. (2005) Fast noninvasive activation and inhibition of neural and network activity by vertebrate rhodopsin and green algae channelrhodopsin, *Proc Natl Acad Sci USA 102*, 17816–17821.

51. Han, X., Qian, X., Bernstein, J. G., Zhou, H. H., Franzesi, G. T., Stern, P., Bronson, R. T., Graybiel, A. M., Desimone, R., and Boyden, E. S. (2009) Millisecond-timescale optical control of neural dynamics in the nonhuman primate brain, *Neuron 62*, 191–198.

Index

Alexei Morozov (ed.), *Controlled Genetic Manipulations*, Neuromethods, vol. 65,
DOI 10.1007/978-1-61779-533-6, © Springer Science+Business Media, LLC 2012

S

T

U

V

W

MIX
Papier aus verantwortungsvollen Quellen
Paper from responsible sources
FSC® C105338

If you have any concerns about our products,
you can contact us on
ProductSafety@springernature.com

In case Publisher is established outside the EU,
the EU authorized representative is:
Springer Nature Customer Service Center GmbH
Europaplatz 3, 69115 Heidelberg, Germany

Printed by Libri Plureos GmbH
in Hamburg, Germany